Concrete Structures: Latest Advances and Prospects for a Sustainable Future

Concrete Structures: Latest Advances and Prospects for a Sustainable Future

Editors

Mariella Diaferio
Francisco B. Varona

Basel • Beijing • Wuhan • Barcelona • Belgrade • Novi Sad • Cluj • Manchester

Editors

Mariella Diaferio	Francisco B. Varona
Polytechnic University of Bari	University of Alicante
Bari	San Vicente del Raspeig
Italy	Spain

Editorial Office
MDPI
St. Alban-Anlage 66
4052 Basel, Switzerland

This is a reprint of articles from the Special Issue published online in the open access journal *Applied Sciences* (ISSN 2076-3417) (available at: https://www.mdpi.com/journal/applsci/special_issues/Sustainable_Future_Concrete).

For citation purposes, cite each article independently as indicated on the article page online and as indicated below:

Lastname, A.A.; Lastname, B.B. Article Title. *Journal Name* **Year**, *Volume Number*, Page Range.

ISBN 978-3-7258-1271-4 (Hbk)
ISBN 978-3-7258-1272-1 (PDF)
doi.org/10.3390/books978-3-7258-1272-1

Contents

Editorial

Concrete Structures: Latest Advances and Prospects for a Sustainable Future

Mariella Diaferio [1,*] **and Francisco B. Varona** [2]

1 Department of Civil, Environmental, Land, Building Engineering and Chemistry, Polytechnic University of Bari, 70125 Bari, Italy
2 Department of Civil Engineering, University of Alicante, 03690 San Vicente del Raspeig, Spain; borja.varona@ua.es
* Correspondence: mariella.diaferio@poliba.it

Citation: Diaferio, M.; Varona, F.B. Concrete Structures: Latest Advances and Prospects for a Sustainable Future. *Appl. Sci.* **2024**, *14*, 3803. https://doi.org/10.3390/app14093803

Received: 18 April 2024
Revised: 23 April 2024
Accepted: 24 April 2024
Published: 29 April 2024

1. Introduction

Along with structural steel, structural concrete is probably one of the most widely used construction materials worldwide for building construction and civil engineering infrastructures. Concrete is manufactured from cement and water and a mixture of fine and coarse aggregates that is suitably adapted in a wide range of regional variations. As a result, concrete does not really depend on strategic materials for its use, which has probably been the main reason for its pre-eminence as a building material in industrialized countries from the last quarter of the 19th century up to the present day. While it offers an attractive set of undoubted advantages, such as the one just mentioned, its essential ingredient, cement, has become its Achilles' heel in the recent two decades. The architecture and engineering sectors must face the challenge of anthropogenic climate change.

Cement is responsible for between 5% and 8% of the annual global anthropogenic CO_2 emissions [1], which are mainly concentrated in the production of clinker. Moreover, from 2015 to 2022, the ratio of tonnes of CO_2 emitted per tonne of cement (global average), far from decreasing, has increased by 7%, standing at a historical record of 0.58 tonnes of CO_2 per tonne of cement [2]. It should not be forgotten that structural concrete needs to work together with steel to configure reinforced concrete and pre-stressed concrete structures. Therefore, the targets of the global Sustainable Development Goals [3] challenge the concrete construction sector to enforce a responsible and rational consumption of cement and steel in order to control their environmental impact as well as to maximize recycling and recirculation possibilities. For obvious reasons, this aim must be achieved without compromising the structural safety or resilience of our infrastructures and urban services and dwellings.

In contrast to what happened with the MDGs (Millennium Development Goals) program [4] that preceded the 2030 Agenda, the research and engineering community today enjoys a scenario of great technological progress in information processing, in the training of database processing algorithms, and in artificial intelligence in general [5]. Likewise, the methods and tools for life cycle assessment (LCA) analysis in building and civil engineering projects now have a much more advanced degree of formalization and implementation than two decades ago. The same is true for topics such as the progressive replacement of Portland cement by other solutions that began to be proposed and researched in the late 1990s, such as geopolymer cements [6,7] and recycled concrete [8,9], the knowledge of which has progressed considerably in the recent 25 years. All this represents a powerful support when it comes to managing the challenges faced by the world of structural concrete for a sustainable future.

As noted in Angst [10], the downside of any subject that has been thoroughly and meticulously researched and developed over so many decades is that concepts and ideas have become deeply ingrained in the minds of those involved. Thus, after 150 years of growth in the concrete sector, there is a risk that the scientific community will show a tendency to take certain aspects of concrete for granted, rather than questioning alternative ideas or the quality of their impact on the sustainable development of society, the economy, and technology. This is the main reason that justifies this Special Issue, entitled "Concrete Structures: Latest Advances and Prospects for a Sustainable Future", that we present here.

2. An Overview of Published Articles

Cement-based materials are the most widely adopted construction materials worldwide [11] and their applications date back many centuries. In contribution 1, the authors propose a deep and detailed survey on the chemistry of cement. The paper highlights the fundamental role of water in the hydration reactions in ordinary Portland cement. The evolution of the chemical composition and the formation of pores are examined. The authors particularly discuss the evolution from plastic paste to the set material during the setting time and the development of the mechanical properties after the hardening step. Water plays an important role in these mechanical properties as it affects the porosity of cement. Moreover, a review of the use of recycled materials in the production of new green cement-based materials is also presented with the aim of summarizing new developments and trends in this construction field, such as the adoption of smart inorganic materials, which can represent a possible tool in the challenge of self-healing and health monitoring of structures.

The design of retrofitting interventions on existing structures firstly requires the acquisition of structural materials' mechanical properties. Among these properties, concrete compressive strength plays a key role in the vulnerability analysis of reinforced concrete (r.c.) structures, but the 'classical' approaches require the execution of destructive tests, which are time-consuming, expensive, invasive, and often difficult to execute for logistical reasons. Alternative approaches make use of non-destructive tests, which are more versatile and quicker than destructive tests; as such, contribution 2 investigates the use of the rebound hammer test. Attention is devoted to the evaluation of the accuracy of the strength assessment related to the choice of the conversion model that correlates the rebound index to the concrete compressive strength. The paper compares several conversion models calibrated by means of a great number of combinations of the data acquired in an experimental investigation on an existing r.c. building. The numerical analysis reveals that the values of coefficients of those analytical models manifest a high variability. The constants of the relationships between the coefficients of these models were established by utilizing regression analysis, and the influence of the coefficients of variation of both concrete strength and rebound index on the results of the calibration procedure were estimated.

Another topic involved in the design of r.c. structures is the realisation of foundations. Among such structural elements, the design of pile foundations, retaining walls, and submerged foundations is investigated in contribution 3. In detail, the paper explores the soil–structure interfacial shear strength and the frictional parameters, which represent key parameters in the design procedure for these foundations. The evaluation of these parameters is investigated considering several types of soils, such as silty sand, medium-coarse sand, clay, and sandstone. In detail, the authors investigate the influence of soil moisture content, normal stress, and interface filling material on the shear parameters of the soil–concrete interface. To this end, large-scale direct shear test apparatus is realised, and different soil–concrete conditions are tested. The results of the experimental campaign highlight that, for stable shear stress, the shear displacement of the sandy soil is smaller than that of clayey soil. Moreover, as the soil moisture content increases, the friction angle of the clayey soil–concrete interface decreases rapidly. The authors show that in the case of medium-coarse sand and the concrete interface, the friction coefficient is greater than that of the silty sand–concrete interface. Moreover, for a fixed moisture content, the authors discuss the friction angle when the soil–concrete interface is filled with a thin layer of sandy soil or silt.

The success of cement-based composites in civil structures is mainly due to their versatility, durability, and cost, even if they require rigorous design procedures to overcome their weak tensile strength and low ductility. In view of improving the mechanical behaviour of these materials, fibre-reinforced cement mortar (FRCM) has been developed by adding chopped fibres in cement composites. In detail, two kinds of FRCMs are available: mono fibre-reinforced cement mortar (MFRCM), which is characterised by the use of only one type of fibre, and hybrid fibre-reinforced cement mortar (HyFRCM), which is obtained by adopting two or more different fibres. These materials are commonly used in during reparations. Consequently, a deep knowledge of their mechanical behaviour represents an important objective in the management of r.c. structures. In contribution 4, the authors study the flexural performance, compressive strength, and impact resistance of both MFRCMs, containing only steel fibre (SF) or carbon fibre (CF), and hybrid fibre-reinforced cement mortar (HyFRCM) containing different combinations of SF and CF. The main aim of the study was to select the optimal fibre combination able to guarantee the required mechanical performance compared with plain mortar. The authors highlighted that the use of MFRCM or HyFRCM improves flexural performance in comparison to plain mortar. Moreover, they estimated how to choose the percentages of SF and CF to acquire higher values of impact resistance or flexural toughness, showing that the adoption of HyFRCM can improve the energy absorption capacity compared with MFRCM.

In recent decades, several multi-storey reinforced concrete buildings have been built with pillars on the ground floor for parking purposes, while the upper floors were reserved for residential use and were realised by inserting unreinforced masonry infills. However, earthquake events highlighted that this configuration is responsible for a different distribution of stiffness along the height of the structure, which is characterised by high stiffness on the upper floors and a "soft-storey" on the ground level. Therefore, during earthquakes, the ground floor columns and/or shear walls are subjected to an increased earthquake demand. However, usually, these structural elements are not designed to withstand such forces; thus, plastic hinges become damaged and in these scenarios, which can lead to a partial or total collapse of the building. This behaviour can also be observed when there is an irregular distribution of infills in plan, which can be responsible for torsional responses and an increase in vulnerability. In contribution 5, the authors investigate a retrofitting intervention aimed to increase the stiffness, shear capacity, and ductility of ground floors in irregular buildings. The intervention involves the realisation of reinforced concrete infills within existing r.c. frames located on the ground level after the execution of r.c. jacketing of the chosen frames. The retrofitting intervention is investigated by performing experimental tests on six 1/3-scaled single-story one-bay r.c. frames subjected to cyclic horizontal forces. The study highlights the key role played by the connections between the r.c. infill and the surrounding jacketed frame realised by means of steel ties. The paper also discusses a numerical procedure for the estimation of the behaviour of retrofitted frames, considering the nonlinearities related to the RC infill, RC frame, steel ties, and mechanisms at the interface. The numerical procedure is also validated by means of the experimental results and may be utilized for design purposes.

In the design of r.c. structures in seismic prone zones, building codes [12–14] introduce a factor whose symbol and name vary for each code but whose introduction is aimed to reduce the full elastic demand in a structure to bring the structure closer to the inelastic range. The evaluation of such a factor represents an important step in the seismic design of structures and conditions as well as energy dissipation requirements. The selection of this factor depends on the main features of the structure and may be quite difficult for buildings with irregularities. On the other hand, recently, due to architectural purposes, several buildings have been built with irregular distributions in their plans and/or elevations. Therefore, the selection of an appropriate reduction factor is quite difficult due to the torsional effects connected to the complexity of a structure. This topic is investigated in contribution 6, which proposed a method for the selection of this factor based on resilience aspects. The method was tested by selecting an appropriate factor value for a high-rise

L-shaped building, considering both unidirectional and bidirectional loading at two design levels. The performed analysis considered several factors, such as building performance level and ductility demand, and its findings can be applied to irregular buildings.

Several structures need to be strengthened for adoption to actual performance requirements. Several approaches have been proposed, but this topic is still highly debated due to the development of new materials and associated design procedures. The actual codes do not provide clear prescriptions for the design of these interventions. This circumstance inevitably affects their wide application in practice. In contribution 7, the authors investigated this research topic by studying the bearing capacity of reinforced concrete beams on shear without internal shear reinforcement. They proposed a method for the design of a strengthening intervention with a fibre-reinforced cementitious matrix (FRCM). The design procedure considered, as a new factor, the loading level at which strengthening occurs. In this regard, the authors discussed the test results of six samples and proposed a procedure for the assessment of a reinforcement system's additional carrying capacity. A new coefficient was introduced, which was able to consider the change in the load level and the related variation in the reinforcement effect. The proposed method has good accuracy and may be easily adopted in practical design.

Contribution 8 investigates the use of a machine learning approach for assessing the subsurface tensile strength of cementitious composites containing waste granite powder. The latest one derived from the crushing or cutting of granite rocks and the use of this material to produce cement composites may contribute to a reduction in the effects of these waste products on the environment and a reduction in the use of cement. However, the calibration of components for obtaining prescribed mechanical properties remains to be investigated. In this field, the availability of non-destructive procedures may represent a powerful tool to predict the effects of components on a product's properties. Indeed, the most commonly adopted procedures make use of destructive tests, which are time-consuming and costly, not only for the execution of these tests but due to the effect they have on the environment. In this view, the adoption of non-destructive tests may represent a good alternative, such as the Schmidt hammer. However, to improve the accuracy of their prediction, artificial neural networks were adopted by the authors. In detail, the authors proposed to combine the results of non-destructive tests with artificial neural networks, allowing them to estimate subsurface tensile strength for different compositions of cementitious composite elements containing granite powder. Different artificial neural networks were tested by the authors, and their performances were compared.

Concrete is one of the most adopted construction materials all over the world due to its several advantages. However, it requires the use of natural resources for its realisation. It has been estimated that the concrete industry produces 14 billion cubic meters of concrete every year [15], to which approximately 8% of the global emissions of carbon dioxide are connected. Thus, a new challenge of the research is the development of strategies capable of reducing the level of harmful emissions for environmental purposes. A promising approach is to replace conventional cement ingredients with waste materials that are also advantageous for reducing waste and improving their management. Moreover, they may also represent a quite promising approach for developing countries, and contributions 9 and 10 deal with this topic.

Contribution 9 discusses a procedure to prepare so-called 'green concrete', i.e., a geopolymer, and cement mortars using mineral wool waste. In the study, durability tests were performed to estimate the relationship between the amount of mineral wool and the flexural and compressive strength of mortars with the aim of determining if the immobilisation of mineral wool in the geopolymer allows to reduce the leaching of phenol and formaldehyde emissions to the environment. Moreover, the authors highlighted that the use of the highest cement-to-wool content ratio increases compressive strength. Moreover, geopolymer mortars show better flexural strength compared to cement mortars. The authors proved that immobilization of the wool in the geopolymer allowed them to significantly reduce phenol and formaldehyde leaching.

Contribution 10 investigated the effects of a partial replacement of ordinary Portland cement with cow dung ash (CDA) in mortar mixes. The authors evaluated the changes on workability, water absorption, bulk density, compressive strength, homogeneity, surface attack resistance, thermal decomposition, and mineralogical composition. The study highlighted that as the percentage of CDA content increased over the 5%, the mortar's workability reduced and affected its density and compressive strength.

An important topic in the field of the maintenance and management of r.c. structures is the knowledge of the effects of some commonly adopted maintenance tasks on the durability of structures and on their safety level. In contribution 11, the authors dealt with this topic and investigated the effects of the use of de-icing salts in bridge decks on the corrosion of steel rebars and on post-tensioning elements. The paper highlighted that in concrete sensitive to an aggregate–alkali reaction, internal stresses may occur, thus causing a high number of cracks. These cracks, in conjunction with the lack of an adequate waterproof treatment, may cause chlorides to enter from de-icing salts. Therefore, the macroscopic appearance of a brittle fracture of transverse reinforcements can be observed. These pathologies may affect the safety level of bridges and the costs of structural maintenance in a bridge deck, along with the use of de-icing salts. The issue is also explored by analysing a bridge, which was repaired in 2020. The execution of high-performance waterproof treatments during the construction phases can play a key role in avoiding the evolution of the previously mentioned pathologies and can reduce their effects on structural safety. Existing structures that are not waterproof are also discussed.

Contemporary architecture is characterised by complex geometrical features, whose realisation requires structural materials able to be adapted to such geometries and with high strengths. Based on these demands, self-compacting high-strength cement-based materials can represent an ideal solution. In contribution 12, this topic is investigated with the aim of identifying models able to optimise the performance of self-compacting mortars. The study focuses on fresh state properties and strength development. The authors used a statistical response surface methodology, which was adopted for the analysis of a database that included a total of thirty formulations of self-compacting mortars with four quantitative input variables. The authors tested the performance of the models by assessing the coefficient of determination related to the D-Flow, T-Funnel, compressive strength development after 24 h, and compressive strength development after 28 h. The estimated coefficients confirmed that the models displayed good performance, proving that this approach has practical potential.

Several authors [16,17] investigated the application of artificial intelligence techniques on the assessment of concrete mixtures' strength, a goal which may have immediate and significant returns in practical applications. In contribution 13, the authors developed a new procedural binary particle swarm optimization algorithm to predict the density and compressive strength values of concrete mixtures. The procedure used within the study outlined various fresh state properties, such as slump, temperature, and grade of cement, of several ready-mix concrete plants to predict the density and compressive strength of concrete mixtures. The results showed that the proposed algorithm achieved better performance in speed and accuracy in comparison to the binary particle swarm algorithm.

Contribution 14 dealt with the assessment of r.c. structures' seismic vulnerability. Several actual building codes [13,18,19] prescribe, among other procedures, the execution of experimental tests to estimate the mechanical properties of concrete and steel, whose number depends on the desired knowledge levels. Three different knowledge levels are proposed by the codes, and three different values of a confidence factor are provided. This factor is utilized to reduce the mean values of experimentally evaluated mechanical strengths, which are then adopted for the numerical assessment of structural safety. Even if the number of prescribed tests is fixed for each knowledge level, different technicians in charge of structural evaluations can make different choices regarding the structural elements that need to be tested for obtaining a prescribed level of knowledge. The authors, using a Monte Carlo approach, numerically simulated the acquisition of experimental strengths related to different choices of test locations, which were then utilized to assess the structural safety level of a reinforced concrete framed structure built in the 1960s in Italy. A huge number of possible choices for each knowledge level were considered. It is worth pointing out that different testing locations may lead researchers to assess different values of mechanical properties, which is based on what safety indexes are obtained. Therefore, the probability distributions of the estimated safety levels were evaluated, and the probability of unsuccessful safety estimations was discussed for the three knowledge levels considered in the Italian technical codes and the Eurocodes.

3. Conclusions

The contributions included in this Special Issue can be classified into four groups. Contributions 1, 9, 10, and 11 deal with advances in the knowledge of cement and concrete chemistry. They delve deeper into their understanding and improve techniques for predicting the working life of infrastructures. Moreover, they incorporate new materials that have a positive impact on sustainable development. Contributions 2, 3, 4, 7, and 14 present experimental studies for characterizing the mechanical behaviour of structural concrete and present quantitative results. Similarly, contributions 5 and 6 also describe experimental and numerical tests for the evaluation of the structural behaviour of concrete, but they focus specifically on the effect of seismic action, with the aim of providing conclusions that will make it possible to achieve resilient building designs that do not compromise the safety of users. Finally, contributions 8, 12, and 13 make use of the latest advances in technology for the treatment of large databases. Additionally, they utilize the application of artificial intelligence algorithms to improve the design of experiments or to optimize concrete dosage parameters and estimate mechanical properties.

To summarize the conclusions drawn from these four groups of papers, we would like to highlight the following:

- Despite being responsible for a remarkable proportion of anthropogenic CO_2 emissions, cement (as a construction material) is backed by a vast knowledge and technology. We cannot ignore the amount of effort that has been put into its research over several decades. This research is essential not only for the design of new and environmentally innovative civil engineering infrastructures, but also for the maintenance and eventual repair of an enormous heritage of buildings, bridges, waterworks, etc., that were built during the 20th century and are still in use. Cement shows great potential and synergy for use with recycled materials for sustainable development. In fact, this Special Issue presents two examples of the use of mineral wool waste as a component for geopolymer and cement-based materials and the use of cow dung ash as an additive to cement. In the first study, resin-impregnated mineral wool fibres were successfully immobilized, significantly reducing formaldehyde leaching into the environment and demonstrating the potential of wool in geopolymer mortars for ecological building materials. In the second example, it was found that while cow dung ash could potentially meet the ASTM requirements for cement, its incorporation affects the workability, water absorption, and compressive strength of mortar mixes. Nevertheless, substituting up to 10% of OPC with CDA shows promise for sustainable mortar production. These research efforts underscore the complexity of building materials and the need for sustainable and environmentally friendly practices in the construction industry. They advocate innovative approaches such as the use of recycled materials, supplementary cementitious materials, and effective preservation strategies to address environmental concerns and ensure the longevity and safety of infrastructure.
- The evaluation of the structural safety of concrete buildings and infrastructure involves a wide range of disciplines and techniques that are constantly evolving. These include an adequate understanding of the soil–foundation interaction, which sometimes requires in-service intervention to improve the interface properties of varying soil moisture contents; the use of fibre-reinforced polymers and fibre-reinforced cementi-

tious matrices for various purposes, such as structural retrofitting of existing structures; the importance of using appropriate assessment analyses to validate the safety indexes of existing buildings, which may require a non-linear approach in combination with a sensitivity analysis to deal with the inherent uncertainties present in the input data; and, finally, the importance of the different locations of non-destructive test results within the same existing structure when using them for structural analysis, with an example that estimates the safety levels in a reinforced concrete moment resisting frame built in the 1960s by means of Monte Carlo simulations, considering different sets of experimental data.

- In the specific case of the seismic assessment of existing reinforced concrete buildings, the papers in this Special Issue demonstrate the power of the numerical analysis tools and methods available today as well as the development of innovative retrofit solutions. On the one hand, structural concrete offers the possibility of seismically retrofitting vulnerable existing structures by using reinforced concrete auxiliary structures and infills to enclose and strengthen them. On the other hand, aesthetic considerations often compromise the seismic safety of new structures, which lose symmetry and may exhibit torsional movements during an earthquake, thus challenging the structural assessment via response reduction factors. This Special Issue contains two papers that successfully address these issues, considering the required resilience of the designs, i.e., whether they are new or retrofitted. However, the papers also note that the standards applied in different countries make it difficult to generalize conclusions due to the different regional parameters that characterize earthquakes.

- Finally, machine learning tools are being applied to classical problems in concrete technology, such as predicting the properties of fresh and hardened conventional and self-compacting concrete mixes as well as the effects of adding waste materials. Artificial intelligence algorithms can be developed (such as the new Procedural Binary Particle Swarm Optimization method presented in one of the papers) and subsequently trained to provide accurate estimates of concrete properties, making them highly useful for their speed, accuracy, and low cost. This pioneering research advances our understanding of statistical methods in the field of civil engineering and construction materials. Artificial intelligence is therefore particularly relevant to the development of advanced, sustainable, and environmentally friendly cement-based products. There are still some limitations that need to be addressed before machine learning tools can be generalized for common practice—such as field tests, experimental verification, more comprehensive sets of mechanical properties—but the successful exploration of the proposed models further clarifies a roadmap for future explorations, with the potential to reduce production costs, limit the environmental impact, and improve the technical performance of mortars and concretes.

Conflicts of Interest: The authors declare no conflicts of interest.

List of Contributions:

1. Lavagna, L.; Nisticò, R. An Insight into the Chemistry of Cement—A Review. *Appl. Sci.* **2023**, *13*, 203. https://doi.org/10.3390/app13010203.
2. Diaferio, M.; Varona, F.B. The Performance of Empirical Laws for Rebound Hammer Tests on Concrete Structures. *Appl. Sci.* **2022**, *12*, 5631. https://doi.org/10.3390/app12115631.
3. Li, D.; Shi, C.; Ruan, H.; Li, B. Shear Characteristics of Soil—Concrete Structure Interaction Interfaces. *Appl. Sci.* **2022**, *12*, 9145. https://doi.org/10.3390/app12189145.
4. Park, J.-G.; Seo, D.-J.; Heo, G.-H. Impact Resistance and Flexural Performance Properties of Hybrid Fiber-Reinforced Cement Mortar Containing Steel and Carbon Fibers. *Appl. Sci.* **2022**, *12*, 9439. https://doi.org/10.3390/app12199439.
5. Manos, G.C.; Katakalos, K.; Soulis, V.; Melidis, L. Earthquake Retrofitting of "Soft-Story" RC Frame Structures with RC Infills. *Appl. Sci.* **2022**, *12*, 11597. https://doi.org/10.3390/app12221 1597.

6.   Prasanth, S.; Ghosh, G.; Gupta, P.K.; Casapulla, C.; Giresini, L. Accounting for Resilience in the Selection of R Factors for a RC Unsymmetrical Building. *Appl. Sci.* **2023**, *13*, 1316. https://doi.org/10.3390/app13031316.

7.   Kos, Z.; Blikharskyi, Z.; Vegera, P.; Grynyova, I. A Calculation Model for Determining the Bearing Capacity of Strengthened Reinforced Concrete Beams on the Shear. *Appl. Sci.* **2023**, *13*, 4658. https://doi.org/10.3390/app13084658.

8.   Czarnecki, S.; Moj, M. Comparative Analyses of Selected Neural Networks for Prediction of Sustainable Cementitious Composite Subsurface Tensile Strength. *Appl. Sci.* **2023**, *13*, 4817. https://doi.org/10.3390/app13084817.

9.   Łaźniewska-Piekarczyk, B.; Smyczek, D.; Czop, M. Comparison of the Effectiveness of Reducing the Leaching of Formaldehyde from Immobilized Wool in Geopolymer and Cement Mortar. *Appl. Sci.* **2023**, *13*, 4895. https://doi.org/10.3390/app13084895.

10.   Worku, M.A.; Taffese, W.Z.; Hailemariam, B.Z.; Yehualaw, M.D. Cow Dung Ash in Mortar: An Experimental Study. *Appl. Sci.* **2023**, *13*, 6218. https://doi.org/10.3390/app13106218.

11.   Ripa Alonso, T.L.; Corral Moraleda, N.; García Alberti, M.; Pavón, R.M.; Gálvez, J.C. The Use of De-Icing Salts in Post-Tensioned Concrete Slabs and Their Effects on the Life of the Structure. *Appl. Sci.* **2023**, *13*, 6961. https://doi.org/10.3390/app13126961.

12.   Rocha, S.; Ascensão, G.; Maia, L. Exploring Design Optimization of Self-Compacting Mortars with Response Surface Methodology. *Appl. Sci.* **2023**, *13*, 10428. https://doi.org/10.3390/app131810428.

13.   Alsaleh, F.; Hammami, M.B.; Wardeh, G.; Al Adday, F. Developing a New Procedural Binary Particle Swarm Optimization Algorithm to Estimate Some Properties of Local Concrete Mixtures. *Appl. Sci.* **2023**, *13*, 10588. https://doi.org/10.3390/app131910588.

14.   Sepe, V.; Diaferio, M.; Caraccio, R. Safety Evaluation of Existing R.C. Buildings: Uncertainties Due to the Location of In Situ Tests. *Appl. Sci.* **2024**, *14*, 2749. https://doi.org/10.3390/app14072749.

References

1.   Cheng, D.; Reiner, D.M.; Yang, F.; Cui, C.; Meng, J.; Shan, Y.; Liu, Y.; Tao, S.; Guan, D. Projecting future carbon emissions from cement production in developing countries. *Nat. Commun.* **2023**, *14*, 8213. [CrossRef] [PubMed]

2.   International Energy Agency (IEA). *Tracking Clean Energy Progress 2023*; International Energy Agency: Paris, France, 2023. Available online: https://www.iea.org/energy-system/industry/cement (accessed on 11 April 2024).

3.   United Nations. *Sustainable Development Goals*; United Nations Organization: New York, NY, USA. Available online: https://www.un.org/sustainabledevelopment/ (accessed on 11 April 2024).

4.   United Nations. *Buildings and Climate Change: Summary for Decision Makers*; United Nations Environment Programme: Nairobi, Kenya, 2009. Available online: https://wedocs.unep.org/20.500.11822/32152 (accessed on 11 April 2024).

5.   Van Damme, H. Concrete material science: Past, present, and future innovations. *Cem. Con. Res.* **2018**, *112*, 5–24. [CrossRef]

6.   Davidovits, E.J. Properties of geopolymer cements. In Proceedings of the 1st International Conference on Alkaline Cements and Concretes, Kiev, Ukraine, 11–14 October 1994; pp. 131–149.

7.   Akbarnezhad, A.; Huan, M.; Mesgari, S.; Castel, A. Recycling of geopolymer concrete. *Cons. Build. Mat.* **2015**, *101*, 152–158. [CrossRef]

8.   Oikonomou, N.D. Recycled Concrete Aggregates. *Cem. Conc. Comp.* **2005**, *27*, 315–318. [CrossRef]

9.   Varona, F.B.; Baeza-Brotons, F.; Tenza-Abril, A.J.; Baeza, F.J.; Bañón, L. Residual Compressive Strength of Recycled Aggregate Concretes after High Temperature Exposure. *Materials* **2020**, *13*, 1981. [CrossRef] [PubMed]

10.   Angst, U.M. Challenges and Opportunities in Corrosion of Steel in Concrete. *Mater. Struct.* **2018**, *51*, 4. [CrossRef]

11.   Gagg, C.R. Cement and Concrete as an Engineering Material: An Historic Appraisal and Case Study Analysis. *Eng. Fail. Anal.* **2014**, *40*, 114–140. [CrossRef]

12.   *ASCE/SEI 7-22*; Minimum Design Loads and Associated Criteria for Buildings and Other Structures. American Society of Civil Engineers: Reston, VA, USA, 2005.

13.   *EN 1998-1*; Eurocode 8: Design of Structures for Earthquake Resistance. BSi: Brussels, Belgium, 2004.

14.   *IS: 1893*; Part-1. Indian Standard Criteria for Earthquake Resistance Design of Structures. Bureau of Indian Standards: New Delhi, India, 2016.

15.   Global Cement and Concrete Association (GCCA). *Global Cement and Concrete Industry Announces Roadmap to Achieve Groundbreaking 'Net Zero' CO2 Emissions by 2050*; Global Cement and Concrete Association (GCCA): London, UK, 2021.

16.   Golafshani, E.M.; Behnood, A.; Arashpour, M. Predicting the compressive strength of eco-friendly and normal concretes using hybridized fuzzy inference system and particle swarm optimization algorithm. *Artif. Intell. Rev.* **2022**, *56*, 7965–7984. [CrossRef]

17.   Lyngdoh, G.A.; Zaki, M.; Krishnan, N.A.; Das, S. Prediction of concrete strengths enabled by missing data imputation and interpretable machine learning. *Cem. Concr. Compos.* **2022**, *128*, 104414. [CrossRef]

18. *ASCE/SEI 41-17*; Seismic Evaluation and Upgrade of Existing Buildings. American Society of Civil Engineers: Reston, VA, USA, 2017.
19. Italian Ministry for Infrastructures and Transportation. *Aggiornamento delle Norme Tecniche per le Costruzioni (Italian Building Code)*; Gazzetta Ufficiale: Rome, Italy, 2018. (In Italian)

applied
sciences

Review

An Insight into the Chemistry of Cement—A Review

Luca Lavagna [1,*] and **Roberto Nisticò** [2,*]

1   Department of Applied Science and Technology, Politecnico di Torino, C.so Duca degli Abruzzi 24, 10129 Torino, Italy
2   Department of Materials Science, University of Milano-Bicocca, INSTM, Via R. Cozzi 55, 20125 Milano, Italy
*   Correspondence: luca.lavagna@polito.it (L.L.); roberto.nistico@unimib.it (R.N.)

Abstract: Even if cement is a well-consolidated material, the chemistry of cement (and the chemistry inside cement) remains very complex and still non-obvious. What is sure is that the hydration mechanism plays a pivotal role in the development of cements with specific final chemical compositions, mechanical properties, and porosities. This document provides a survey of the chemistry behind such inorganic material. The text has been organized into five parts describing: (i) the manufacture process of Portland cement, (ii) the chemical composition and hydration reactions involving a Portland cement, (iii) the mechanisms of setting, (iv) the classification of the different types of porosities available in a cement, with particular attention given to the role of water in driving the formation of pores, and (v) the recent findings on the use of recycled waste materials in cementitious matrices, with a particular focus on the sustainable development of cementitious formulations. From this study, the influence of water on the main relevant chemical transformations occurring in cement clearly emerged, with the formation of specific intermediates/products that might affect the final chemical composition of cements. Within the text, a clear distinction between setting and hardening has been provided. The physical/structural role of water in influencing the porosities in cements has been analyzed, making a correlation between types of bound water and porosities. Lastly, some considerations on the recent trends in the sustainable reuse of waste materials to form "green" cementitious composites has been discussed and future considerations proposed.

Keywords: ceramic materials; composites; inorganic materials; oxides; Portland cement; porous materials

Citation: Lavagna, L.; Nisticò, R. An Insight into the Chemistry of Cement—A Review. *Appl. Sci.* **2023**, *13*, 203. https://doi.org/10.3390/app13010203

Academic Editors: Mariella Diaferio and Francisco B. Varona Moya

Received: 23 November 2022
Revised: 15 December 2022
Accepted: 21 December 2022
Published: 23 December 2022

1. Introduction

Cement is a hydraulic binder; it consists of a finely ground inorganic material which forms a paste when mixed with water, is able to set and harden because of numerous exothermic hydration reactions (and processes), and is thus capable of binding fragments of solid matter to form a compact whole solid [1–3]. After hardening, cement retains its strength and stability, even under the effect of water. Cement forms a composite defined as mortar when mixed with water and fine aggregate (i.e., sand), whereas it forms concrete when mixed with water, sand and gravel (i.e., small stones) [4]. Cement-based materials, such as concrete, have been used for many centuries, mostly in the construction and civil engineering fields, thus becoming the most widely used material, and the second most consumed resource on Earth [5]. Among the different types of cement, ordinary Portland cement (OPC) is the most widely used one [6].

Even if the technology of cement seems quite well established, recent discoveries in nanotechnology and materials science (e.g., the discovery of graphene) opened the possibility of inducing novel smart functionalities in cement and concrete, allowing their use in advanced technological applications. Examples of such systems are self-healing systems [7,8], health-monitoring systems [9,10], conductive materials [11], water permeable materials [12], thermal energy storage materials [13], rubberized concrete [14–17], and sustainable composites containing bio-based fillers [18–20].

From the economic viewpoint, even if cement-based materials are low-cost products (with a maximum cost of 155 EUR ton^{-1}), cements still represent raw materials of remarkable interest for the global market (i.e., cement accounted for 0.074% of total world trade in 2020). Countries leading in the worldwide exportation of cement in 2020 were Vietnam (USD 1.5 billion, covering almost 12% of the global market), Turkey (USD 1.3 billion, 10.3%), United Arab Emirates (USD 578 million, 4.7%), Thailand (USD 574 million, 4.6%), and Germany (USD 528 million, 4.3%) [21]. On the other hand, the largest global importers of cement-based materials are the US (USD 2.3 billion, 10.5%), China (USD 1.2 billion, 9.7%), Bangladesh (USD 513 million, 4.1%), Philippines (USD 464 million, 3.7%), and France (USD 452 million, 3.7%) [21].

This document aims to provide a useful survey of the chemistry behind cement and cement-based materials. Particular attention has been dedicated to the key role played by water in the hydration reactions occurring in OPC, and the different types of compounds generated during this step. In this context, the recent literature [22–41] clearly highlights the pivotal interest still present in the development of new advanced cement and cementitious materials. Even if cement is a well-consolidated material, the chemistry of cement (and the chemistry inside cement) remains very complex and not obvious. Therefore, in this study, the different chemical compounds generated during hydration and setting are clearly investigated, keeping an eye on their mechanisms of formation. Furthermore, the physical and structural role of water in driving the entire process of pores formation has been considered. Lastly, a dedicated section regarding the latest developments in the field of sustainable recycling for the construction industry has been proposed, with a special focus on the use of recycled (waste) materials to produce a new generation of "green" concretes and mortars.

2. The Manufacture of OPC

OPC is a fine grey/white powder consisting of a mixture of calcium silicates, aluminates, and aluminoferrites. These raw materials can be classified into four distinct groups: calcareous, siliceous, argillaceous, and ferriferous. Such raw materials are thermally treated via pyrolysis and mechanically processed in order to obtain the desired product with specific compositions and defined mechanical properties. Cement manufacturing implies four different steps: quarrying, raw material preparation, clinkering, and cement preparation [42]. Figure 1 schematically represents the production line of an OPC [43]. The entire production process begins with the extraction of the previously described raw materials that are rock mixes mainly constituted by ca. 80% of limestone (i.e., primarily calcium carbonate mineral $CaCO_3$) and 20% of either clays or shale (i.e., a source of silica SiO_2, alumina Al_2O_3, and hematite Fe_2O_3) [44].

Once raw materials are extracted, they are pre-crushed inside a quarry. Subsequently, crushed raw material is prepared by applying a variety of blending and sizing operations to confer appropriate chemical and physical properties to the feed. These processing steps can be divided into two different approaches, namely, either dry or wet. Concerning the dry route, both limestone and clays are independently crushed, and then fed together inside a mill. The wet route, instead, requires clays to be mixed forming a paste inside a wash mill (i.e., a tank where clays are grinded in the presence of water), and crushed limestone is added only at the end. Finely ground materials are dried, thermally treated inside a kiln, and then cooled down. This series of steps is defined as clinkering and it confers the main relevant features of the OPC manufacturing. In fact, clinkering reshapes the starting raw materials mixture into clinkers, which are grey, spherically shaped nodules whose diameter is 5–25 mm in size [45]. The chemical reactions occurring inside the kiln, kept heated by the spontaneous ignition of the pulverized coal dusts introduced inside the system due to the high temperatures, is still not completely understood. These chemical reactions are influenced by several factors, such as the large variety in terms of starting chemical compositions, the operating parameters of processing, and the difficulties in performing an efficient in situ sampling under the processing conditions (i.e., high temperatures). However, the chemical reactions and physical processes involved in this step can be rationalized as follow:

1. Evaporation of the physically sorbed water molecules from the raw mix (20–100 °C).
2. Dehydration (100–430 °C) with the production of oxides, such as silica, alumina, and hematite.
3. Calcination (800–1100 °C) with the development of calcium oxide, according to the carbonate decomposition reaction: $CaCO_3 \rightarrow CaO + CO_2$
4. Exothermic reactions (1100–1300 °C) with the formation of secondary silicate phases: $2CaO + SiO_2 \rightarrow 2CaO \cdot SiO_2$
5. Sintering and reactions occurring inside the melt (1300–1450 °C) with conversion of secondary silicate phases into both ternary silicates and tetracalcium aluminoferrites: $2CaO \cdot SiO_2 + CaO \rightarrow 3CaO \cdot SiO_2 \; 3CaO \cdot Al_2O_3 + CaO + Fe_2O_3 \rightarrow 4CaO \cdot Al_2O_3 \cdot Fe_2O_3$
6. Cooling of the system, and the crystallization of the other mineral phases.

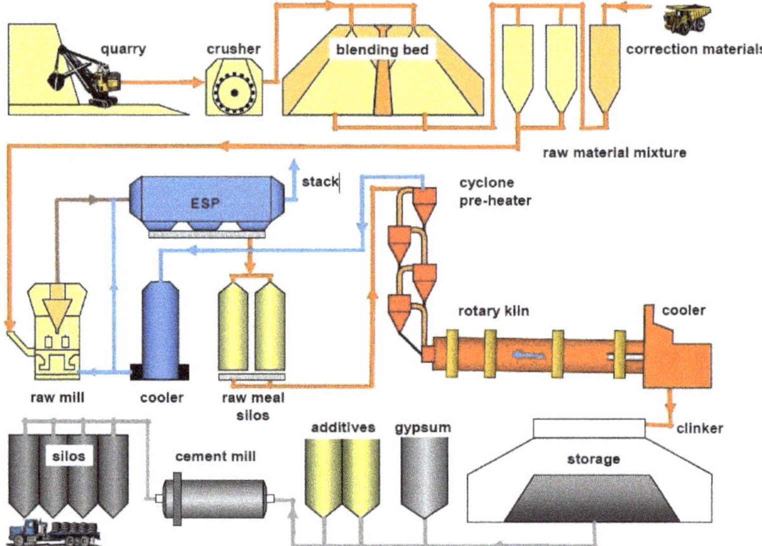

Figure 1. Schematic representation of the process of cement manufacturing. Reprinted with permission from [43].

Finally, the resulting clinker is further united with gypsum and then ground. From the energetic viewpoint, the grinding process consumes ca. 60% of the total electrical energy involved in a standard cement plant (high cost). In a conventional cement making process, the electrical energy consumed is approx. 110 kWh/tons, with ca. 30% used for raw material preparation and ca. 40% for the grinding step. The resulting cement is pneumatically pumped toward specific silos for storage and is subsequently drawn either for packing inside paper bags or for storage inside bulk vessels [46].

3. Hydration Step Involving OPC

Hydration refers to chemical reactions between anhydrous compounds with water molecules, giving a plethora of hydrated compounds. The hydration phenomenon occurring in cement consists of the chemical reaction between non-hydrated cement and water that allows for important physicochemical transformations and mechanical changes. In this context, a distinction should be made between the partial hydration by means of the moisture contained in an air atmosphere, and the hydration achieved by the direct mixing of the cement with water [47]. In this last case (i.e., direct mixing), several chemical reactions take place, inducing the conversion from a workable paste with plastic features to a hard solid.

Furthermore, since the hydration of OPC consists of a multitude of simultaneous single chemical reactions involving the different main components forming the cement powder with water, the resulting pathway is rather complex and non-obvious, and its degree of complexity is strongly influenced by the starting composition of the clinker that directly affects the transformation occurring within the cementitious matrices.

3.1. Typical Starting Composition of Cement and Mineral Phases

The starting composition of clinker is generally indicated by the oxide content (in wt.%). The main relevant oxides forming an OPC clinker are CaO, SiO_2, Al_2O_3 and Fe_2O_3. These oxides form different mineral phases when forming the clinker. Since clinker primarily contains these four chemical species, it is possible to represent these phases with the Bogue formulae, thus indicating CaO with C, Al_2O_3 with A, Fe_2O_3 with F, SiO_2 with S, and water with H [26]. Furthermore, there are also several further minor components deriving from natural clays, e.g., sodium (Na), potassium (K), magnesium (Mg), and other metal ions [48,49]. Table 1 summarizes the chemical composition of OPC, expressing the single component content in wt.%. Additionally, Table 2 describes the approximate composition of the cement clinker [50].

Table 1. Main components of an OPC.

Compounds	Chemical Formula	Bogue Formula	Amount (wt.%)
Alite, or Tricalcium silicate	Ca_3SiO_5 [$3CaO \cdot SiO_2$]	C_3S	30–50
Belite, or Dicalcium silicate	Ca_2SiO_4 [$2CaO \cdot SiO_2$]	C_2S	20–45
Celite, or Tricalcium aluminate	$Ca_3Al_2O_6$ [$3CaO \cdot Al_2O_3$]	C_3A	8–12
Brownmillerite, or Tetracalcium aluminoferrite	$Ca_4Al_2Fe_2O_{10}$ [$4CaO \cdot Al_2O_3 \cdot Fe_2O_3$]	C_4AF	6–10
Gypsum, or Calcium sulphate dihydrated	$CaSO_4 \cdot 2H_2O$	-	4–8
Potassium oxide	K_2O	K	<2
Sodium oxide	Na_2O	N	<2

Table 2. Chemical composition of OPC clinkers.

Component	Amount (wt.%)
CaO	58.0–68.0
SiO_2	16.0–26.0
Al_2O_3	4.0–8.0
Fe_2O_3	2.0–5.0
MgO	1.0–4.0
SO_3	0.1–2.5

3.2. Chemical Reactions Occurring during the Hydration Process

A series of separate/independent (parallel and/or sequential) reactions involving water molecules and the principal mineral phases forming the cement led to the hydration of OPC. In general, the hydration process is strongly driven by the starting chemical composition, the dimensions and size (e.g., specific surface areas, roughness, and particle size distribution), the quantity of water added, the water-to-cement ratio (w/c), the curing temperature, and the presence of additives [51]. In order to simplify the comprehension of the process, the following hydration reactions are discussed separately by considering minerals individually. Figure 2 reports the principal hydrated products in OPC [52].

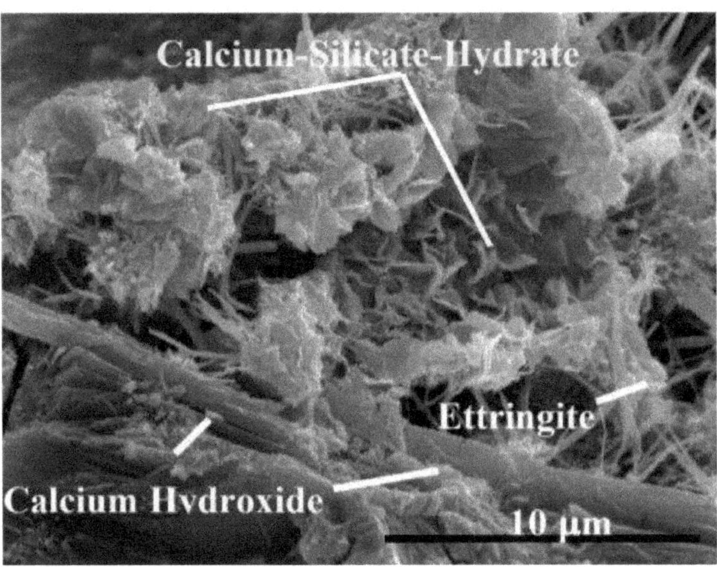

Figure 2. SEM micrograph of the fracture surface of hardened OPC paste after seven days of hydration. Reprinted with permission from [52].

3.2.1. Reaction Involving C_3S

The principal constituent of the OPC is the tricalcium silicate (C_3S, or alite). Alite largely affects the cement mechanical properties (strength) and the hardening process. During the alite hydration, four stages may be distinguished. The first one is the pre-induction period (duration time: a few minutes, just after being exposed to water) where the rapid hydration of C_3S occurs. When C_3S is exposed to water, at the surface level, ions start to a rapidly dissolve, with release of hydroxyl and silicate ions in the liquid phase. As a result, the positive charge of Ca^{2+} is counterbalanced by the negative ones due to silicate and hydroxyl anions. Since the dissolution rate of C_3S is faster than the diffusion rate of ions at the surface level, the liquid phase rapidly turns oversaturated considering the silicate hydrated phase. The direct consequence of this phenomenon is the precipitation of a first layer of C-S-H (i.e., tobermorite) nearby the surface. A high heat of hydration is produced in this first step of reaction [53]. Immediately after, the rate of hydration slows down and then increases, with a second heat release. Finally, the hydration process slows down again, entering the deceleration step [54]. The explanation of this specific behavior is the object of several theories, and some considerations regarding kinetic aspects are reported in the literature [55].

C-S-H is the main relevant binding phase, and is chemically defined as $CaO \cdot SiO_2 \cdot H_2O$. The stoichiometry of C-S-H in the cement paste is variable (and this is particularly true for the bound water). Additionally, a secondary product, calcium hydroxide or portlandite (i.e., $Ca(OH)_2$, or CH), is generated in the case of a CaO/SiO_2 molar ratio which is smaller than the alite one (namely, 3:1).

3.2.2. Reaction Involving C_2S

In analogy with the previously described alite, the hydration of dicalcium silicate (C_2S, or belite) also requires different reaction steps. In general, the induction time is significant, with a low hydration rate. Then, a second slight increment in the hydration rate occurs. Finally, the hydration rate reduces again in the third step. Belite evolves into both C-S-H and CH [56].

3.2.3. Reaction Involving C_3A

Tricalcium aluminate (C_3A, or celite), the most reactive component present in cement, is the third species involved in the hydration mechanism. Its role in the first step of hydration is of paramount relevance as it influences both the mechanical and rheological properties. Furthermore, gypsum can determine the chemistry of the products obtainable through the hydration of celite. Without gypsum, the formation of a gel-like material can be registered at the surface of cement as the first hydration product. Proceeding with the hydration, this material transforms itself into hexagonal crystals of C_2AH_8 and C_4AH_{13} (or C_4AH_{19}). These products convert into cubic C_3AH_6. It is important to note that both C_2AH_8 and C_4AH_{19} are metastable products, whereas the cubic C_3AH_6 is the calcium aluminate hydrate form stable at RT [57]. The fast precipitation of hexagonal platelets causes the phenomenon defined as "flash set". A high temperature of hydration corresponds with a high rate of conversion towards the most stable hydration products. Additionally, at temperatures higher than 80 °C, C_3AH_6 directly forms from the hydration of C_3A. Both C_2AH_8 and C_4AH_{19} are considered as Afm, which indicates a subfamily of mono-hydrated calcium aluminate phases structurally related to hydrocalumite, with the representative formula of $[Ca_2(Al,Fe)(OH)_6] \cdot X \cdot xH_2O$, where X corresponds to an exchangeable anion.

To prevent the fast setting of the paste, and elongate the period of workability, gypsum is usually added to the clinker. With gypsum, the principal hydration phase is ettringite (or $C_6AS_3H_{32}$), which belongs to the Aft group, which indicates a subfamily of three-hydrated calcium aluminate phases with the representative formula $[Ca_3(Al,Fe)(OH)_6] \cdot X_3 \cdot xH_2O$, where X equals a doubly charged anion [58,59]. If sulfate ions are available in the system, the ettringite phase remains stable. When gypsum is fully consumed, ettringite chemically reacts with further C_3A, forming Afm. The gradual consumption of ettringite, instead, favors the growth of hexagonal calcium aluminate hydrate, C_4AH_{13} [60].

3.2.4. Reaction Involving C_4AF

The last component that participates in the hydration reaction is calcium aluminoferrite (C_4AF, or brownmillerite). This compound is the only variable phase present in cement with an unfixed composition. Brownmillerite constitutes a wide category of solid solution $Ca_2(Al_yFe_{(2-y)})O_5$, where the values of y are in the 0.00–1.33 range. Analogous to the previous component C_3A, the hydration of C_4AF is affected by the presence/absence of gypsum. The absence of gypsum favors the reaction of C_4AF with water, forming metastable C(A,F)H hydrates (defined as hydroxy-Afm phase) onto the surface of the non-hydrated grains, and eventually the stable phase $C_3(A,F)H_6$. Experimentally, it has been demonstrated that the Al^{3+}/Fe^{3+} ratio is usually lower in the products formed by the reaction with water with respect to the original C_4AF [61]. The reduction in Al^{3+}/Fe^{3+} in the hydrated products can be clarified by means of the production of a secondary by-product, namely, an amorphous iron-rich gel or iron hydroxide [62].

Conversely, in the presence of gypsum, C_4AF produces hydration phases analogous to the case of C_3A. However, the reaction rates of both components are quite different. In fact, in the presence of gypsum, C_4AF reacts much slower than C_3A, i.e., gypsum retards the hydration of C_4AF more efficiently than in the case of C_3A. However, the rate of hydration of C_4AF is affected by the composition of the ferrite phase: the high amount of Fe corresponds to a low rate of hydration [63]. Therefore, gypsum favors the AFt phase as the main product of reaction, whereas in a successive reaction step, the AFt phase evolves into AFm by reaction with further C_4AF.

3.3. Mechanism and Heat of Hydration of OPC

The kinetics of hydration defines both the cement microstructure and its final properties. The cement–water reaction causes both the dissolution of the anhydrous phases and the precipitation of the hydrated ones. Hence, its evolution with time is subject to the kinetics of dissolution [64], and the growth of the hydrated crystals with their rate of nucleation [65]. Figure 3 reports the kinetics of hydration for an OPC [1]. The hydration process of cement can be organized into four stages: pre-induction, induction (dormant), acceleration, and a post-acceleration period. Every step is discussed in the following paragraphs [66].

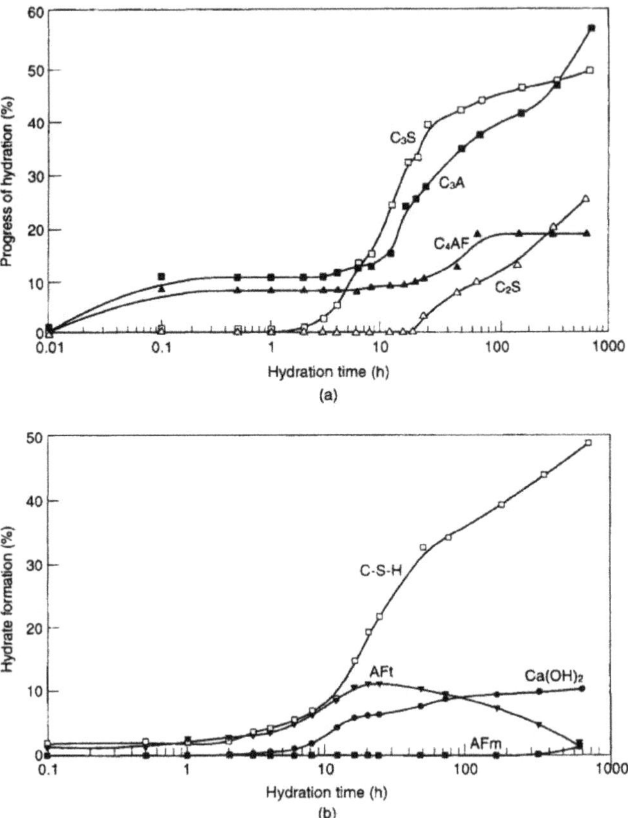

Figure 3. Graph reporting the progress of hydration as a function of the hydration time for the principal components of an OPC: consumption of clinker phases ((**a**), top), and formation of hydrate phases ((**b**), bottom). Reprinted with permission from [1].

3.3.1. Pre-Induction Period

Once cement interacts with water, ionic species rapidly go in the solution allowing the formation of hydrates. From the dissolution of alkali sulphates, there is the release of free potassium, sodium, and sulphate ions. Gypsum dissolves until saturation, releasing Ca^{2+} and SO_4^{2-}.

The literature evidences the creation of a coating growing over the surface of the cement grains immediately after contact with water [67,68]. This product is recognized as being C-S-H. The layer of C-S-H grows, covering most of the grain surface, until the exhaustion of reactants available at the surface. According to Thomas et al. [69], when the content in the C_3S product is higher than the CaO/SiO_2 ratio, during this first step of hydration, there is an increment in the concentration of Ca^{2+} and OH^- in the liquid. Furthermore, adding water to the cement leads to the rapid dissolution of both gypsum and clinker minerals, producing Ca^{2+}, OH^-, SO_4^{2-} ions, and ionic concentration differences through steps in the liquid phase, and causing the supersaturation phenomenon necessary for the formation of the ettringite phase.

Moreover, Jiahui et al. [70] pointed out that the formation of octahedral $[Al(OH)_6]^{3-}$ is the time-determining step in controlling the AFt growth. Furthermore, it has been demonstrated that the ionic concentration affects the kinetic rate of the reaction. Among the different ions forming ettringite, $[AlO_2]^-$ is the one with the lowest concentrated one, and is consequently the limiting species. There are only a small quantity of C_2S hydrates within this first step. Furthermore, the belite (i.e., C_2S) reaction rate is similar, but slower, in respect to that of C_3S, and this is probably attributable to both belite's high thermodynamic stability in respect to alite, and its different crystalline structure. Lastly, C_2S presents high density, and the presence of holes/pores/channels in the structure of C_3S facilitates the reaction with water [71].

3.3.2. Induction (Dormant) Period

After the fast hydration step of cement, the general hydration rate slows down for some hours. The explanation of this sharp slowdown is probably attributable to two main theories, considering both the apparent low reactivity of C_3S and its high solubility calculated from the enthalpy of formation. The first theory is named as the theory of the protective membrane, and it considers the decrease in the hydration rate attributable to the formation of metastable C-S-H layers covering the reacting surface of cement particles, thus preventing the fast dissolution ability of C_3S [72]. Several studies reported in the literature evidenced the tendency of C-S-H to rapidly form once it has interacted with water [73]. In principle, the formation of a low-permeable layer of hydrates, with enough density and coverage, should be able to slow down the ion diffusion rate from the anhydrous cement grains. However, experimental evidence confirms that C-S-H does not form a continuous layer around the grains. In this way, there is still contact with the pore solution, thus weakening this theory [74].

The second theory, the theory of dissolution, instead suggests that the slowdown rate is principally due to the decrease in the kinetic of dissolution of both C-S-H and alite when the system is evolving towards the equilibrium condition. In this context, it should be highlighted that the dissolution of C_3S has been mostly simplistically considered in all models addressing the hydration kinetics of cementitious matrices. This is probably caused by the dissolution step, which in most cases has been studied independently from the other mechanisms that characterize the hydration step. As dissolution occurs together with the precipitation of hydrate phases at the surface, this might interfere with the dissolution process involving the cement components. Hence, this theory remains valid, but the role of the kinetics of dissolution should be highlighted [75].

3.3.3. Acceleration Period

During this step, hydration accelerates once again. The C_3S hydration favors the formation of the "outer" C-S-H, which is different from the "inner" C-S-H. Tennis et al. [76] proposed a model describing the formation of calcium silicate hydrate, making a distinction between low-density (LD) C-S-H ("outer" product), and high-density (HD) C-S-H ("inner" product). Figure 4 shows a C_3S grain after 96 h of hydration, where both C-S-H products are present [77]. In particular, the "outer" C-S-H forms during the early hydration step. It is easily recognizable as it remains away from the cement particle surface, and it is highly porous. On the other hand, the "inner" C-S-H forms during the late hydration step, and it has low porosity. Furthermore, the amount of "inner" C-S-H increases as the w/c ratio decreases [78].

During this step, a noticeable hydration of C_2S is also registered. Moreover, Portlandite (i.e., CH) precipitates from the liquid, thus decreasing the amount of Ca^{2+} ions in the liquid. Gypsum, instead, completely dissolves. Quite surprisingly, the concentration of SO_4^{2-} anions in the liquid decreases, probably because of the occurrence of two phenomena: (i) the formation of the AFt phase, and (ii) the adsorption of SO_4^{2-} ions at the surface of the C-S-H phase.

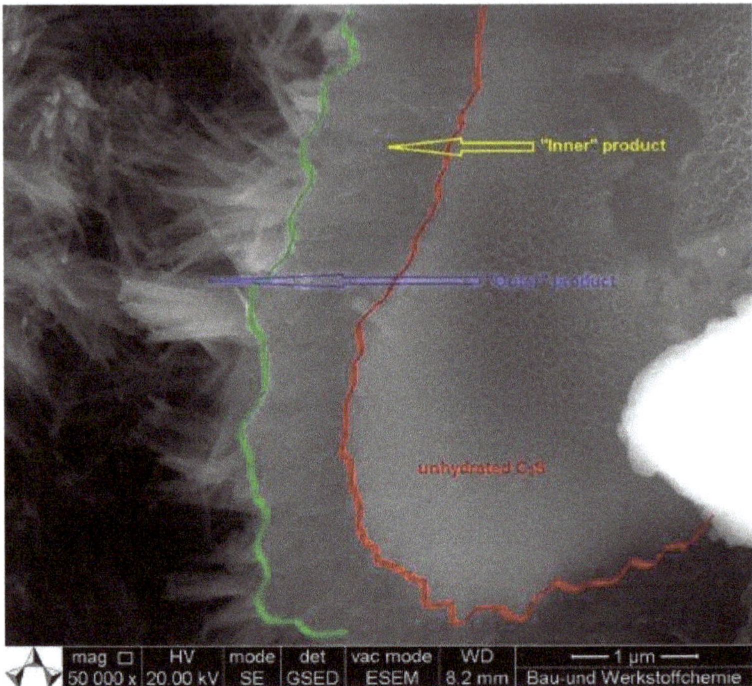

Figure 4. SEM micrograph showing the two types of C-S-H produced during hydration; namely, the "outer" product away from the cement particle surface, occupying the water-filled space and characterized by high porosity, and the "inner" C-S-H formed during later steps of hydration and characterized by low porosity. Reprinted with permission from [77].

3.3.4. Post-Acceleration Period

During this step, the hydration rate decelerates, and the "inner" C-S-H starts forming from hydration reactions involving both C_3S and C_2S. According to Bazzoni et al. [79], the acceleration period involves the nucleation (and growth) of C-S-H clusters onto the surfaces of the cement grains. The deceleration period, instead, involves a substantial decrease in the growth rate of C-S-H that, at this point of the process, covers much of the grains surface. In fact, this final period of hydration involves a low hydration rate, probably caused by the less available space. In particular, the formation of hydrates occurs only in space filled with water. Such a lack of available space is attributable to two phenomena: (i) the depletion of the water volume (necessary for hydrates' precipitation), and (ii) the sequestration of the remaining water into pores whose critical pore sizes are smaller for precipitation [80].

Moreover, together with the formation of "inner" C-S-H, there is a continuous consumption of gypsum, until it is fully consumed. In this way, the concentration of the sulphate ions in the liquid decreases, and the AFt phase (produced during the earlier steps of hydration) begins to chemically react with further C_3A and $C_2(A,F)$, thus forming the AFm phase [81].

3.3.5. Consideration over the Heat of Hydration

Figure 5 reports the heat of hydration of an OPC over the hydration time [1]. The curve profile evidences six different thermal phenomena. The first one is an initial sharp endothermic peak soon after mixing, probably due to the dissolution of potassium sulphate in water (i.e., this contribution is present only in cement containing potassium sulphate, as seen in Figure 5, step 1). The second one is an intense exothermic peak (with a maximum centered in the very first minutes) attributable to the initial hydration reactions involving

C_3S, C_3A, and gypsum (Figure 5, step 2). The third one corresponds to the induction (dormant) period, and it is associated with a minimum of the heat of hydration (Figure 5, step 3). The fourth one is an intense exothermic peak attributable to the hydration of C_3S and the consequent conversion into C-S-H and CH (Figure 5, step 4). The fifth one corresponds to a small descending branch of the principal peak, probably attributable to the AFt formation (Figure 5, step 5), whereas the sixth one corresponds to another small descending branch of the principal peak attributable to the Aft–AFm conversion (Figure 5, step 6) [1].

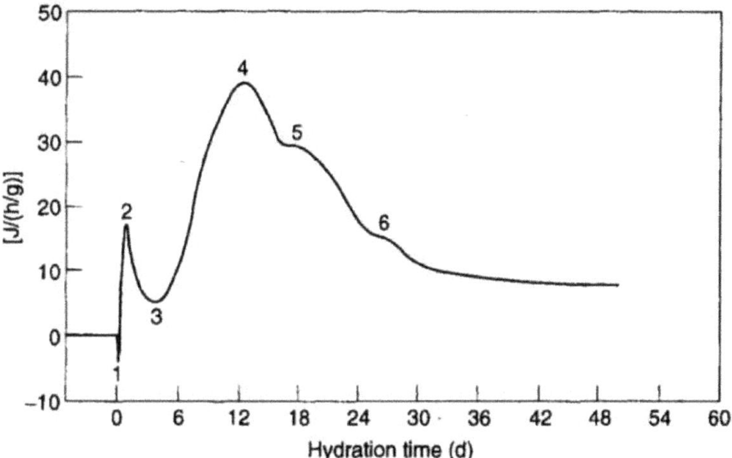

Figure 5. Hydration heat evolution of an OPC vs. hydration time. Step 1: K_2SO_4 dissolution. Step 2: Early-stage period. Step 3: Dormant period. Step 4: Middle-stage period (C-S-H formation). Step 5: AFt formation. Step 6: Aft–AFm conversion. Reprinted with permission from [1].

4. Setting of OPC

The setting time is the elapsed time from the addition of water to a cement mixture until it reaches a specified level of rigidity (measured by a specific technical procedure). Setting consists of the conversion of a plastic cementitious paste into a set material, which is no longer deformable [1]. This conversion is gradual and continuous, and the setting time is usually quantified by measuring the penetration resistance by means of a Vicat needle (i.e., ASTM C191-21) [82]. In general, setting is usually preceded by a stiffening of the paste (with an increment in the material viscosity, even if the mixture does not lose its plastic behavior). Furthermore, it is important to distinguish between the term "setting" and "hardening". Hardening occurs after the setting step, and it refers to an increase in terms of mechanical properties (i.e., Young's modulus, strength, and hardness) until the material reaches the final value of these ones.

4.1. Mechanism of Setting

The mechanism of setting passes through the contact between water and the OPC grains. Quite soon, formed particles of cement undertake flocculation during mixing, thus increasing the viscosity of the cement. After a few minutes of mixing, only coarse (approx. 10 μm) and fine (approx. 3 μm) particles are dispersed in the medium (i.e., water), with the formation of few aggregates of coarse particles, which entrap a fraction of the water. During the induction (dormant) period (i.e., hydration of C_3S), flocculation is reversible, and aggregates are re-dispersed by remixing the paste. Once the acceleration period starts, C-S-H starts to precipitate from the AFt phase, and it is registered as an increment in the fraction of hydrated material and a consequent decrement in the volume of liquid. Thus, particles agglomerate with the fine ones. The principal hydrated compound is C-S-H, which is strongly 3D-connected, and agglomeration cannot be re-dispersed by simply mixing. By continuing with the hydration process, the quantity of products continues growing and the bonding between particles strengthens, thus providing a gradual increment in the strength of the cementitious set paste. Figure 6 reports the mechanism of the flocculation of the cement paste [83,84].

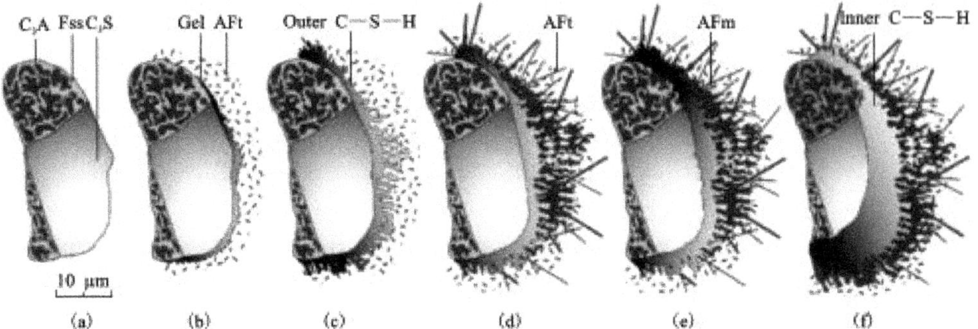

Figure 6. Image showing the mechanism of flocculation: (**a**) unhydrated section of polymineralic grain (scale of interstitial phase is slightly exaggerated), (**b**) after 10 min, (**c**) after 10 h, (**d**) after 18 h, (**e**) after 1–3 days, (**f**) after 14 days [83].

4.2. Flash Setting vs. False Setting

At this point, it is mandatory to clarify the difference existing between flash setting and false setting. "Flash setting" consists of the fast loss of plasticity of the pastes once mixed with water, thus reducing the effective time of workability of the cement. Such a rapid transformation is caused by the increment in the early reactions involving aluminate and ferrite chemical species, with the production of plates of AFm all over the material, and the release of a high heat of hydration [65]. The meshwork of AFm plates, responsible for minimizing the possibilities of remixing, is overcome by the more compact coatings of AFt phases formed in presence of gypsum, which is a set regulator. Gypsum dissolves in water, releasing Ca^{2+} and SO_4^{2-} ions. As previously discussed, the quantity of hydrated C_3A and C_4AF in the pre-induction period is reduced, with the consequent formation of the AFt phase during hydration. This AFt phase precipitates at the surface of the cement grain forming a microcrystalline layer. In this way, the flowability and plasticity of the cement paste is maintained until the formation of the hydrated phases, such as C-S-H that induces the "normal setting".

On the other hand, the term "false setting" refers to the formation of secondary gypsum (i.e., calcium sulphate dehydrated) crystals, in the presence of either a very low amount or an absence of C_3A. In this case, this "false setting" is caused by the interlocking of gypsum crystals, and, simply, intensive mixing can restore the plasticity of the mixture. "False setting" is typical for cements containing high concentrations of either K_2O (i.e., precipitation of $K_2Ca(SO_4)_2$), or C_3A (i.e., excessive formation of the AFt phase).

5. Pore Structure in OPC

5.1. Type of Porosities

In hardened cementitious materials, there are four types of pores, namely: gel pores, capillary pores, hallow-shell pores, and entrapped/entrained air voids [85].

Gel pores (size of approx. a few nm) form within the hydrated phases at the interface between the cement grains with the liquid. Due to their small size and the great affinity between gel surfaces with water molecules, the movement of the liquid water within gel pores does not contribute to the cement paste permeability.

Capillary pores (sized between a few nm to a few microns) form from the spaces in the fresh paste initially filled with water. During hardening, such spaces evolve into either interconnected channels or gel-pore interconnected cavities. Capillary pores are highly irregular, and their presence is due to the volume of water used during the hydration process. Since a high w/c ratio causes high porosity in the hardened paste, capillary pores are typically formed in formulation with w/c ratios larger than 0.42 [86].

Hollow-shell pores (size of several microns) are closed, ink-bottle pores that form within the void spaces at the borders of cement grains as they move back during hydration, whose shape is a relict of the cement particles.

Entrapped air voids form during mixing due to the high viscosity of the paste. These voids are irregular and typically isolated from each other. Entrained air voids, instead, are intentionally formed during the cement mixing. These voids are uniformly distributed, spherical in shape, and not interconnected with each other [86].

5.2. Type of Water

The introduction of water within the initial mixing of the cementitious paste allows for the formation of hardened cement through hydration. Moreover, there are three different categories of water: (i) chemically bound water, (ii) physically bound (gel) water, and (iii) free (evaporable) water. Chemically bound water forms a solid cementitious paste. Gel water, instead, is the water physically bound to C-S-H gel. The sum of chemically and physically bound water provides the minimum amount of water to fully hydrate a given quantity of cement, thus fixing the w/c ratio at 0.42 [86,87]. Lastly, free (evaporable) water is the water contained in the capillary network of pores.

5.3. Bleeding Phenomenon Involcing the Action of Water in Cementitious Paste

Bleeding is a particular type of sedimentation, where a definite volume of mixing water remains separated at the surface of the cementitious matrix. The implication of this is the reduction in the final volume of the hardened paste soon after the placement, resulting in interparticle distance reduction and variation in the effective w/c ratio, which results being smaller than the initial one. The properties of cement paste showing bleeding are defined by the sedimentation rate and the volume of bleeding water, calculated as the difference between the initial mixing water and the water effectively inside the cement. The bleeding-induced effect increases with the w/c ratio, thus for a high w/c ratio, it is mandatory to add anti-settling agents. Examples of anti-settling agents are bentonite, hydro-soluble polymers, and inorganic salts. Such anti-settling agents act in different ways, namely: adsorbing large quantities of water to preserve the homogeneity of the slurry (i.e., bentonite) [88], increasing the cohesiveness between cement hydrates and the viscosity of the medium (i.e., hydro-soluble polymers) [89], or entrapping water within weak-bounded hydroxide structures formed within the slurry volumes (i.e., inorganic salts) [90].

6. Future Perspective in the Use of Recycled (Waste) Materials in Cementitious Matrices

The use of recycled (waste) materials in the construction sector can achieve significant benefits in terms of both environmental preservation (i.e., saving natural resources, reducing the amount of greenhouse gas emissions and energy consumption) and processing costs. Quite recently, several studies report the use of waste materials as a substitute counterpart for the aggregate fraction in the preparations of concrete and mortar, trying to either maintain or (better) improve the final mechanical properties. The use of materials deriving from the demolition of structures (i.e., construction and demolition waste, CDW) is one of the major studied waste substrates that attract the interests of scientific literature. For example, Villoria Sàez et al. [91] proposed to examine and compare CDW generation in all EU member states in correlation with their respective national construction businesses, gross domestic product, and capital, together with an assessment of the policy framework and CDW recovery performance of each member state against the recovery target of the waste framework directive. The results show that Austria, Germany, the Netherlands, Belgium, and France are the highest CDW producers, whereas Croatia, Slovenia, Slovakia, Poland, Portugal, and Spain were the lowest ones. Liikanen et al. [92], instead, evaluated the role played by raw materials for wood-plastic composites (WPC) to achieve the CDW recovery target in Finland. Specifically, the objective of this analysis was to assess the environmental impacts of WPC production using specific CDW fractions (i.e., wood, plastic, gypsum board, and mineral wool) as raw materials, and to compare these impacts with the traditional situation in which these CDW fractions are treated by conventional methods. The results indicate that, compared with the traditional situation, the environmental impacts of CDW management can be reduced when CDW fractions are used in WPC production. Moreover, Coelho et al. [93] evaluated the economic implications of traditional demolition and selective demolition, analyzing a case study in Portugal. Several scenarios are considered, based on possible waste management options, some of which favor selective demolition over a conventional one. By considering the Italian situation, Borghi et al. [94] applied life-cycle assessment (LCA) methodology to assess the environmental performance of construction and demolition materials in the current context of the management of the Lombardy region, identifying critical aspects related to the management system of non-hazardous CDW and possible actions for improvement (e.g., increment and refinement of the quality of the recycled aggregate). Moving away from Europe, Contreras et al. [95] analyzed a case study in Brazil where CDW was used to produce new building materials, replacing the natural aggregate fraction to produce novel (sustainable) bricks showing superior average compression strength than standard bricks. This finding shows that it is possible to produce low-cost bricks with excellent physical properties using CDW as an aggregate and lime/cement as an additive. Ossa et al. [96] conducted a study regarding the use of recycled CDW aggregate to create asphalt mixtures for urban roads. Numerous tests were conducted to evaluate the susceptibility of the asphalt concrete samples to moisture damage and plastic deformation (typical for this specific application). The results indicate that it is possible to use CDW aggregates (up to 20%) to pave urban roads. In a recent study by Coelho et al. [97], the technological, economic, and environmental aspects related to operational CDW recycling facilities that produce medium- to high-quality recycled concrete aggregates have been reported (for details please refers to [97]). Interestingly, Marzouk et al. [98] evaluate the impact of two alternatives for CDW management, namely: either recycling or disposal. The results show that CDW recycling leads to significant reductions in emissions, energy consumption, global warming potential, and conserves landfill space compared to the disposal of waste in landfills. Gálvez-Martos et al. [99] summarize the key principles and best practices for CDW management across the entire construction value chain. Systematic implementation of these best practices could dramatically improve resource efficiency and reduce environmental impacts: reducing waste generation, minimizing transportation impacts and reuse/recycling, improving the quality of secondary materials, and optimizing the environmental performance of treatment methods. Lastly, Jesus et al. [100] studied the behavior of cementitious renderings incorporating very fine recycled aggregates from two types of CDW: recycled concrete aggregate (RCA, smaller one) and mixed recycled aggregate (MRA). Results pointed out that the modified mortars in most of the tests showed superior performance than the reference mortar (without CDW). An interesting and alternative use of recycled waste materials in concrete is reported by Ma et al. [101], who propose a novel route for utilizing recycled coarse aggregate in high-quality recycled manufactured sand, and the use of such high-quality sand to prepare recycled mortar with good mechanical strength and durability, thus further increasing the value of recycling concrete waste.

Furthermore, the use of alternative aggregates, besides those deriving from CDW, is another frequently discussed topic from the literature [102]. Among the possible recycled substrates, plastic waste that would be destined for landfills is attractive not only from the sustainability and economy viewpoint, but also for their very peculiar mechanical properties [103]. In this context, the literature proposes a multitude of different polymeric substrates to be exploited as aggregate fractions, fillers or fibers in the preparation of novel (green) functional concrete, such as polyethylene terephthalate (PET) [104–112], poly vinyl chloride (PVC) [113,114], high density polyethylene (HDPE) [115], shredded and recycled plastic waste [102,116,117], expanded polystyrene foams (EPS) [118,119], polycarbonate [120], or polyurethane foams [121,122]. Even if the use of polymeric materials in concrete often leads to a decrease in the mechanical properties due to several factors, such as the poor adherence between the plastic aggregates and the cementitious matrix, and the granular characteristics of the polymeric waste (which in most cases is not sufficient to achieve the optimum packing leading to the minimum of porosity) [117], the use of plastic waste can also be properly exploited for the introduction of some specific properties, unconventional for ordinary concrete. Currently, there is a growing interest in the use of recycled rubber (mostly derived from tire waste) in concrete as a partial replacement of the aggregate fraction [16,17,123,124]. In fact, the addition of rubber particles into concrete can introduce novel properties in cement, such as a limitation of the water absorption and a consequent improvement against corrosion, a reduction in noise propagation, and a further improvement in fire resistance [125]. Furthermore, in the literature, the use of rubber concrete in non-structural applications is also suggested [14,126]. In this context, an important point that should be considered is that the introduction of rubber fragments as aggregate fraction in cementitious matrices significantly affects the final mechanical properties of the cement/concrete in a proportional way in respect to the rubber content [16,17].

7. Conclusions

Due to their remarkable mechanical properties, cementitious materials are among the most largely exploited substrates used by the construction industry for building cities, and linking them with highways, roads, and bridges. However, even if cement is a well-consolidated material, its chemistry remains very complex and not obvious. In order to try to improve the comprehension of the chemistry of cement, the present document has been organized into five parts.

Part I (Paragraph 2) describes the manufacture process for obtaining ordinary Portland cement (OPC).

Part II (Paragraph 3) provides information on the chemical composition of OPC, which is a mixture of different inorganic oxides (i.e., mainly CaO, SiO_2, Al_2O_3, Fe_2O_3), and deepens the understanding of the hydration mechanisms of every component forming the cement. Particular attention has been dedicated towards the evolution of the chemical composition in the cement system with time, after the reaction with water.

Part III (Paragraph 4) is focused on the mechanism of setting, fundamental for improving structural integrity in the final material.

Part IV (Paragraph 5) deals with the different types of porosities available in a cement. Additionally, a paragraph dedicated to the pivotal role of water in driving the pore formation has been provided.

Part V (Paragraph 6) reports the recent findings of the alternative use of recycled (waste) materials in cementitious matrices, thus highlighting the future trends for the sustainable development of cementitious formulations.

Based on the analysis of the literature, the following important statements can be delivered:

(1) Water plays a fundamental role in the evolution of the cement matrix. Since the hydration process consists of a series of different chemical reactions involving the main components of the cement powder with water, the reaction products deriving from the hydration mechanism strongly affect the final chemical composition of the cement.

(2) During the setting time, there is a conversion from a plastic paste towards a set material, whereas the mechanical properties of the formulation are defined after the hardening step.

(3) It is possible to distinguish between three different types of water: (i) chemically bound water, (ii) physically bound (gel) water, and (iii) free (evaporable) water.

(4) Every type of water influences the porosity of the cement, and consequently the mechanical performance.

(5) It is possible to use recycled (waste) materials and reuse them in cementitious matrices to form sustainable (advanced) cementitious composites.

(6) The recent interest in smart inorganic materials for advanced technological applications opens the possibility of using cement as a matrix for novel nanoscopic composites with improved properties (e.g., self-healing and health-monitoring materials).

Author Contributions: Conceptualization, L.L. and R.N.; writing—original draft preparation, L.L.; writing—review and editing, L.L. and R.N. All authors have read and agreed to the published version of the manuscript.

Funding: This research received no external funding.

Institutional Review Board Statement: Not applicable.

Informed Consent Statement: Not applicable.

Data Availability Statement: Not applicable.

Acknowledgments: Authors would like to thank M. Pavese, and E. Chiavazzo (Polytechnic of Torino, Italy) for fruitful discussions, and D. Burlon, and M.L. Aghemo for their precious help.

Conflicts of Interest: The authors declare no conflict of interest.

References

1. Odler, I. 6-Hydration, Setting and Hardening of Portland Cement. In *Lea's Chemistry of Cement and Concrete*, 4th ed.; Hewlett, P.C., Ed.; Butterworth-Heinemann: Oxford, UK, 1998; pp. 241–297, ISBN 978-0-7506-6256-7.
2. Cadix, A.; James, S. Chapter 5-Cementing Additives. In *Fluid Chemistry, Drilling and Completion*; Wang, Q., Ed.; Gulf Professional Publishing: Dhahram, Saudi Arabia, 2022; pp. 187–254, ISBN 978-0-12-822721-3.
3. Ludwig, H.-M.; Zhang, W. Research Review of Cement Clinker Chemistry. *Cem. Concr. Res.* **2015**, *78*, 24–37. [CrossRef]
4. Krishnya, S.; Elakneswaran, Y.; Yoda, Y. Proposing a Three-Phase Model for Predicting the Mechanical Properties of Mortar and Concrete. *Mater. Today Commun.* **2021**, *29*, 102858. [CrossRef]
5. Gagg, C.R. Cement and Concrete as an Engineering Material: An Historic Appraisal and Case Study Analysis. *Eng. Fail. Anal.* **2014**, *40*, 114–140. [CrossRef]
6. Liu, M.; Xia, Y.; Zhao, Y.; Cao, Z. Immobilization of Cu (II), Ni (II) and Zn (II) in Silica Fume Blended Portland Cement: Role of Silica Fume. *Constr. Build. Mater.* **2022**, *341*, 127772. [CrossRef]
7. Xue, C. Performance and Mechanisms of Stimulated Self-Healing in Cement-Based Composites Exposed to Saline Environments. *Cem. Concr. Compos.* **2022**, *129*, 104470. [CrossRef]
8. Papaioannou, S.; Amenta, M.; Kilikoglou, V.; Gournis, D.; Karatasios, I. Synthesis and Integration of Cement-Based Capsules Modified with Sodium Silicate for Developing Self-Healing Cements. *Constr. Build. Mater.* **2022**, *316*, 125803. [CrossRef]
9. Sun, M.; Liew, R.J.Y.; Zhang, M.-H.; Li, W. Development of Cement-Based Strain Sensor for Health Monitoring of Ultra High Strength Concrete. *Constr. Build. Mater.* **2014**, *65*, 630–637. [CrossRef]
10. Cassese, P.; Rainieri, C.; Occhiuzzi, A. Applications of Cement-Based Smart Composites to Civil Structural Health Monitoring: A Review. *Appl. Sci.* **2021**, *11*, 8530. [CrossRef]
11. Lavagna, L.; Bartoli, M.; Suarez-Riera, D.; Cagliero, D.; Musso, S.; Pavese, M. Oxidation of Carbon Nanotubes for Improving the Mechanical and Electrical Properties of Oil-Well Cement-Based Composites. *ACS Appl. Nano Mater.* **2022**, *5*, 6671–6678. [CrossRef]
12. Jia, L.; Shi, C.; Pan, X.; Zhang, J.; Wu, L. Effects of Inorganic Surface Treatment on Water Permeability of Cement-Based Materials. *Cem. Concr. Compos.* **2016**, *67*, 85–92. [CrossRef]
13. Lavagna, L.; Burlon, D.; Nisticò, R.; Brancato, V.; Frazzica, A.; Pavese, M.; Chiavazzo, E. Cementitious Composite Materials for Thermal Energy Storage Applications: A Preliminary Characterization and Theoretical Analysis. *Sci. Rep.* **2020**, *10*, 12833. [CrossRef] [PubMed]
14. Nisticò, R.; Lavagna, L.; Boot, E.A.; Ivanchenko, P.; Lorusso, M.; Bosia, F.; Pugno, N.M.; D'Angelo, D.; Pavese, M. Improving Rubber Concrete Strength and Toughness by Plasma-induced End-of-life Tire Rubber Surface Modification. *Plasma Process Polym.* **2021**, *18*, 2100081. [CrossRef]
15. Siddika, A.; Mamun, M.A.A.; Alyousef, R.; Amran, Y.H.M.; Aslani, F.; Alabduljabbar, H. Properties and Utilizations of Waste Tire Rubber in Concrete: A Review. *Constr. Build. Mater.* **2019**, *224*, 711–731. [CrossRef]
16. Lavagna, L.; Nisticò, R.; Sarasso, M.; Pavese, M. An Analytical Mini-Review on the Compression Strength of Rubberized Concrete as a Function of the Amount of Recycled Tires Crumb Rubber. *Materials* **2020**, *13*, 1234. [CrossRef] [PubMed]
17. Roychand, R.; Gravina, R.J.; Zhuge, Y.; Ma, X.; Youssf, O.; Mills, J.E. A Comprehensive Review on the Mechanical Properties of Waste Tire Rubber Concrete. *Constr. Build. Mater.* **2020**, *237*, 117651. [CrossRef]
18. Chen, Y.K.; Sun, Y.; Wang, K.Q.; Kuang, W.Y.; Yan, S.R.; Wang, Z.H.; Lee, H.S. Utilization of Bio-Waste Eggshell Powder as a Potential Filler Material for Cement: Analyses of Zeta Potential, Hydration and Sustainability. *Constr. Build. Mater.* **2022**, *325*, 126220. [CrossRef]

19. Nisticò, R.; Lavagna, L.; Versaci, D.; Ivanchenko, P.; Benzi, P. Chitosan and Its Char as Fillers in Cement-Base Composites: A Case Study. *Boletín Soc. Española Cerámica Vidr.* **2020**, *59*, 186–192. [CrossRef]
20. Suarez-Riera, D.; Merlo, A.; Lavagna, L.; Nisticò, R.; Pavese, M. Mechanical Properties of Mortar Containing Recycled Acanthocardia Tuberculata Seashells as Aggregate Partial Replacement. *Boletín Soc. Española Cerámica Vidr.* **2021**, *60*, 206–210. [CrossRef]
21. Cement | OEC. Available online: https://oec.world/en/profile/hs/cement (accessed on 12 December 2022).
22. Xu, Z.; Guo, Z.; Zhao, Y.; Li, S.; Luo, X.; Chen, G.; Liu, C.; Gao, J. Hydration of Blended Cement with High-Volume Slag and Nano-Silica. *J. Build. Eng.* **2023**, *64*, 105657. [CrossRef]
23. Balachandran, C.; Muñoz, J.F.; Peethamparan, S.; Arnold, T.S. Alkali -Silica Reaction and Its Dynamic Relationship with Cement Pore Solution in Highly Reactive Systems. *Constr. Build. Mater.* **2023**, *362*, 129702. [CrossRef]
24. Xia, Y.; Liu, M.; Zhao, Y.; Chi, X.; Guo, J.; Du, D.; Du, J. Hydration Mechanism and Phase Assemblage of Blended Cement with Iron-Rich Sewage Sludge Ash. *J. Build. Eng.* **2023**, *63*, 105579. [CrossRef]
25. Zheng, X.; Liu, K.; Gao, S.; Wang, F.; Wu, Z. Effect of Pozzolanic Reaction of Zeolite on Its Internal Curing Performance in Cement-Based Materials. *J. Build. Eng.* **2023**, *63*, 105503. [CrossRef]
26. Deng, X.; Guo, H.; Tan, H.; Zhang, J.; Zheng, Z.; Li, M.; Chen, P.; He, X.; Yang, J.; Wang, J. Comparison on Early Hydration of Portland Cement and Sulphoaluminate Cement in the Presence of Nano Ettringite. *Constr. Build. Mater.* **2022**, *360*, 129516. [CrossRef]
27. Fang, Z.; Wang, C.; Hu, H.; Zhou, S.; Luo, Y. Effect of Electrical Field on the Stability of Hydration Products of Cement Paste in Different Liquid Media. *Constr. Build. Mater.* **2022**, *359*, 129489. [CrossRef]
28. Hou, J.; He, X.; Ni, X. Hydration Mechanism and Thermodynamic Simulation of Ecological Ternary Cements Containing Phosphogypsum. *Mater. Today Commun.* **2022**, *33*, 104621. [CrossRef]
29. Zhang, Y.; Yang, W.; Wang, Y.; Gong, Z.; Zhang, K. Molecular Dynamics Simulation of C-S-H Corrosion in Chloride Environment. *Mater. Today Commun.* **2022**, *33*, 104568. [CrossRef]
30. Aretxabaleta, X.M.; López-Zorrilla, J.; Labbez, C.; Etxebarria, I.; Manzano, H. A Potential C-S-H Nucleation Mechanism: Atomistic Simulations of the Portlandite to C-S-H Transformation. *Cem. Concr. Res.* **2022**, *162*, 106965. [CrossRef]
31. Yi, Y.; Ma, W.; Sidike, A.; Ma, Z.; Fang, M.; Lin, Y.; Bai, S.; Chen, Y. Synergistic Effect of Hydration and Carbonation of Ladle Furnace Aslag on Cementitious Substances. *Sci. Rep.* **2022**, *12*, 14526. [CrossRef]
32. Zunino, F.; Dhandapani, Y.; Ben Haha, M.; Skibsted, J.; Joseph, S.; Krishnan, S.; Parashar, A.; Juenger, M.C.G.; Hanein, T.; Bernal, S.A.; et al. Hydration and Mixture Design of Calcined Clay Blended Cements: Review by the RILEM TC 282-CCL. *Mater. Struct.* **2022**, *55*, 234. [CrossRef]
33. Maruyama, I.; Sugimoto, H.; Umeki, S.; Kurihara, R. Effect of Fineness of Cement on Drying Shrinkage. *Cem. Concr. Res.* **2022**, *161*, 106961. [CrossRef]
34. Li, H.-W.; Wang, R.; Wei, M.-W.; Lei, N.-Z.; Sun, H.-X.; Fan, J.-J. Mechanical Properties and Hydration Mechanism of High-Volume Ultra-Fine Iron Ore Tailings Cementitious Materials. *Constr. Build. Mater.* **2022**, *353*, 129100. [CrossRef]
35. Ma, Y.; Li, W.; Jin, M.; Liu, J.; Zhang, J.; Huang, J.; Lu, C.; Zeng, H.; Wang, J.; Zhao, H.; et al. Influences of Leaching on the Composition, Structure and Morphology of Calcium Silicate Hydrate (C–S–H) with Different Ca/Si Ratios. *J. Build. Eng.* **2022**, *58*, 105017. [CrossRef]
36. Liu, R.; Fang, B.; Zhang, G.; Guo, J.; Yang, Y. Investigation of Sodium Alginate as a Candidate Retarder of Magnesium Phosphate Cement: Hydration Properties and Its Retarding Mechanism. *Ceram. Int.* **2022**, *48*, 30846–30852. [CrossRef]
37. Wang, X.; Zhao, Q.; Zhang, S.; Lu, N. Pre-Curing Time Effect on Reactive Powder Concrete Impact Resistance. *AIP Adv.* **2022**, *12*, 105025. [CrossRef]
38. Li, B.; Tang, Z.; Huo, B.; Liu, Z.; Cheng, Y.; Ding, B.; Zhang, P. The Early Age Hydration Products and Mechanical Properties of Cement Paste Containing GBFS under Steam Curing Condition. *Buildings* **2022**, *12*, 1746. [CrossRef]
39. Yan, Z.; Zhang, H.; Zhu, Y. Hydration Kinetics of Sulfoaluminate Cement with Different Water/Cement Ratios as Grouting Material Used for Coal Mines. *Mag. Concr. Res.* **2022**, *74*, 1056–1064. [CrossRef]
40. Dong, P.; Allahverdi, A.; Andrei, C.M.; Bassim, N.D. The Effects of Nano-Silica on Early-Age Hydration Reactions of Nano Portland Cement. *Cem. Concr. Compos.* **2022**, *133*, 104698. [CrossRef]
41. Ma, J.; Liu, X.; Yu, Z.; Shi, H.; Wu, Q.; Shen, X. Effects of Limestone Powder on the Early Hydration of Tricalcium Aluminate. *J. Therm. Anal. Calorim.* **2022**, *147*, 10351–10362. [CrossRef]
42. Achternbosch, M.; Bräutigam, K.R.; Hartlieb, N.; Kupsch, C.; Richers, U.; Stemmermann, P.; Gleis, M. *Heavy Metals in Cement and Concrete Resulting from the Co-Incineration of Wastes in Cement Kilns with Regard to the Legitimacy of Waste Utilisation*; Forschungszentrum Karlsruhe GmbH: Karlsruhe, Germany, 2003.
43. de Queiroz Lamas, W.; Palau, J.C.F.; de Camargo, J.R. Waste Materials Co-Processing in Cement Industry: Ecological Efficiency of Waste Reuse. *Renew. Sustain. Energy Rev.* **2013**, *19*, 200–207. [CrossRef]
44. Hwidi, R.S.; Tengku Izhar, T.N.; Mohd Saad, F.N. Characterization of Limestone as Raw Material to Hydrated Lime. *E3S Web Conf.* **2018**, *34*, 02042. [CrossRef]
45. Schumacher, G.; Juniper, L. 15-Coal Utilisation in the Cement and Concrete Industries. In *The Coal Handbook: Towards Cleaner Production*; Osborne, D., Ed.; Woodhead Publishing: Sawston, UK, 2013; Volume 2, pp. 387–426, ISBN 978-1-78242-116-0.
46. Lea, F.M.; Mason, T.O. *Cement-Extraction and Processing*; Encyclopædia Britannica, Inc.: Chicago, IL, USA, 2019.

47. Whittaker, M.; Dubina, E.; Plank, J.; Black, L. The Effects of Prehydration on Cement Performance. In Proceedings of the 30th Cement and Concrete Science Conference, Birmingham, UK, 13–15 September 2010.

48. Nisticò, R. A Comprehensive Study on the Applications of Clays into Advanced Technologies, with a Particular Attention on Biomedicine and Environmental Remediation. *Inorganics* **2022**, *10*, 40. [CrossRef]

49. Nisticò, R. The Importance of Surfaces and Interfaces in Clays for Water Remediation Processes. *Surf. Topogr. Metrol. Prop.* **2018**, *6*, 043001. [CrossRef]

50. Powers, T.C.; Brownyard, T.L. Studies of the Physical Properties of Hardened Portland Cement Paste. *ACI J. Proc.* **1946**, *43*, 249–336. [CrossRef]

51. Alesiani, M.; Pirazzoli, I.; Maraviglia, B. Factors Affecting Early-Age Hydration of Ordinary Portland Cement Studied by NMR: Fineness, Water-to-Cement Ratio and Curing Temperature. *Appl. Magn. Reson.* **2007**, *32*, 385–394. [CrossRef]

52. Soler, J.M. *Thermodynamic Description of the Solubility of C-S-H Gels in Hydrated Portland Cement Literature Review*; Posiva Oy: Helsinki, Finland, 2007. Available online: http://www.posiva.fi/publications/WR2007-88web.pdf (accessed on 12 December 2022).

53. Wang, L.; Yang, H.Q.; Zhou, S.H.; Chen, E.; Tang, S.W. Hydration, Mechanical Property and C-S-H Structure of Early-Strength Low-Heat Cement-Based Materials. *Mater. Lett.* **2018**, *217*, 151–154. [CrossRef]

54. de Jong, J.G.M.; Stein, H.N.; Stevels, J.M. Hydration of Tricalcium Silicate. *J. Appl. Chem.* **2007**, *17*, 246–250. [CrossRef]

55. Cuesta, A.; Zea-Garcia, J.D.; Londono-Zuluaga, D.; De La Torre, A.G.; Santacruz, I.; Vallcorba, O.; Dapiaggi, M.; Sanfélix, S.G.; Aranda, M.A.G. Multiscale Understanding of Tricalcium Silicate Hydration Reactions. *Sci. Rep.* **2018**, *8*, 8544. [CrossRef]

56. Thomas, J.J.; Ghazizadeh, S.; Masoero, E. Kinetic Mechanisms and Activation Energies for Hydration of Standard and Highly Reactive Forms of β-Dicalcium Silicate (C2S). *Cem. Concr. Res.* **2017**, *100*, 322–328. [CrossRef]

57. Park, S.M.; Jang, J.G.; Son, H.M.; Lee, H.K. Stable Conversion of Metastable Hydrates in Calcium Aluminate Cement by Early Carbonation Curing. *J. CO2 Util.* **2017**, *21*, 224–226. [CrossRef]

58. Lea, F.M.; Hewlett, P.C.; Liška, M. (Eds.) *Lea's Chemistry of Cement and Concrete*, 5th ed.; Butterworth-Heinemann: Oxford, UK, 2019; ISBN 978-0-08-100773-0.

59. Taylor, H.F.W.; Famy, C.; Scrivener, K.L. Delayed Ettringite Formation. *Cem. Concr. Res.* **2001**, *31*, 683–693. [CrossRef]

60. Marchon, D.; Flatt, R.J. 8-Mechanisms of Cement Hydration. In *Science and Technology of Concrete Admixtures*; Aïtcin, P.-C., Flatt, R.J., Eds.; Woodhead Publishing: Sawston, UK, 2016; pp. 129–145, ISBN 978-0-08-100693-1.

61. Meller, N.; Hall, C.; Crawshaw, J. ESEM Evidence for Through-Solution Transport during Brownmillerite Hydration. *J. Mater. Sci.* **2004**, *39*, 6611–6614. [CrossRef]

62. Rose, J.; Bénard, A.; El Mrabet, S.; Masion, A.; Moulin, I.; Briois, V.; Olivi, L.; Bottero, J.Y. Evolution of Iron Speciation during Hydration of C4AF. *Waste Manag.* **2006**, *26*, 720–724. [CrossRef]

63. Flick, E.W. *Handbook of Adhesive Raw Materials*; William Andrew: Norwich, NY, USA, 1989; ISBN 0-8155-1185-X.

64. Nicoleau, L.; Schreiner, E.; Nonat, A. Ion-Specific Effects Influencing the Dissolution of Tricalcium Silicate. *Cem. Concr. Res.* **2014**, *59*, 118–138. [CrossRef]

65. Scherer, G.W.; Zhang, J.; Thomas, J.J. Nucleation and Growth Models for Hydration of Cement. *Cem. Concr. Res.* **2012**, *42*, 982–993. [CrossRef]

66. Scrivener, K.L.; Juilland, P.; Monteiro, P.J.M. Advances in Understanding Hydration of Portland Cement. *Cem. Concr. Res.* **2015**, *78*, 38–56. [CrossRef]

67. Double, D.D.; Hellawell, A.; Perry, S.J.; Hirsch, P.B. The Hydration of Portland Cement. *Proc. R. Soc. Lond. A Math. Phys. Sci.* **1978**, *359*, 435–451. [CrossRef]

68. Meredith, P.; Donald, A.M.; Luke, K. Pre-Induction and Induction Hydration of Tricalcium Silicate: An Environmental Scanning Electron Microscopy Study. *J. Mater. Sci.* **1995**, *30*, 1921–1930. [CrossRef]

69. Thomas, N.L.; Double, D.D. Calcium and Silicon Concentrations in Solution during the Early Hydration of Portland Cement and Tricalcium Silicate. *Cem. Concr. Res.* **1981**, *11*, 675–687. [CrossRef]

70. Jiahui, P.; Jianxin, Z.; Jindong, Q. The Mechanism of the Formation and Transformation of Ettringite. *J. Wuhan Univ. Technol. -Mater. Sci. Ed.* **2006**, *21*, 158–161. [CrossRef]

71. El-Didamony, H.; Sharara, A.M.; Helmy, I.M.; Abd El-Aleem, S. Hydration Characteristics of β-C2S in the Presence of Some Accelerators. *Cem. Concr. Res.* **1996**, *26*, 1179–1187. [CrossRef]

72. Jennings, H.M.; Pratt, P.L. An Experimental Argument for the Existence of a Protective Membrane Surrounding Portland Cement during the Induction Period. *Cem. Concr. Res.* **1979**, *9*, 501–506. [CrossRef]

73. Jia, F.; Yao, Y.; Wang, J. Influence and Mechanism Research of Hydration Heat Inhibitor on Low-Heat Portland Cement. *Front. Mater.* **2021**, *8*, 697380. [CrossRef]

74. Ouzia, A.R.C.W.C. *Modeling the Kinetics of the Main Peak and Later Age of Alite Hydration*; EPFL: Lausanne, Switzerland, 2019; p. 288. [CrossRef]

75. Juilland, P.; Nicoleau, L.; Arvidson, R.S.; Gallucci, E. Advances in Dissolution Understanding and Their Implications for Cement Hydration. *RILEM Tech. Lett.* **2017**, *2*, 90–98. [CrossRef]

76. Tennis, P.D.; Jennings, H.M. Model for Two Types of Calcium Silicate Hydrate in the Microstructure of Portland Cement Pastes. *Cem. Concr. Res.* **2000**, *30*, 855–863. [CrossRef]

77. Sakalli, Y.; Trettin, R. Investigation of C3S Hydration by Environmental Scanning Electron Microscope. *J. Microsc.* **2015**, *259*, 53–58. [CrossRef]

78. Zhutovsky, S.; Kovler, K. Chemical Shrinkage of High-Strength/High-Performance Cementitious Materials. *Int. Rev. Chem. Eng.* **2018**, *10*, 110–118. [CrossRef]

79. Bazzoni, A.; Ma, S.; Wang, Q.; Shen, X.; Cantoni, M.; Scrivener, K.L. The Effect of Magnesium and Zinc Ions on the Hydration Kinetics of C3S. *J. Am. Ceram. Soc.* **2014**, *97*, 3684–3693. [CrossRef]

80. Barnes, P.; Bensted, J. *Structure and Performance of Cements*; CRC Press: Boca Raton, FL, USA, 2002; ISBN 0-429-17842-5.

81. Ben Haha, M.; Winnefeld, F.; Pisch, A. Advances in Understanding Ye'elimite-Rich Cements. *Cem. Concr. Res.* **2019**, *123*, 105778. [CrossRef]

82. C01 Committee. *Test Methods for Time of Setting of Hydraulic Cement by Vicat Needle*; ASTM International: West Conshohocken, PA, USA, 2006.

83. Yuan, Q.; Liu, Z.; Zheng, K.; Ma, C. Chapter 2-Inorganic Cementing Materials. In *Civil Engineering Materials*; Yuan, Q., Liu, Z., Zheng, K., Ma, C., Eds.; Elsevier: Amsterdam, The Netherlands, 2021; pp. 17–57, ISBN 978-0-12-822865-4.

84. Taylor, H.F.W. 4 Properties of Portland Clinker and Cement. In *Cement Chemistry*; Thomas Telford Publishing: London, UK, 1997; pp. 89–112.

85. Wang, L.; Jin, M.; Wu, Y.; Zhou, Y.; Tang, S. Hydration, Shrinkage, Pore Structure and Fractal Dimension of Silica Fume Modified Low Heat Portland Cement-Based Materials. *Constr. Build. Mater.* **2021**, *272*, 121952. [CrossRef]

86. Aïtcin, P.-C.; Lessard, J.-M. 8-The Composition and Design of High-Strength Concrete and Ultrahigh-Strength Concrete. In *Developments in the Formulation and Reinforcement of Concrete*, 2nd ed.; Mindess, S., Ed.; Woodhead Publishing: Sawston, UK, 2019; pp. 171–192, ISBN 978-0-08-102616-8.

87. Ley-Hernandez, A.M.; Lapeyre, J.; Cook, R.; Kumar, A.; Feys, D. Elucidating the Effect of Water-To-Cement Ratio on the Hydration Mechanisms of Cement. *ACS Omega* **2018**, *3*, 5092–5105. [CrossRef]

88. Mesboua, N.; Benyounes, K.; Benmounah, A. Study of the Impact of Bentonite on the Physico-Mechanical and Flow Properties of Cement Grout. *Cogent Eng.* **2018**, *5*, 1446252. [CrossRef]

89. Nguyen, D.D.; Devlin, L.P.; Koshy, P.; Sorrell, C.C. Impact of Water-Soluble Cellulose Ethers on Polymer-Modified Mortars. *J. Mater. Sci.* **2014**, *49*, 923–951. [CrossRef]

90. Xie, N.; Dang, Y.; Shi, X. New Insights into How MgCl2 Deteriorates Portland Cement Concrete. *Cem. Concr. Res.* **2019**, *120*, 244–255. [CrossRef]

91. Villoria Sáez, P.; Osmani, M. A Diagnosis of Construction and Demolition Waste Generation and Recovery Practice in the European Union. *J. Clean. Prod.* **2019**, *241*, 118400. [CrossRef]

92. Liikanen, M.; Grönman, K.; Deviatkin, I.; Havukainen, J.; Hyvärinen, M.; Kärki, T.; Varis, J.; Soukka, R.; Horttanainen, M. Construction and Demolition Waste as a Raw Material for Wood Polymer Composites—Assessment of Environmental Impacts. *J. Clean. Prod.* **2019**, *225*, 716–727. [CrossRef]

93. Coelho, A.; de Brito, J. Economic Analysis of Conventional versus Selective Demolition—A Case Study. *Resour. Conserv. Recycl.* **2011**, *55*, 382–392. [CrossRef]

94. Borghi, G.; Pantini, S.; Rigamonti, L. Life Cycle Assessment of Non-Hazardous Construction and Demolition Waste (CDW) Management in Lombardy Region (Italy). *J. Clean. Prod.* **2018**, *184*, 815–825. [CrossRef]

95. Contreras, M.; Teixeira, S.R.; Lucas, M.C.; Lima, L.C.N.; Cardoso, D.S.L.; da Silva, G.A.C.; Gregório, G.C.; de Souza, A.E.; dos Santos, A. Recycling of Construction and Demolition Waste for Producing New Construction Material (Brazil Case-Study). *Constr. Build. Mater.* **2016**, *123*, 594–600. [CrossRef]

96. Ossa, A.; García, J.L.; Botero, E. Use of Recycled Construction and Demolition Waste (CDW) Aggregates: A Sustainable Alternative for the Pavement Construction Industry. *J. Clean. Prod.* **2016**, *135*, 379–386. [CrossRef]

97. Coelho, A.; De Brito, J. 9-Preparation of Concrete Aggregates from Construction and Demolition Waste (CDW). In *Handbook of Recycled Concrete and Demolition Waste*; Pacheco-Torgal, F., Tam, V.W.Y., Labrincha, J.A., Ding, Y., de Brito, J., Eds.; Woodhead Publishing: Sawston, UK, 2013; pp. 210–245, ISBN 978-0-85709-682-1.

98. Marzouk, M.; Azab, S. Environmental and Economic Impact Assessment of Construction and Demolition Waste Disposal Using System Dynamics. *Resour. Conserv. Recycl.* **2014**, *82*, 41–49. [CrossRef]

99. Gálvez-Martos, J.-L.; Styles, D.; Schoenberger, H.; Zeschmar-Lahl, B. Construction and Demolition Waste Best Management Practice in Europe. *Resour. Conserv. Recycl.* **2018**, *136*, 166–178. [CrossRef]

100. Jesus, S.; Maia, C.; Brazão Farinha, C.; de Brito, J.; Veiga, R. Rendering Mortars with Incorporation of Very Fine Aggregates from Construction and Demolition Waste. *Constr. Build. Mater.* **2019**, *229*, 116844. [CrossRef]

101. Ma, Z.; Shen, J.; Wang, C.; Wu, H. Characterization of Sustainable Mortar Containing High-Quality Recycled Manufactured Sand Crushed from Recycled Coarse Aggregate. *Cem. Concr. Compos.* **2022**, *132*, 104629. [CrossRef]

102. Saikia, N.; de Brito, J. Use of Plastic Waste as Aggregate in Cement Mortar and Concrete Preparation: A Review. *Constr. Build. Mater.* **2012**, *34*, 385–401. [CrossRef]

103. El Bitouri, Y.; Perrin, D. Compressive and Flexural Strengths of Mortars Containing ABS and WEEE Based Plastic Aggregates. *Polymers* **2022**, *14*, 3914. [CrossRef]

104. Albano, C.; Camacho, N.; Hernández, M.; Matheus, A.; Gutiérrez, A. Influence of Content and Particle Size of Waste Pet Bottles on Concrete Behavior at Different w/c Ratios. *Waste Manag.* **2009**, *29*, 2707–2716. [CrossRef]

105. Choi, Y.W.; Moon, D.J.; Kim, Y.J.; Lachemi, M. Characteristics of Mortar and Concrete Containing Fine Aggregate Manufactured from Recycled Waste Polyethylene Terephthalate Bottles. *Constr. Build. Mater.* **2009**, *23*, 2829–2835. [CrossRef]
106. Choi, Y.-W.; Moon, D.-J.; Chung, J.-S.; Cho, S.-K. Effects of Waste PET Bottles Aggregate on the Properties of Concrete. *Cem. Concr. Res.* **2005**, *35*, 776–781. [CrossRef]
107. Kim, S.B.; Yi, N.H.; Kim, H.Y.; Kim, J.-H.J.; Song, Y.-C. Material and Structural Performance Evaluation of Recycled PET Fiber Reinforced Concrete. *Cem. Concr. Compos.* **2010**, *32*, 232–240. [CrossRef]
108. Marzouk, O.Y.; Dheilly, R.M.; Queneudec, M. Valorization of Post-Consumer Waste Plastic in Cementitious Concrete Composites. *Waste Manag.* **2007**, *27*, 310–318. [CrossRef]
109. Silva, D.A.; Betioli, A.M.; Gleize, P.J.P.; Roman, H.R.; Gómez, L.A.; Ribeiro, J.L.D. Degradation of Recycled PET Fibers in Portland Cement-Based Materials. *Cem. Concr. Res.* **2005**, *35*, 1741–1746. [CrossRef]
110. Yesilata, B.; Isıker, Y.; Turgut, P. Thermal Insulation Enhancement in Concretes by Adding Waste PET and Rubber Pieces. *Constr. Build. Mater.* **2009**, *23*, 1878–1882. [CrossRef]
111. Akçaözoğlu, S.; Atiş, C.D.; Akçaözoğlu, K. An Investigation on the Use of Shredded Waste PET Bottles as Aggregate in Lightweight Concrete. *Waste Manag.* **2010**, *30*, 285–290. [CrossRef] [PubMed]
112. Nisticò, R. Polyethylene Terephthalate (PET) in the Packaging Industry. *Polym. Test.* **2020**, *90*, 106707. [CrossRef]
113. Kou, S.C.; Lee, G.; Poon, C.S.; Lai, W.L. Properties of Lightweight Aggregate Concrete Prepared with PVC Granules Derived from Scraped PVC Pipes. *Waste Manag.* **2009**, *29*, 621–628. [CrossRef]
114. Merlo, A.; Lavagna, L.; Suarez-Riera, D.; Pavese, M. Mechanical Properties of Mortar Containing Waste Plastic (PVC) as Aggregate Partial Replacement. *Case Stud. Constr. Mater.* **2020**, *13*, e00467. [CrossRef]
115. Naik, T.R.; Singh, S.S.; Huber, C.O.; Brodersen, B.S. Use of Post-Consumer Waste Plastics in Cement-Based Composites. *Cem. Concr. Res.* **1996**, *26*, 1489–1492. [CrossRef]
116. Ismail, Z.Z.; AL-Hashmi, E.A. Use of Waste Plastic in Concrete Mixture as Aggregate Replacement. *Waste Manag.* **2008**, *28*, 2041–2047. [CrossRef]
117. Merlo, A.; Lavagna, L.; Suarez-Riera, D.; Pavese, M. Recycling of WEEE Plastics Waste in Mortar: The Effects on Mechanical Properties. *Recycling* **2021**, *6*, 70. [CrossRef]
118. Kan, A.; Demirboğa, R. A Novel Material for Lightweight Concrete Production. *Cem. Concr. Compos.* **2009**, *31*, 489–495. [CrossRef]
119. Kan, A.; Demirboğa, R. A New Technique of Processing for Waste-Expanded Polystyrene Foams as Aggregates. *J. Mater. Process. Technol.* **2009**, *209*, 2994–3000. [CrossRef]
120. Hannawi, K.; Kamali-Bernard, S.; Prince, W. Physical and Mechanical Properties of Mortars Containing PET and PC Waste Aggregates. *Waste Manag.* **2010**, *30*, 2312–2320. [CrossRef] [PubMed]
121. Mounanga, P.; Gbongbon, W.; Poullain, P.; Turcry, P. Proportioning and Characterization of Lightweight Concrete Mixtures Made with Rigid Polyurethane Foam Wastes. *Cem. Concr. Compos.* **2008**, *30*, 806–814. [CrossRef]
122. Ben Fraj, A.; Kismi, M.; Mounanga, P. Valorization of Coarse Rigid Polyurethane Foam Waste in Lightweight Aggregate Concrete. *Constr. Build. Mater.* **2010**, *24*, 1069–1077. [CrossRef]
123. Sgobba, S.; Borsa, M.; Molfetta, M.; Marano, G.C. Mechanical Performance and Medium-Term Degradation of Rubberised Concrete. *Constr. Build. Mater.* **2015**, *98*, 820–831. [CrossRef]
124. Nacif, G.L.; Panzera, T.H.; Strecker, K.; Christoforo, A.L.; Paine, K. Investigations on Cementitious Composites Based on Rubber Particle Waste Additions. *Mater. Res.* **2012**, *16*, 259–268. [CrossRef]
125. Asdrubali, F.; D'Alessandro, F.; Schiavoni, S.; Baldinelli, G. *Lightweight Screeds Made of Concrete and Recycled Polymers: Acoustic, Thermal, Mechanical and Chemical Characterization*; Forum Acusticum: Aalborg, Denmark, 2011; pp. 821–826.
126. Abaza, O.A.; Shtayeh, S.M. Crumbed Rubber for Non-Structural Portland Cement Concrete Applications. *Eng. Sci.* **2010**, *37*, 214–225.

Article

The Performance of Empirical Laws for Rebound Hammer Tests on Concrete Structures

Mariella Diaferio [1,*] **and Francisco B. Varona** [2]

[1] Department of Civil, Environmental, Land, Building Engineering and Chemistry (DICATECh), Polytechnic University of Bari, 70126 Bari, Italy

[2] Department of Civil Engineering, University of Alicante, 03690 San Vicente del Raspeig, Spain; borja.varona@ua.es

* Correspondence: mariella.diaferio@poliba.it

Abstract: The assessment of concrete compressive strength plays a key role in the analysis of the seismic vulnerability of existing buildings. However, the adoption of classical destructive tests is usually limited by their invasiveness, cost and time needed for the execution. Thus, in order to overcome these limits and allow investigations to be extended to a large number of points, the use of the rebound hammer test is investigated here with a detailed analysis of the effects on the accuracy of the strength assessment related to the choice of the conversion model relating rebound index to compressive strength. The analysis has been performed by comparing several empirical laws calibrated with data acquired in an experimental investigation of an existing concrete building. The relationships between the coefficients of the examined conversion models are then established, with the aim of reducing the unknowns in the calibration procedure. Furthermore, the influence of the coefficients of variation of concrete strength and rebound index on the results of the calibration procedure has been analyzed, thereby supporting the assessment of the accuracy of the concrete strength.

Keywords: rebound hammer test; concrete compressive strength; regression analysis; conversion models

Citation: Diaferio, M.; Varona, F.B. The Performance of Empirical Laws for Rebound Hammer Tests on Concrete Structures. *Appl. Sci.* **2022**, *12*, 5631. https://doi.org/10.3390/app12115631

Academic Editor: Dario De Domenico

Received: 29 April 2022
Accepted: 29 May 2022
Published: 1 June 2022

Publisher's Note: MDPI stays neutral with regard to jurisdictional claims in published maps and institutional affiliations.

1. Introduction

In the last decade, the focus of structural building projects has been on interventions in existing structures rather than on the design and construction of new buildings. Such intervention projects aimed to extend, change and/or adapt the building to new performance requirements. Thus, strengthening and retrofitting activities actually represent an important amount of the activities of the construction sector.

In addition, the occurrence of several seismic events in the last decades has also highlighted the key role played by adequate maintenance of existing construction patrimony, while also drawing attention to the assessment of the seismic vulnerability of such structures and their retrofitting. The approach to these issues is quite different from that of designing new constructions from scratch, and this is also confirmed by the presence of ad hoc parts of the building codes devoted to the interventions on existing buildings, as in the Italian Building Code [1,2], and in the CEN [3].

Regardless of the final objective of the interventions on existing buildings, the first step is the acquisition of an adequate level of knowledge of the structure, which can be conducted through the analysis of available documentation and through a detailed study that involves experimental tests. However, even if thorough documentation is available, the uncertainties associated with the mechanical properties of the materials still remain, because these can be strongly influenced by the construction process itself, degradation that occurred during the service life, etc.

In other cases, uncertainties about the actual strength of the construction materials arise because of additional factors, such as lack of documentation, numerous previous interventions during the service life and environmental events. These uncertainties significantly affect the assessment of the safety level of the structure.

The adoption of in-situ investigations to assess the strength of construction materials is highly recommended, particularly in the case of concrete structures, as in many cases it has been found that design specifications are not reliable for an accurate estimation of their safety level [4–6].

However, direct determination of the material strength requires experimental tests whereby samples of the material are loaded to failure. This approach is especially recommended for reinforced concrete (r.c.) structures because the concrete strength is highly influenced by: the curing phase; the casting procedure, which may produce aggregate segregation and affect homogeneity; subsequent degradation processes depending on environmental exposure conditions; loading conditions and types of stress to which concrete is subjected to; building imperfections; damage, etc. Hence, building codes suggest extracting specimens—typically, cylindrical concrete cores—which are then cut and have their surface prepared for uniaxial compression tests.

The disadvantages of this approach are mainly associated with the characteristics of the test, which requires damaging the structural integrity. Furthermore, before drilling and extracting the concrete cores, it is necessary to use specific monitoring tools to locate the reinforcing bars and avoid their accidental cutting during the core exaction. Finally, it is also necessary to repair the investigated element. In some instances, e.g., in already damaged structures, it may be appropriate to limit the number of tested elements for stability reasons.

Moreover, the difficulty in selecting the structural element to be tested is also connected to the variability of concrete properties in the whole structure, due to the peculiarities of such a construction material, which may potentially have an impact on the accuracy of the experimental measurement of the concrete compressive strength. Indeed, in practical applications, the criteria for choosing the points on which to perform the tests are strictly related to subjective considerations and the experience of the technician, and thus, may represent a limit of the procedure. In [7] the authors highlight that the concrete strength exhibits a wide variability within a single building.

To overcome these limits, non-destructive tests have been proposed that allow the estimation of the concrete compressive strength indirectly, i.e., starting from the measurement of another property that is strictly connected to the targeted strength. These tests also have the advantage of allowing tests to be extended to a large number of elements, as they require limited preparatory work, reduced testing times and less expensive equipment [8–11].

One of the most adopted procedures is the Rebound Hammer (RH) test which proceeds by impacting a spring driven hammer mass on the surface of the investigated r.c. element and measuring the rebound of the hammer mass. Indeed, the rebound depends on the energy absorbed by the impacted material which may be linked to its strength [12].

It is worth noting that to guarantee reliable measurements, the surface of the investigated element must be smooth. Therefore, to this aim, the surface is prepared by means of an abrasive stone in order to minimize the effects of smoothness and carbonation of the impacted surface on the measurements. Moreover, the technician must control that the mass impacts perpendicularly to the surface and, to improve the accuracy, several readings must be acquired by repeating the measure at each point of a grid fixed on the chosen element.

However, as with any non-destructive test, even if the correct procedure is followed, there is still a number of factors affecting the rebound index (see [13–15]).

The effects of variability of the concrete strength on non-destructive tests results have been pointed out for several structural typologies and environmental exposures: in [16] the authors analyzed and discussed the results of an extensive experimental campaign on a viaduct, while in [17] the authors analyzed the data acquired on a huge number of buildings by means of the Sonreb method. In the latter work, the effects on the structural safety related to the assessment of concrete strength are investigated by considering different analytical formulations of the laws that correlated the concrete strength to the results derived from Sonreb tests and by proposing new empirical laws.

In addition, the reliability of the estimated concrete compressive strength depends on the conversion model that correlates the measured rebound index with it.

Consequently, the main challenge is the calibration of the conversion model to estimate the compressive strength of concrete.

As previously cited, many factors influence concrete strength and are mostly related to the execution phase. This variability is well known and accepted in practical applications, as the influence of some factors may be canceled or significantly decreased only by adopting the precast process. Thus, usually, the aim of the tests for the assessment of concrete strength is not to acquire a local evaluation of this parameter, but to establish a value of the concrete strength that may lead to an accurate evaluation of the structural response of the investigated building to any kind of actions. For this purpose, the present study aims to assess this strength value. It should be noted that the aforementioned variability does not raise significant concerns when the sought mechanical property is directly measured, as it occurs when it is possible to drill cores and perform compression tests. However, when the choice is made to proceed by means of rebound hammer tests, this variability has an impact on the measurements, which cannot be precisely quantified. In fact, it has an effect on the variability of the coefficients of the conversion model that correlates the rebound index to the concrete strength. Therefore, a topic is related to the analysis of the variation of the coefficients of the conversion model and, thus, the accuracy of the assessed strength.

Many Authors have underlined that in practical applications the concrete strength varies due to uneven maturation of concrete, effects of corrosion of the steel bars, and vibration puddling, which may be responsible for non-uniform distribution of the aggregates, and temperature during the curing phase, etc. To limit the influence of these factors in the following analysis, it has been chosen to perform the analysis on a single building.

Another topic is the choice of the analytical formulation that correlates the concrete strength and rebound index and, in detail, if by varying the adopted formulation, the range of variability of the corresponding coefficients varies. Indeed, the influencing factors on the rebound index make it quite difficult to define a conversion model which can be applied to all testing conditions, as has been pointed out by several authors [9,15].

This paper deals with these topics by analyzing the data acquired during an experimental campaign in an existing building on which both compression tests on drilled cores and rebound hammer tests have been performed. The acquired data have been assembled in a huge number of combinations, as will be described later in Section 2, in this way the data refer to the same building, have been measured in the same experimental conditions and with the same instruments and technicians; therefore, the effects of some factors, such as age, humidity, etc., may be considered equal for all the considered combinations, thus, the variations of coefficients of the conversions models are only related to other factors, i.e., those that usually affect the variability of the concrete strength and the structural response. The main aim is to verify the variability of these coefficients of the conversion model for different choices of its analytical formulation and to establish a relationship that correlates these coefficients. To this end, all the collected combinations of data have been utilized to calibrate conversion models, in this way, even if the data have been acquired for the same structure, it is verified that if the number and location of the investigated elements change, also the coefficients of the conversion model vary even if they are related to the same structure. In the present paper, several conversion models are calibrated and empirical regression laws between the coefficients of the conversion models are determined. In detail, the relationships between the coefficients of the conversion model have been established by varying the chosen analytical formulation, the measurement positions, the number of points to be tested, the coefficients of variation of the concrete strength and the rebound index. The results show that for all the investigated conversion models, a different law between their coefficients can be evaluated, which is strictly linked to the adopted conversion model and is independent of the coefficients of variation of the concrete strength and the rebound index.

2. Experimental Data

In the present paper, the assessment of the in situ compressive strengths of the concrete structural elements of an existing building is discussed. In [7] the authors showed that the concrete strength may alter from one floor level to another and/or even from one structural element to another of the same typology within the same structure. In the case of using conventional destructive testing to evaluate existing conventional buildings, the extraction of concrete cores and the subsequent calibration of empirical models may be difficult because, on the one hand, a high value for the coefficient of variation of the concrete strength (Cov_{fc}) may be recorded, and on the other hand, extending the destructive tests to a sufficient number of locations may be difficult or even impossible due to economic and logistic reasons. Therefore, to acquire detailed information on the structure, the use of non-destructive tests (NDTs) is quite useful.

This is also confirmed by the experimental tests here investigated which have been conducted on a building located in Bari (Italy). The building was built in the 1970s and it is composed of five floors. The experimental tests have been performed on some walls and repeated on the same walls on each floor; the walls have been chosen, as often happens in practical applications, based on some logistic constraints mainly due to the possibility of drilling a core on each investigated wall. Although the transversal section of the tested walls varies on each floor, there is one dimension that remains constant and equal at 35 cm, which ultimately determined the direction of the performed tests. The structure does not manifest cracks or damage.

The rebound tests have been performed by means of a Schmidt concrete test hammer type N, oriented orthogonally to the investigated elements and thus laying in a horizontal direction, in accordance with the manufacturer's instructions. The testing method consists in provoking the impact of a standardized mass against the surface of the structural element under testing and in measuring the height of the rebound, the measurement is expressed in terms of the percentage of the rebound height compared to the distance traveled by the moving mass between the instant it is released and when it hits the concrete surface, this percentage is called the rebound index and is measured on a scale numbered from 10 to 100. It is worth noting that the presence of internal voids, cracks, aggregates, etc., in correspondence with the point on which the hammer is positioned for the execution of the test, may be responsible for the overestimation or underestimation of the concrete strength. Therefore, the rebound index related to each investigated element has been evaluated as the mean value of the measurements acquired in the nine points of a grid fixed upon the element's surface, in this way the impact of possible "local" conditions on the assessed strength may be attenuated. The RH tests have been performed in a single day, to reduce the influence of environmental conditions on the acquired measurements. The surface had been rubbed off with an abrasive stone before testing, in order to guarantee the smoothness of the test surface and to minimize the influence of carbonatation. Moreover, the points of impact have been chosen as far as possible from edges and corners, and the tests have been performed in dry-air conditions inside the building.

After the rebound hammer tests, a core has been drilled on each investigated element, and subjected in a laboratory to compression tests, in accordance with prescriptions of [18,19]. In detail, the cylindrical cores have been extracted with a diameter of 100 mm and a ratio height/diameter equal to 2; afterward, the load-bearing surfaces of the cores have been prepared to improve the contact with the loading machine and ensure uniaxial compression conditions. In detail, 28 cores have been drilled and subjected to compression laboratory tests. In Section 3 the calibration procedure of the conversion models based on the cited experimental data is described.

The mean value of the compression strength evaluated on the 28 drilled cores is equal to 25.26 MPa, the standard deviation is 3.4 MPa, thus, the coefficient of variation of the core compressive strength is equal to 0.13.

Regarding the RH tests, as has been previously described, the rebound index has been measured on each one of the 28 investigated walls. It has been evaluated that the mean value of the 28 measured rebound indexes is equal to 42.55 and the standard deviation is 1.9, thus, the coefficient of variation of the rebound index is equal to 0.04, showing that the dispersion of the acquired rebound indexes around the mean value is quite low compared to that of the core strength. The coefficient of variation and the standard deviation of the rebound index, which are representative of the repeatability of the test results, exhibit rather low values compared to the usual accepted ones (see [9,20]) and make the author confident about the representativeness of the acquired rebound values.

Indeed, if a Gaussian distribution is assumed for the rebound measurements and the adopted values are assumed equal to the average value of the nine readings, the obtained value of the coefficient of variation [9] enables the true measurement to be estimated with an accuracy level lower than 6% and with a 95% confidence level.

The high value of the coefficient of variation of core strength highlights the high variability of the measurements. Based on this observation, it is possible to create a database of several possible combinations of the available data, for each of which a different value of the coefficient of variation of concrete strength would be expected. This approach enables the impact of the chosen points on the calibration procedure of the conversion model for the same building to be investigated. It would also shed light on the advantages and limitations of the RH tests with reference to ordinary practical applications, where the number of extracted cores may be highly limited by economic and logistic reasons. The objective is to verify the influence of the properties of the dataset on the calibration procedure also varying the chosen conversion model.

Thus, the datasets here are considered to have been obtained by collecting a huge number of combinations of the available data, which have been obtained by varying the number of experimental results that compose the dataset considered in the calibration procedure.

It is worth noting that the core strength f_{core} must be converted in the in-situ concrete strength f_c in order to take into account the influence of the height/diameter ratio of the core. For the present investigations, the ratio is equal to 2, the core diameter is 100 mm and a correction term, which considers the damage due to the drilling procedure, in accordance with [21] and the in-situ concrete strength, is assessed in accordance with the following relationship:

$$f_c = 1.06\, f_{core} \tag{1}$$

3. Conversion Models and Calibration Procedures

The reliability of the assessed concrete strength depends on the assumption of a relationship between the acquired rebound number and the actual concrete strength. The available codes [12] suggest performing a calibration procedure of the conversion model that correlates the RH measurement with the strength of concrete cores extracted at the same position of the RH tests.

Many empirical formulations have been proposed which make use of several models and a high number of coefficients for the same model. The main empirical models are the power law [9]:

$$f_c = a_r\, R^{b_r} \tag{2}$$

the linear law [22–24]:

$$f_c = b_l\, R + a_l \tag{3}$$

the exponential law [9]:

$$f_c = a_e\, e^{b_e R} \tag{4}$$

and the polynomial law [25–27]:

$$f_c = a_p R^2 + b_p R + c_p \tag{5}$$

Several authors proposed favorable values for the coefficients of Equations (2)–(5) based on the results of experimental studies, which were performed in many cases in laboratory conditions or in existing buildings. However, the proposed values arguably vary in wide ranges and a conversion model suitable for any kind of application has not yet been successfully established.

Indeed, to improve the accuracy of the adopted model, a calibration procedure should be performed by minimizing the estimation error of the concrete strength on a set of in-situ strength data acquired by means of compression tests on drilled cores $f_{c,\,s}$. In the present paper, the laws of Equations (3)–(5) are investigated.

In detail, the coefficients of the linear empirical law are obtained by minimizing the following mean square error:

$$\epsilon_l = \frac{\sum_{i=1}^{m} [Ln(f_{c,s_i}) - (b_l Ln(R_i) + a_l)]^2}{m} \tag{6}$$

where f_{c,s_i} is the in-situ strength evaluated on the basis of the strength of the i-th core, R_i is the rebound index registered at the same position of the i-th core and m is the number of drilled cores considered in the calibration procedure.

The exponential law may be written using logarithmic notation as follows:

$$Ln\ (f_c) = Ln(a_e) + b_e\ R_i \tag{7}$$

where the coefficients are calibrated based on the minimization procedure of the following mean square error:

$$\epsilon_e = \frac{\sum_{i=1}^{m} [Ln(f_{c,s_i}) - (Ln(a_e) + b_e R_i)]^2}{m} \tag{8}$$

In the case of the polynomial law the mean square error can be written as:

$$\epsilon_p = \frac{\sum_{i=1}^{m} [Ln(f_{c,s_i}) - (a_p R_i^2 + b_p R_i + c_p)]^2}{m} \tag{9}$$

The influencing factors on the rebound index make it quite difficult to define a conversion model which can be applied to all testing conditions, as has been reported by several authors [9,15]. The present work aims to highlight that, even if the same experimental conditions are ensured, the effects of some influencing factors are minimized and variability in the coefficients of conversion models still remains, which can be predicted by means of ad hoc empirical laws that have been proposed for different analytical formulations of the conversion model.

In the experimental campaign examined here, a significant number of cores have been extracted with the final aim of acquiring a deep knowledge of the concrete compressive strength. However, in the present paper, in order to consider the general practical conditions, which are usually characterized by a limited number of drilled cores due to logistic and economic reasons, the available N data have not been utilized to calibrate a single empirical conversion model. The data have been rearranged in order to build several datasets of possible results of the experimental investigations, which have been utilized to calibrate a huge number of conversion models.

In detail, the data acquired during the experimental tests have been utilized to calibrate the coefficients of three different analytical formulations, i.e., the linear law, the exponential law and the polynomial law. The data have been arranged in pairs of values: the first referring to in-situ strength, evaluated according to Equation (1) by inserting the value of core strength for the considered wall, and the second one related to the rebound index measured for the same wall on which the core was drilled. These 28 data have been combined to define a wide database, which was built as follows: firstly, the number m ($m < 28$) of pairs of values was set and several combinations of the available 28 pairs of values are defined. In this way, each combination corresponds to a new possible experimental setup that does not consider all the effective investigated walls. Each defined combination is considered as the acquired data and utilized to calibrate the three laws in Equations (3)–(5), by minimizing Equations (6), (8) and (9), respectively.

To take into consideration the accuracy of each conversion model, the compressive strength derived through the conversion model and the in-situ compressive strength based on the destructive tests of the extracted cores (see Equation (1)) have been compared and the Root Mean Square Error (RMSE) has been evaluated for each conversion model as follows:

$$RMSE = \sqrt{\frac{\sum_{i=1}^{N}\left(f_{c,s_i} - f_{c,emp_i}\right)^2}{N}} \tag{10}$$

where f_{c,emp_i} is the concrete strength related to the i-th investigated point of the structure and assessed by means of the chosen empirical law.

4. Linear Law

Based on the procedure described in Sections 2 and 3, more than 10^7 linear conversion models have been calibrated by varying the number m and, thus, the combinations of data considered in the calibration process. It is worth noting that the coefficient of variation related both to the concrete strength and the rebound index, varies from each combination to another even if they are acquired on the same structure.

The main objective is to study the variation of the coefficients of the linear empirical law. Thus, in Figure 1 the terms b_l and a_l are plotted for several values of the number m of data utilized for the calibration. Each point of the plots corresponds to a different combination of data and, thus, to a different linear model. It is worth mentioning that the values of the coefficients vary significantly. However, a clear relationship can be observed between such coefficients, which have been evaluated by means of a linear regression analysis. The results of such analyses have been reported in each plot with the related coefficient of determination. The values are similar and do not depend on the number m of considered data. Moreover, considering other values of m—whose results have not been plotted for the sake of brevity—it is possible to recognize the following relationships between the coefficients:

$$b_l = -0.0237a_l + 0.5920 \tag{11}$$

With a coefficient of determination equal to 0.9977, the excellent accuracy of the linear relationship in fitting the numerical values is shown.

In Figure 2 the relationship between the coefficients of the linear models has been plotted, also considering the values of the coefficient of variation of the concrete strength and of the rebound index evaluated for each combination of data composed by $m = 15$ measurements. The plots show that for increasing values of the coefficient of variation of the concrete strength, the values of b_l tend to increase while the values of a_l decrease, but the relationship between such coefficients remains almost the same. Comparable behavior is observed for increasing values of the coefficient of variation of the rebound index, but the latter coefficient is characterized by a very low variation. The described trends are also detected for other values of m but are not plotted for the sake of brevity.

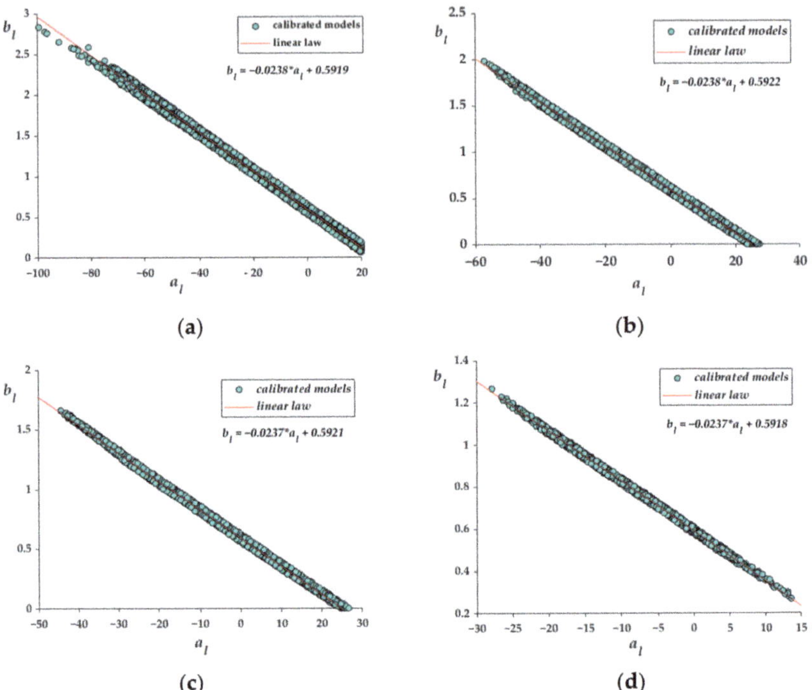

Figure 1. Relationship between the coefficients of the calibrated linear empirical laws b_l vs. a_l for different value of m considered data: (**a**) $m = 11$; (**b**) $m = 15$; (**c**) $m = 18$; (**d**) $m = 22$.

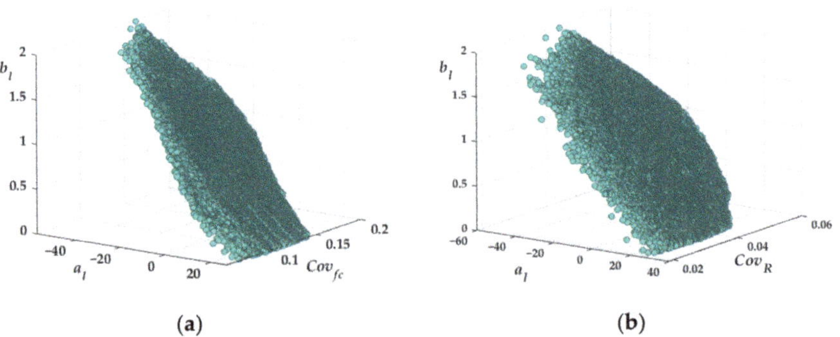

Figure 2. Calibration of the empirical model based on the data acquired in $m = 15$ points of the investigated structure: (**a**) Plot of the coefficients of the calibrated linear empirical laws and of the coefficient of variation of the concrete strength; (**b**) Plot of the coefficients of the calibrated linear empirical laws and of the coefficient of variation of the rebound index.

Figure 3 shows the variation of the RMSE with respect to the coefficient of variation of the concrete strength and of the rebound index for two different values of m. It is observed that even for a higher given value of m, the range of variability of RMSE decreased, and it would not be concluded for the examined case that a direct relationship exists between the RMSE and the considered coefficients of variation.

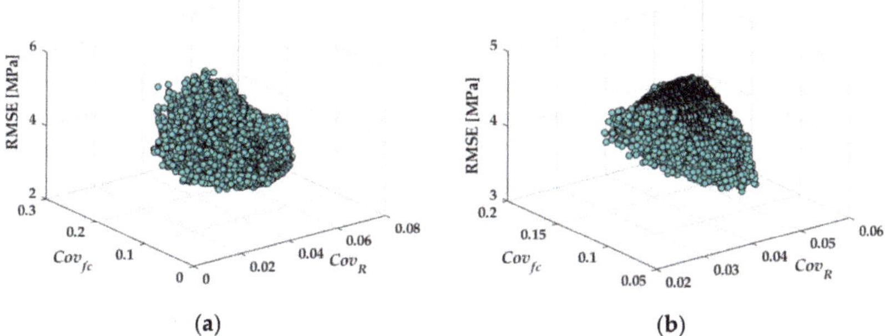

(a)

(b)

Figure 3. RMSE (Equation (10)) evaluated for the linear conversion models vs. coefficient of variation of the in-situ concrete strength Cov_{fc} and coefficient of variation of the rebound index Cov_R: (**a**) $m = 11$; (**b**) $m = 18$.

5. Exponential Law

A total number of more than 10^6 datasets have been defined and, consequently, the same number of exponential conversion models (see Equation (4)) have been established by varying the number m of the considered data. In Figure 4 the evaluated terms of the empirical exponential model for several values of the number m are plotted. The analysis of the plot would show that a logarithmic relationship might exist between the terms b_e and a_e, which has been evaluated by carrying out a regression analysis. This analysis has been conducted for each value of m ranging between 3 and 24, thus, 21 laws have been evaluated. It is found that the terms of such laws are similar; hence, the following relationship has been obtained:

$$b_e = -0.0237\, Ln(a_e) + 0.07605,$$ (12)

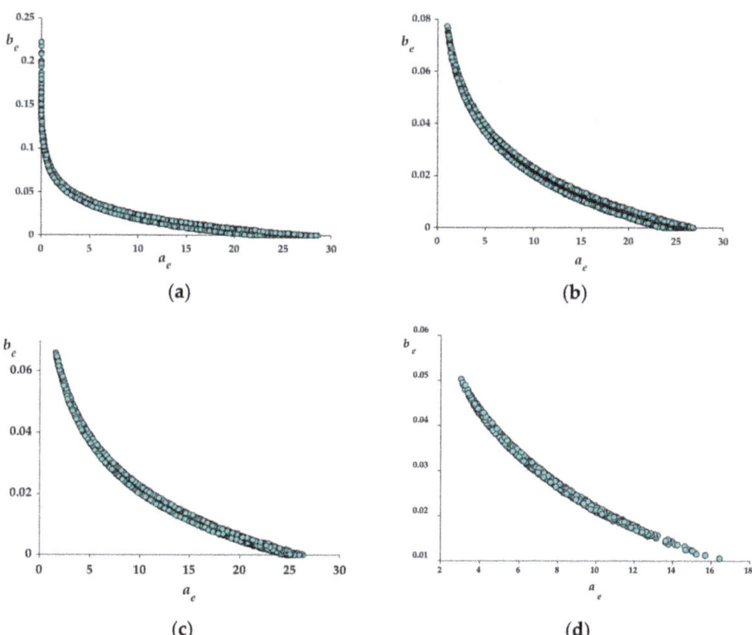

(a)

(b)

(c)

(d)

Figure 4. Relationship between the coefficients of the calibrated exponential empirical laws b_e vs. a_e: (**a**) $m = 8$; (**b**) $m = 15$; (**c**) $m = 18$; (**d**) $m = 22$.

With a coefficient of determination equal to 0.9976. In detail, the first coefficient of Equation (12) shows a maximum variation of 0.2%, while the second coefficient has a maximum variation of 0.14%.

The datasets have been analyzed to ascertain the influence of the coefficient of variation related to the concrete strength Cov_{fc}, on the relationship between the coefficients of the exponential law. Figure 5 shows that the relationship between the coefficients of the exponential law does not change significantly by varying the coefficient of variation of the concrete strength.

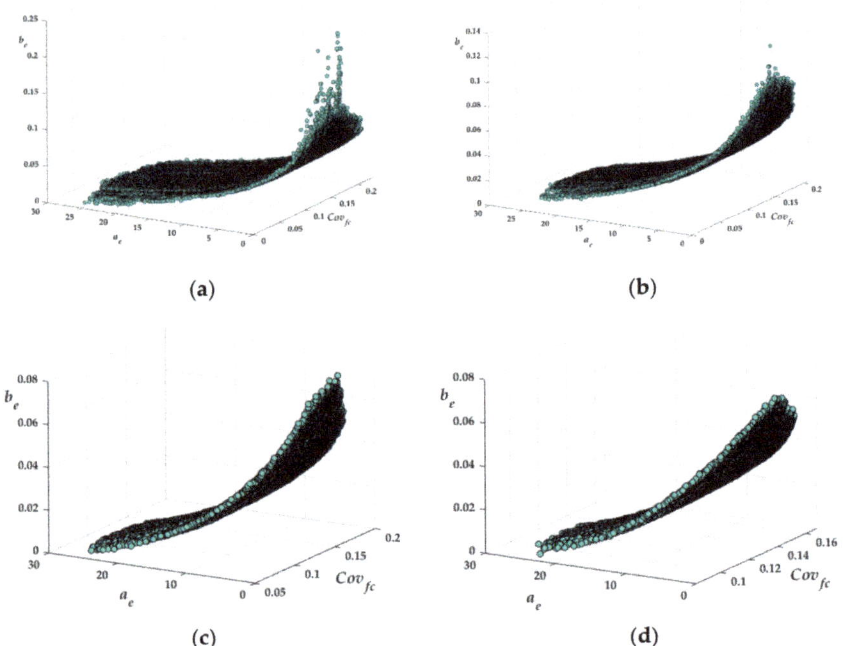

Figure 5. Relationship between the coefficients of the calibrated exponential empirical laws (a_e, b_e) and the coefficient of variation of the concrete strength Cov_{fc}: (**a**) $m = 8$; (**b**) $m = 11$; (**c**) $m = 15$, (**d**) $m = 18$.

A similar trend is observed when analyzing the influence of the coefficient of variation of the rebound index on the law in Equation (12) (see Figure 6).

The analysis of the RMSE for a different number, m, of considered data reveals that the coefficient of variation of the rebound index does not seem to be significantly relevant for the accuracy of the calibrated exponential empirical law, while the latter would be fundamentally dependent on the coefficient of variation of the concrete strength. The increase of the number m of considered data does have an impact on the range of variation of RMSE (see Figure 7).

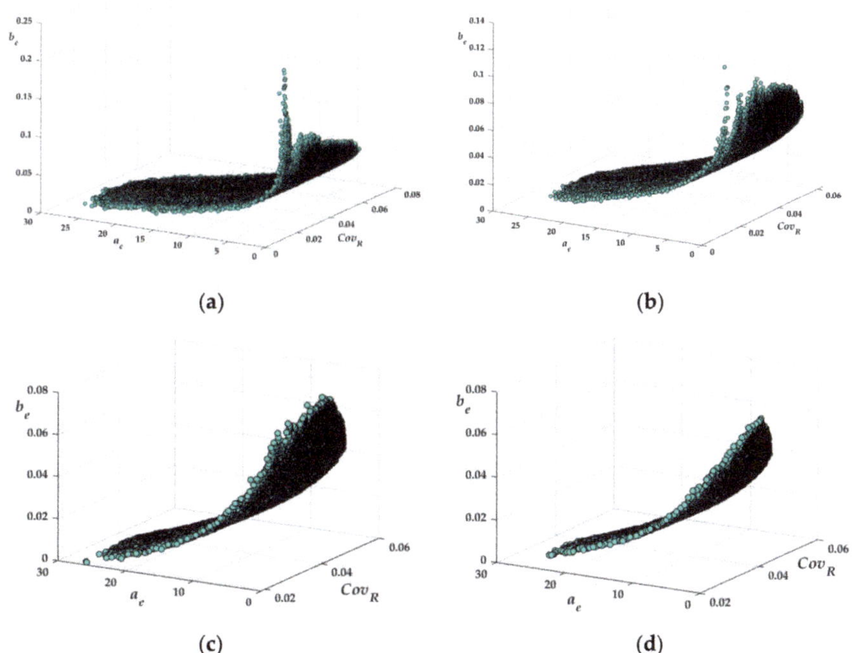

(a)

(b)

(c)

(d)

Figure 6. Relationship between the coefficients of the calibrated exponential empirical laws (a_e, b_e) and the coefficient of variation of rebound index Cov_R: (**a**) $m = 8$; (**b**) $m = 11$; (**c**) $m = 15$, (**d**) $m = 18$.

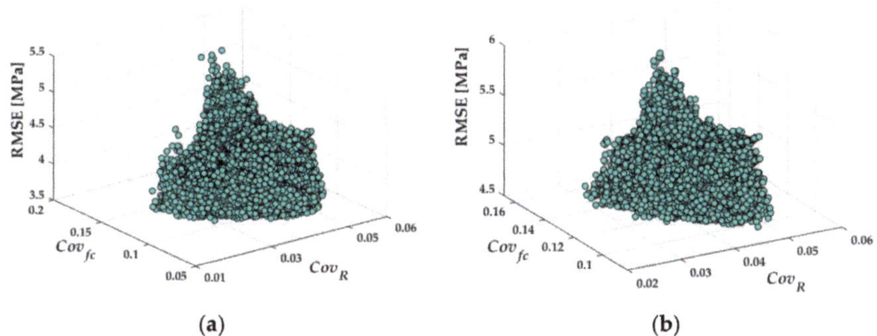

(a)

(b)

Figure 7. RMSE (Equation (10)) evaluated for the exponential conversion models vs. coefficient of variation of the in-situ concrete strength Cov_{fc} and coefficient of variation of the rebound index Cov_R: (**a**) $m = 15$; (**b**) $m = 18$.

6. Polynomial Law

This section discusses the calibration of polynomial empirical laws (see Equation (5)) in accordance with the procedure described in Sections 2 and 3. Additionally, for this law more than 10^6 conversion models have been established by minimizing the value of the Equation (9) and by varying the number m of the considered data between 3 and 24. In Figure 8 the coefficients of the calibrated polynomial laws are plotted for four different values of m (i.e., $m = 8, 11, 15, 18$). The plots would indicate that all the coefficients lie on a plane, and consequently, a relationship between such coefficients may be established.

Therefore, a regression analysis has been carried out for each value of m ranging between 3 and 24. Following, the established equation is written:

$$a_p = -0.0236\, b_p - 0.00055\, c_p + 0.0139 \tag{13}$$

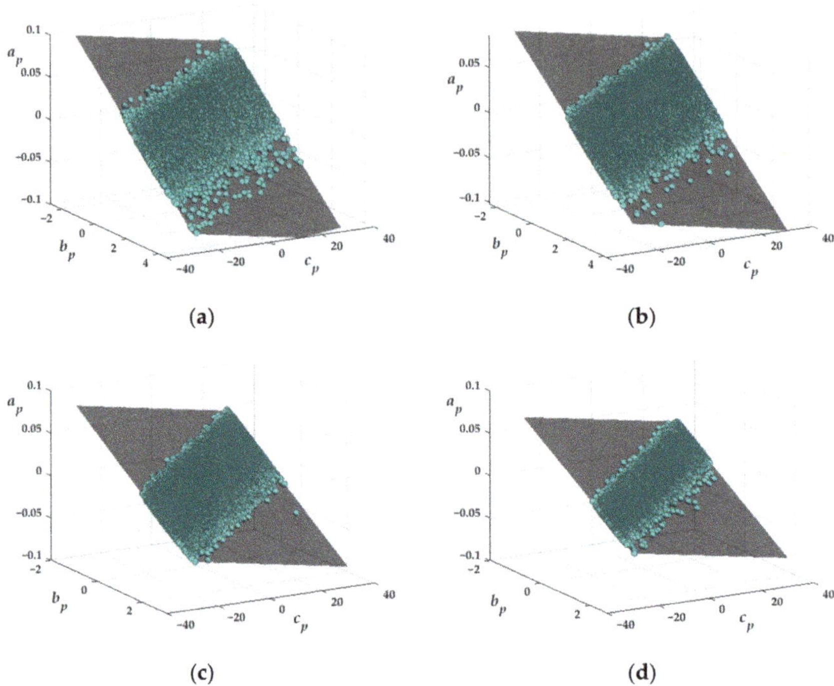

Figure 8. Relationship between the coefficients of the calibrated polynomial empirical laws a_p, b_p, c_p: (**a**) $m = 8$; (**b**) $m = 11$; (**c**) $m = 15$; (**d**) $m = 18$.

With a coefficient of determination equal to 0.9996, showing a very high goodness of fit in the regression model. Moreover, by analyzing all the coefficients of Equation (13) for each value of m, the coefficient of variation of the three coefficients in Equation (13) is about 3.4%, 5.3% and 2.2%, respectively.

In addition, the influence of the coefficient of variation of the concrete strength on the coefficients of the polynomial law has been evaluated. In Figure 9 the coefficients are plotted by including such coefficients of variation related to each one of the calibrated laws. It can be thus observed that the relationship between the coefficients does not seem to depend on the coefficient of variation of both the concrete strength Cov_{fc} and the rebound number Cov_R.

The previously described behavior is not affected by the number m of considered data in the calibration procedure, as it can be verified by comparing Figure 9 with Figure 10 which has been plotted by considering m equal to 18.

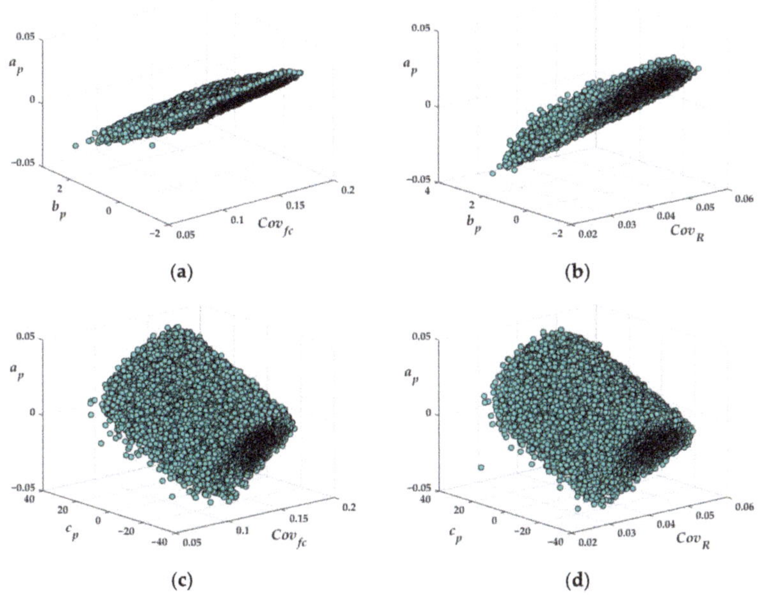

(a)   (b)

(c)   (d)

Figure 9. Relationship between the coefficients of the calibrated polynomial empirical laws a_p, b_p, c_p and the coefficients of variation for $m = 15$: (**a**) plot a_p vs. b_p and the coefficient of variation of the concrete strength Cov_{fc}; (**b**) plot a_p vs. b_p and the coefficient of variation of the rebound index Cov_R; (**c**) plot a_p vs. c_p and the coefficient of variation of the concrete strength Cov_{fc}; (**d**) plot a_p vs. c_p and the coefficient of variation of the rebound index Cov_R.

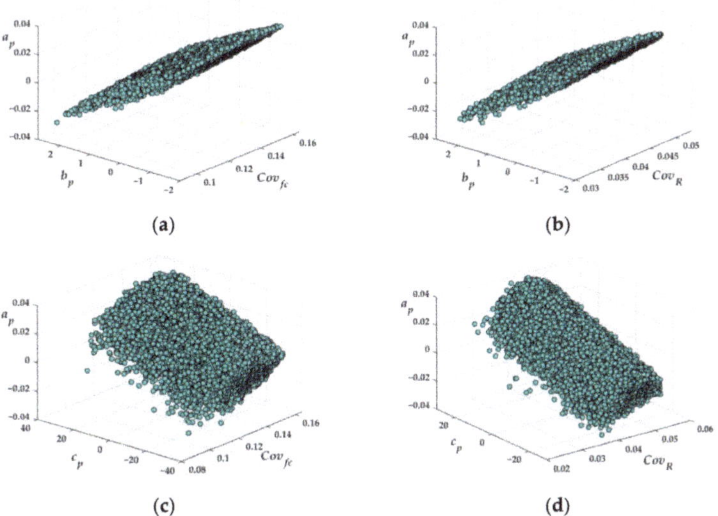

(a)   (b)

(c)   (d)

Figure 10. Relationship between the coefficients of the calibrated polynomial empirical laws a_p, b_p, c_p and the coefficients of variation for $m = 18$: (**a**) plot a_p vs. b_p and the coefficient of variation of the concrete strength Cov_{fc}; (**b**) plot a_p vs. b_p and the coefficient of variation of the rebound index Cov_R; (**c**) plot a_p vs. c_p and the coefficient of variation of the concrete strength Cov_{fc}; (**d**) plot a_p vs. c_p and the coefficient of variation of the rebound index Cov_R.

Figure 11 shows the RMSE evaluated for the polynomial law calibrated for two values of m (i.e., $m = 15$ and 18). Similar trends have been obtained for the other values of m. The graphs show that as the value of m increases, as in the laws discussed above, the range of variability of the RMSE decreases, and although for higher values of the coefficient of variation of the concrete strength, the RMSE reaches high values and no relationship can be established.

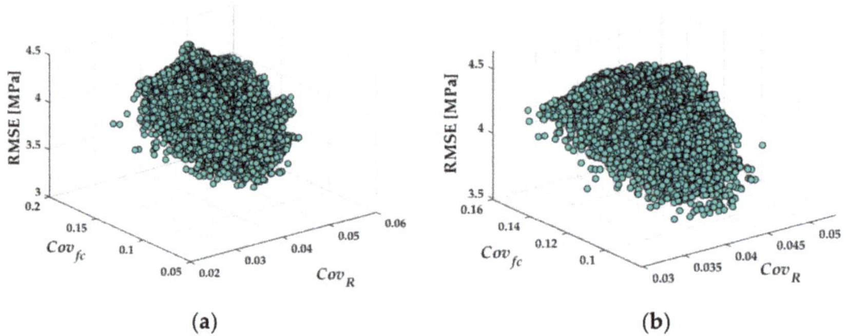

(a)   (b)

Figure 11. RMSE (Equation (10)) evaluated for the polynomial conversion models vs. coefficient of variation of the in-situ concrete strength Cov_{fc} and coefficient of variation of the rebound index Cov_R: (**a**) $m = 15$; (**b**) $m = 18$.

7. Conclusions

The use of non-destructive testing in the field of diagnostics of an existing building is becoming increasingly widespread due to its versatility and swiftness. In this paper, the use of the rebound hammer test has been analyzed with detailed attention given to the choice of the empirical law, correlating the concrete strength with the rebound index. Three conversion models have been investigated to verify the variability of their coefficients, the accuracy of the assessed concrete strength with respect to the coefficient of variation of the concrete strength and of the rebound index, and the possible advantages and/or limitations related to the adopted empirical law.

The use of the linear and exponential laws involved the estimation of two coefficients each, whereas the polynomial law required the calibration of three coefficients. The analysis presented here—based on large numbers of combinations of experimental non-destructive and destructive results from an existing building—would reveal that the coefficients of those analytical models manifest a high variability in their values. Nonetheless, a relationship between such values may be always found by means of a regression analysis. In detail, for the linear law, a linear relationship exists between its two defining coefficients; for the exponential law a logarithmic equation binds the coefficients, and for the polynomial law, the coefficients lie on a plane whose equation was established. Moreover, the influence of the number m of considered data in the calibration procedure of the empirical laws on the established relationships was investigated. It was highlighted that for all the considered conversion models, varying this number in the range from 3 to 24, the highest variation in the factors of the relationships between the coefficients of the conversion models is equal to 5.3% for the polynomial law and 0.2% for the linear and the exponential laws. This result may be due to the weight of the rebound index in the analytical formulation, in fact, as its weight increases, the variation of the coefficients increases.

Furthermore, in all the considered models, the coefficient of variation of the concrete strength and of the rebound index did not affect the previously described relationships. The existence of relationships between the coefficients of the conversion model shows that these coefficients, which are calibrated by a minimization procedure, are not independent from each other. Indeed, the calibration procedure is performed based on a fixed combination of data and, thus, the coefficient of variation of the core strength and of the rebound index interfere directly in the procedure, while the regression analysis carried out on the coefficients of the conversion model has the effect of averaging the influence of these coefficients of variation.

The RMSE was also evaluated for each conversion model. Based on the methodology here examined—involving experimental results from a real existing building—it did not vary significantly when using a different empirical law, but a narrow reduction was observed in the range of variability of RMSE and of its mean value going from the exponential law to the linear one, and from the linear one to the polynomial law.

A future step of the research will be the extension of the procedure and results discussed here, to other buildings. In the first stage, other buildings characterized by comparable age of construction and environmental conditions will be selected to dismiss the possibility that the results may be due to some unconsidered factors. In a second stage, the analysis will be extended to other buildings in order to address these effects, one by one.

Author Contributions: Conceptualization, methodology, validation, data curation, writing—original draft preparation, M.D.; writing—review and editing, F.B.V. All authors have read and agreed to the published version of the manuscript.

Funding: This research was funded by "Studio del comportamento meccanico e del retrofitting di edifici esistenti: approcci innovativi e Life Cycle Assessment" project Politecnico di Bari FRA 2021.

Informed Consent Statement: Not applicable.

Conflicts of Interest: The authors declare no conflict of interest.

References

1. Italian Law. *Minister Decree (D.M.) of the 17/01/2018 "Norme Tecniche per le Costruzioni (NTC 2018)"*; n. 42; Gazzetta Ufficiale: Rome, Italy, 2018. (In Italian)
2. Italian Law. *The Circular of the Ministry of Infrastructures and Transports n. 7 of the 21/01/2019*; Istruzioni per l'Applicazione dell' «Aggiornamento delle "Norme tecniche per le Costruzioni"»; n. 35; Gazzetta Ufficiale: Rome, Italy, 2019. (In Italian)
3. *CEN/TS 17440*; Assessment and Retrofitting of Existing Structures. CEN-CENELEC Management Centre: Brussels, Belgium, 2020.
4. Diaferio, M.; Foti, D.; Ivorra, S.; Giannoccaro, N.I.; Notarangelo, G.; Vitti, M. Monitoring and structural identification of a "Star-shaped" reinforced concrete bell tower. In Proceedings of the 8th International Operational Modal Analysis Conference (IOMAC), Copenhagen, Denmark, 13–15 May 2019; pp. 47–54.
5. Bovo, M.; Tondi, M.; Savoia, M. Infill Modelling Influence on Dynamic Identification and Model Updating of Reinforced Concrete Framed Buildings. *Adv. Civ. Eng.* **2020**, *2020*, 9384080. [CrossRef]
6. Diaferio, M.; Foti, D.; Giannoccaro, N.I.; Sabbà, M.F. Dynamic Identification on an Irregular Structure. *Appl. Sci.* **2022**, *12*, 3445. [CrossRef]
7. Shimizu, Y.; Hirosawa, M.; Zhou, J. Statistical analysis of concrete strength in existing reinforced concrete buildings in Japan. In Proceedings of the 12th World Conference on Earthquake Engineering, Auckland, New Zealand, 30 January–4 February 2000.
8. Diaferio, M.; Vitti, M. Correlation curves to characterize concrete strength by means of UPV tests. *Lect. Notes Civ. Eng.* **2021**, *110*, 209–218.
9. Breysse, D. Nondestructive evaluation of concrete strength: An historical review and a new perspective by combining NDT methods. *Constr. Build. Mater.* **2012**, *33*, 139–163. [CrossRef]
10. Diaferio, M.; Vitti, M. Non-destructive tests for the assessment of r.c. buildings. In Proceedings of the 2021 48th Annual Review of Progress in Quantitative Nondestructive Evaluation, QNDE 2021, Virtual Online, 28–30 July 2021; p. v001t09a002.
11. Machado, M.D.; Shehata, L.C.D. Correlation curves to characterize concretes used in Rio de Janeiro by means of nondestructive tests. *Ibracon Struct. Mater. J.* **2019**, *2*, 100–123.
12. *ASTM C805-13*; Standard Test Method for Rebound Number of Hardened Concrete. American Society for Testing and Materials: West Conshohocken, PA, USA, 2013.
13. Zaky, S. Evaluation of concrete structural members contained different types of coarse aggregate using Schmidt hammer apparatus after 28-day and at later ages. *Inter. Build.* **2001**, 37–50.
14. Olonade, K.A. Influence of water-cement ratio and water reducing admixtures on the rebound number of hardened concrete. *Malays. J. Civ. Eng.* **2020**, *32*. [CrossRef]
15. Kumavata, H.R.; Chandaka, N.R.; Patil, I.T. Factors influencing the performance of rebound hammer used for non-destructive testing of concrete members: A review. *Case Stud. Constr. Mater.* **2021**, *14*, e00491. [CrossRef]
16. De Domenico, D.; Messina, D.; Recupero, A. Quality control and safety assessment of prestressed concrete bridge decks through combined field tests and numerical simulation. *Structures* **2022**, *39*, 1135–1157. [CrossRef]

17. Cristofaro, M.T.; Viti, S.; Tanganelli, M. New predictive models to evaluate concrete compressive strength using the SonReb method. *J. Build. Eng.* **2020**, *27*, 100962. [CrossRef]
18. *EN 12504-1:2019*; Testing Concrete in Structures—Part 1: Cored Specimens—Taking, Examining and Testing in Compression. CEN-CENELEC Management Centre: Brussels, Belgium, 2019.
19. *EN 12390-3:2019*; Testing Hardened Concrete—Part 3: Compressive Strength of Test Specimens. CEN-CENELEC Management Centre: Brussels, Belgium, 2019.
20. American Concrete Institute. *ACI 228.1R-03 In-Place Methods to Estimate Concrete Strength*; American Concrete Institute: Farmington Hills, MI, USA, 2003.
21. Masi, A.; Vona, M. Estimation of the in-situ concrete strength: Provisions of the European and Italian seismic codes and possible improvements. In *Eurocode 8 Perspectives from the Italian Standpoint Workshop*; Doppiavoce: Napoli, Italy, 2009; p. 67.
22. Kim, J.K.; Kim, C.Y.; Yi, S.T.; Lee, Y. Effect of carbonation on the rebound number and compressive strength of concrete. *Cem. Concr. Compos.* **2009**, *31*, 139–144. [CrossRef]
23. Aydin, F.; Saribiyik, M. Correlation between Schmidt hammer and destructive compressions testing for concretes in existing buildings. *Sci. Res. Essays* **2010**, *5*, 1644–1648.
24. Abdulkadir, M.R.; Karim, A.A.; Abdullah, A.H. Destructive and nondestructive strength evaluation of concrete exposed to fire, J. Zankoy Sulaimani. *J. Zankoy Sulaimani. Part A (Pure Appl. Sci.)* **2017**, *19*, 43–55. [CrossRef]
25. Proceq, S.A. *Concrete Test Hammer N/NR, L/LR and Digital Schmidt ND/LD—Rebound Measurement and Carbonation, Information Sheet, ver. 10*; Proceq Europe: Schwerzenbach, Switzerland, 2003.
26. Brozovsky, J.; Bodnarova, L. Contribution to the issue of evaluating the compressive strength of concrete exposed to high temperatures using the Schmidt rebound hammer. *Russ. J. Nondestruct. Test.* **2016**, *52*, 44–52. [CrossRef]
27. Duna, S.; Omoniyi, T.M. Correlation between non-destructive testing (NDT) and destructive testing (DT) of compressive strength of concrete. *Int. J. Eng. Sci. Invent.* **2014**, *3*, 12–17.

applied sciences

Article

Shear Characteristics of Soil—Concrete Structure Interaction Interfaces

Dejie Li [1,2,3,*], Chong Shi [1,3], Huaining Ruan [1,3] and Bingyi Li [4]

1   Key Laboratory of Ministry of Education for Geomechanics and Embankment Engineering, Hohai University, Nanjing 210098, China
2   CCCC Second Harbour Engineering Co., Ltd., Wuhan 430040, China
3   Institute of Geotechnical Research, Hohai University, Nanjing 210098, China
4   School of Civil Engineering, Suzhou University of Science and Technology, Suzhou 215011, China
*   Correspondence: lidejieedu@163.com; Tel.: +86-28-8546-5055

Abstract: The shear characteristics of the interfaces between soil and concrete structures are essential for the safety of the structures. In this study, a large-scale direct shear test apparatus was developed to measure the mechanical parameters of soil–concrete interfaces under conditions with different soil types, soil moisture contents, and interfacial filling materials. The results showed that the shear stress of the soil–concrete interface increased initially and then became stable with the increase in the shear displacement. The shear displacement of the sandy soil when the shear stress became stable was smaller than that of the clayey soil. The silty sand–concrete interface had a smaller friction angle than the interface with the medium-coarse sand. Moreover, with the increase in the soil moisture content, the friction angle of the clayey soil–concrete interface decreased rapidly, whereas the cohesion first increased and then decreased, and the peak cohesion was near the plastic limit of the soil. Under the same moisture content, the friction angle and cohesion of the clay–concrete interface was reduced by filling the interface with a thin layer of sandy soil, while filling the silty sand–concrete interface with a thin layer of silt reduced the friction angle and increased the interfacial cohesion. Nonetheless, the filling had little impact on the overall shear strength of the interface.

Keywords: soil–concrete interface; shear characteristics; large-scale direct shear test; shear strength

Citation: Li, D.; Shi, C.; Ruan, H.; Li, B. Shear Characteristics of Soil—Concrete Structure Interaction Interfaces. *Appl. Sci.* **2022**, *12*, 9145. https://doi.org/10.3390/app12189145

Academic Editors: Francisco B. Varona Moya and Mariella Diaferio

Received: 18 August 2022
Accepted: 10 September 2022
Published: 12 September 2022

Publisher's Note: MDPI stays neutral with regard to jurisdictional claims in published maps and institutional affiliations.

1. Introduction

The interfacial shear strength is a key engineering parameter in the design of pile foundations, retaining walls, and submerged foundations. However, there are currently limited data on the frictional parameters between different soil materials and structures (Konkol et al., 2021) [1]. The shear strength of a soil–concrete interface is affected by various factors, such as the soil moisture content, density, and confining pressure (Abdulghade et al., 2021) [2]. Current designs generally adopt a semi-empirical method based on elasticity theory and experimental data (Johnson et al., 2001) [3]. Thus, it is of great research and practical significance to obtain accurate soil–structure interfacial shear parameters.

Extensive studies have been carried out on the parameters of soil–structure interfaces. For example, in order to obtain reproducible results, Yin et al. (2021) prepared a sand–clay mixture, which showed better uniformity than specimens prepared with traditional soil reconstruction techniques [4]. Canakci et al. (2016) studied the influence of structural materials on the soil–structure interface parameters and found that the friction angle of the soil–concrete interface was larger than those of the soil–steel and soil–wood interfaces [5]. In terms of the value of the friction angle of the soil–concrete interface, Ilori et al. (2017) showed that the normal stress affected the friction coefficient of the soil–concrete interface [6]. Shakir et al. (2008) found that the surface morphology of the structure was the most important factor affecting the friction coefficient of the soil–structure interface [7]. Muszyński et al. (2019) assessed the surface morphology of the concrete pile foundation using a three-dimensional laser scanner and proposed a morphological evaluation method [8]. Konkol et al. (2021) studied the shear softening characteristics and failure load of a soft soil–concrete interface under a constant normal load. Then, a hyperbolic interface model was established based on a direct shear test [1]. Cen et al. (2020) obtained hysteresis and backbone curves under different vertical pressures and shear–displacement amplitudes [9]. Yin et al. (2020) and Ravera et al. (2022) proposed that the thermal cycle is also a potential factor affecting the soil–structure interface response [10,11]. Casagrande et al. (2020) evaluated the mechanical properties of concrete piles under thermal loads, and the results showed that there was an increase in the pile–soil frictional resistance after each thermal cycle [12]. Wang et al. (2022) studied the behavior of a soil–structure interface under a constant normal stiffness using the discrete element method and proposed an algorithm to achieve a constant normal stiffness [13]. Liu et al. (2019) argued that the interface friction angle could change during pile construction, and an on-site test was the best method to obtain the shear parameters of the soil–structure interfaces [14]. In summary, the above studies yielded many important conclusions; yet, there is still a lack of in-depth research on the influence of various factors, such as the soil type, soil moisture content, and interface filling material, on the shear strength of the soil–structure interface.

In this study, large-scale direct shear tests were carried out on different types of soil materials, including silty sand, medium-coarse sand, clay, and sandstone, in order to study the influence of the soil moisture content, normal stress, and interface filling material on the shear parameters of the soil–concrete interface.

2. Materials and Methods

2.1. Test Apparatus

The large-scale direct shear test apparatus DZJ-1 was used in this study. The length, width, and height of the shear box were all 150 mm. A normal force was applied by a rolling diaphragm cylinder, and the horizontal shear force was under servo control. The apparatus had a horizontal shear rate range of 0.002–4.0 mm/min, and the shear rate used in this study was 1 mm/min. During the shear test, the horizontal shear displacement and shear stress were automatically collected via sensors, the accuracies of which were ±1%. The large-scale direct shear apparatus is shown in Figure 1.

Figure 1. DZJ-1 large-scale direct shear test apparatus.

When the large-scale direct shear test apparatus was used for the shear test, C30 concrete blocks with a length of 15 cm, a width of 15 cm, and a height of 5 cm were placed in the lower shear box, and soil samples with a length of 15 cm, a width of 15 cm, and a height of 5 cm were placed in the upper shear box. The schematic diagram of the placement of the shear test samples is shown in Figure 2.

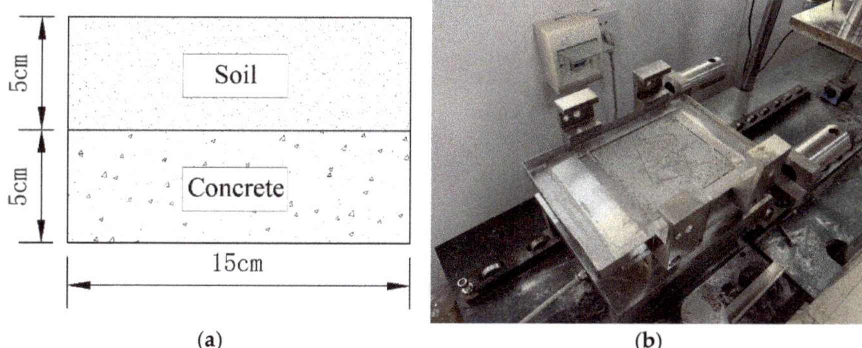

(**a**)　　　　　　　　　　　　　(**b**)

Figure 2. The schematic diagram of the placement of shear test samples: (**a**) sample placement diagram and (**b**) concrete blocks.

2.2. Soil Materials

Silty sand, medium-coarse sand, and sandstone were collected from Wuhan City, Hubei Province, China. The clay was obtained from the Oujiang River estuary in Wenzhou City, Zhejiang Province, China. The soil samples were collected from the same batch, and the same soil samples had good consistency under different experimental conditions. The particle gradation of silty sand and medium-coarse sand was tested, and the gradation curve is shown in Figure 3.

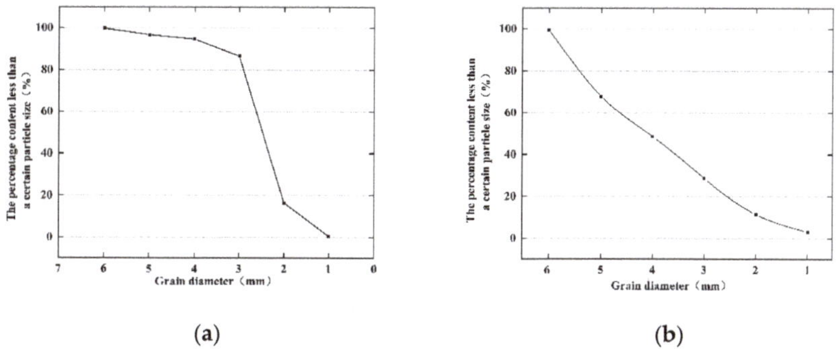

(**a**)　　　　　　　　　　　　　(**b**)

Figure 3. Gradation curve of sandy soil particles: (**a**) silty sand and (**b**) medium-coarse sand.

The liquid limit and the plastic limit of the clay and sandstone were measured by a liquid–plastic limit testing instrument. The liquid limit and the plastic limit of the clay were 37.51% and 19.96%, respectively, and those of the sandstone were 29.64% and 17.04%, respectively. The parameters of the different soil types, such as the dry density, the proportion of the soil particles, the saturated density, and the soil classification, are shown in Table 1.

Table 1. Physical properties of soil samples.

Soil Types	Dry Density (g·cm^{-3})	Proportion of Soil Particles	Saturated Moisture Content (%)	Plastic Limit (%)	Liquid Limit (%)	The Soil Classification
Medium-coarse sand	1.65	2.67	21.47	/	/	
Silty sand	1.51	2.7	24.87	/	/	
Clay	0.88	2.75	44.58	19.96	37.51	HL
Sandstone	1.44	2.68	36.57	17.04	29.64	CL

2.3. Test Conditions

First, four single-soil type specimens were prepared, and shear tests were carried out under four different moisture contents and three normal stress conditions. At high soil moisture contents, the strengths of the clay and sandstone were typically low. Hence, the normal stress levels on the clay were set to 25, 50, and 75 kPa when the soil moisture content was 30% and 50, 100, and 150 kPa when the moisture content was 25%. The clay–concrete interface was filled with thin layers of silty sand and medium-coarse sand, and the silty sand–concrete interface was filled with thin silt. A total of 69 shear tests were carried out, as shown in Table 2.

Table 2. Test conditions.

Soil	Moisture Content (%)	Normal Stress (kPa)
Silty sand	0, 5, 10, 15	100, 200, 300
Medium-coarse sand	0, 5, 10, 15	100, 200, 300
Clay	15, 20, 25	100, 200, 300
	30	25, 50, 75
Sandstone	10, 15, 20	100, 200, 300
	25	50, 100, 150
Clay + thin layer of medium-coarse sand	15 (Clay)	100, 200, 300
Clay + thin layer of silty sand	15 (Clay)	100, 200, 300
Silty sand + thin layer of silt	50 (Silt)	100, 200, 300

In order to ensure the reproducibility of the tests, the soil specimens were dried and ground, and different quantities of water were then added to prepare soil specimens with different moisture contents. Prefabricated concrete blocks were placed in the lower shear box, and then, the soil specimens were loaded into the shear box in layers. A predetermined normal force was applied to consolidate the soil samples. Figure 4 shows the different soil specimens. After each shear test, the soil–concrete interface was observed. Examples of the shear surfaces of the different soil specimens are shown in Figure 5.

(a) (b) (c) (d)

Figure 4. Different types of soil samples: (**a**) silty sand, (**b**) medium-coarse sand, (**c**) clay, and (**d**) sandstone.

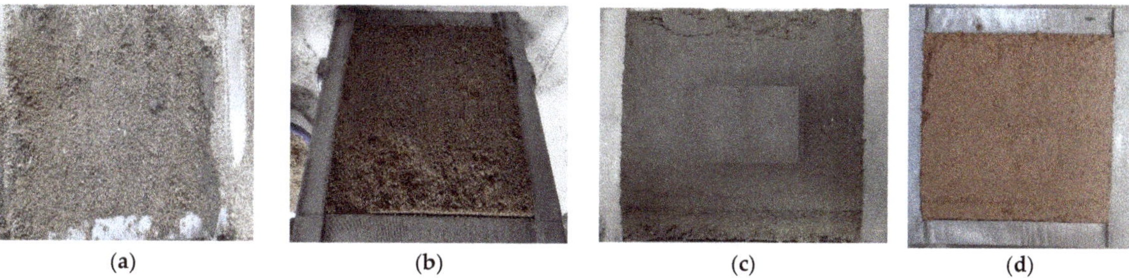

(a)   (b)   (c)   (d)

Figure 5. Examples of interface shear surfaces: (**a**) silty sand, (**b**) medium-coarse sand, (**c**) clay, and (**d**) sandstone.

3. Results and Analysis

3.1. Shear Test of Unfilled Soil–Concrete Interface

The variations of the shear stress with the shear displacement for the specimens with different moisture contents under different normal stress conditions are shown in Figure 6. The shear stresses of all the specimens first increased with the increase in the shear displacement, and when the shear displacement reached a critical value, the shear stress stabilized. The critical value increased with the increase in the normal stress. In the shear test of different samples, the shear stress first increases and then tends to be stable with the increase in the shear displacement. Silty sand and medium-coarse sand require less shear displacement to stabilize than clay and sandstone. This is mainly due to the fact that the silty and the medium-coarse sand are composed of loose particles, which are easily redistributed during the shear process. However, clay and sandstone are flocculated structures, which are damaged during shear slip, and the shear displacement required to achieve stable shear stress is large.

Figure 7 shows the results for different soil specimens with a moisture content of 15%. When the normal stress was 100 kPa, the maximum shear stresses of the different soil specimens in descending order were clay, sandstone, medium-coarse sand, and silty sand. The difference in the maximum shear stress between the clay and the sandstone was small; yet, as the normal stress increased, the difference gradually increased. This was mainly due to the formation of a thin layer of silt at the clayey soil–concrete interface under the action of the normal stress, which reduced the interfacial shear strength. In comparison, the sandstone had more evident water softening properties than the clayey soil. For the silty sand and medium-coarse sand, a water film formed at the interface due to water discharge. However, the water film had little effect on the shear stress of the interface. The shear stress of the silty sand was lower than that of the medium-coarse sand. This was because the particle size of the silty sand was small (0.25–0.5 mm), and compared with the medium-coarse sand, the silty sand particles could more easily fill the voids on the concrete surface, thereby reducing the interface friction coefficient. The particle sizes of the medium-coarse sand were evenly distributed in the range of 0.1–5 mm. The gaps between the particles and between the particles and the concrete surface resulted in an embedding effect, which improved the interfacial shear strength.

The critical values of the shear stress under the conditions of the different soil types, soil moisture contents, and normal stresses were taken as the interfacial shear strength, as shown in Figure 8. With the increase in the moisture content, the shear strengths of the different soil types decreased to varying degrees. The shear strength of the medium-coarse sand and silty sand decreased by 5–22 kPa, but the overall extent of the decrease was small. For the silt and sandstone, under the same normal stress, the interface was greatly affected by the moisture content. When the moisture content was at a low level, the shear strength of the interface decreased slightly with the increase in the moisture content. However, when the moisture content was above a certain threshold, the shear strength of the interface decreased rapidly, and the maximum decrease was 96 kPa. In addition, the shear strength of the silt and sandstone clayey soil interface was more sensitive to the soil moisture content.

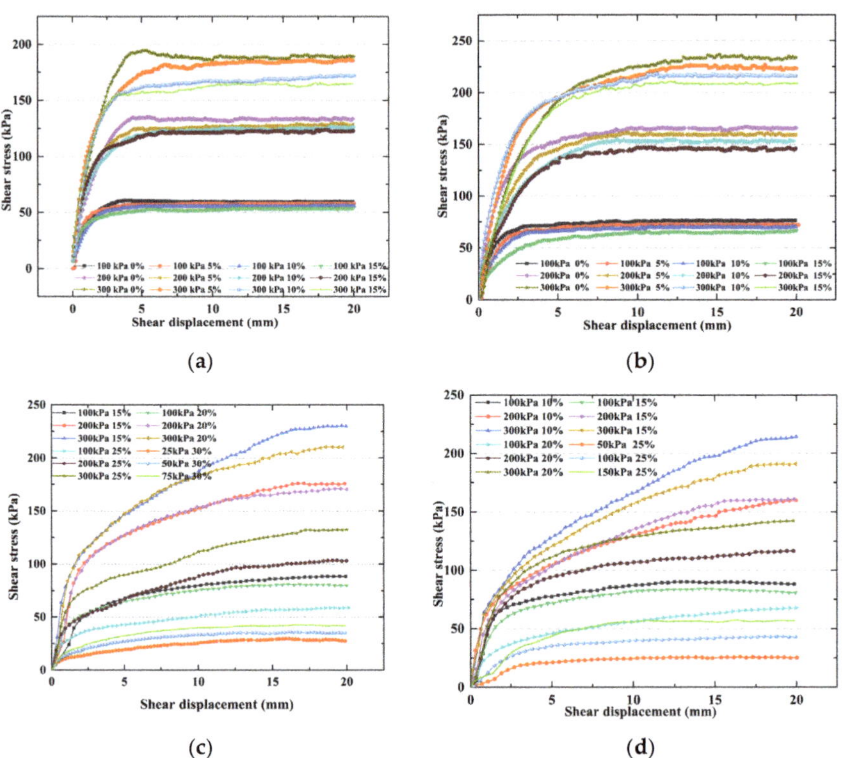

Figure 6. Variation of shear stress with shear displacement for different soil specimens: (**a**) silty sand, (**b**) medium-coarse sand, (**c**) clay, and (**d**) sandstone.

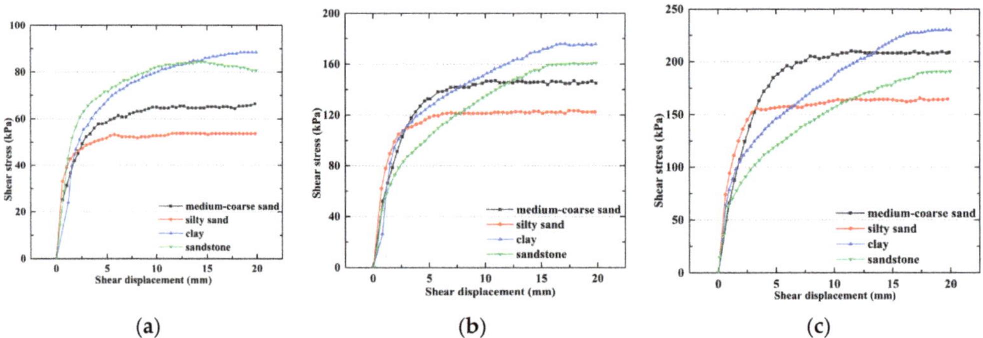

Figure 7. Shear stress–shear displacement curves of different soil specimens with 15% moisture contents: (**a**) normal stress: 100 kPa, (**b**) normal stress: 200 kPa, and (**c**) normal stress: 300 kPa.

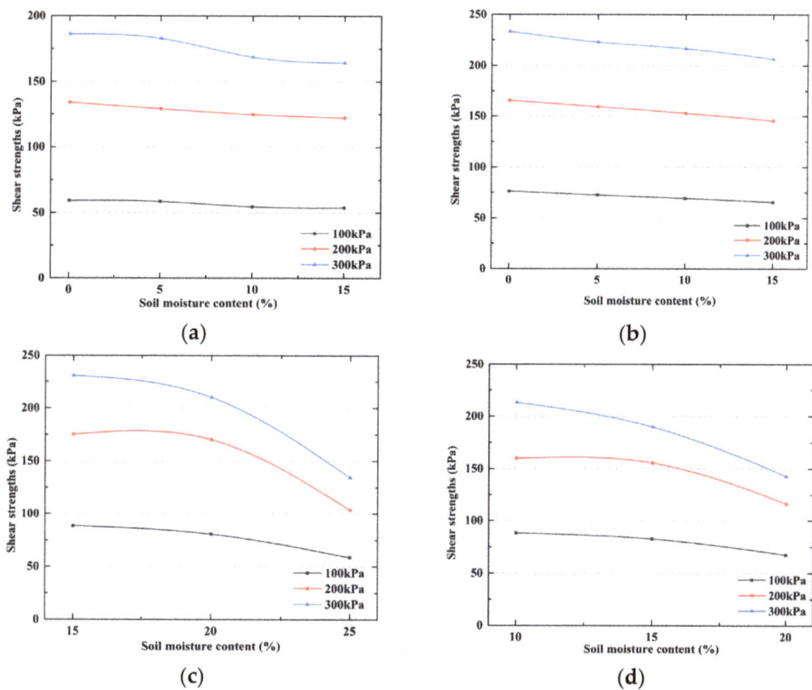

Figure 8. Variation of interfacial shear strength with moisture content: (**a**) silty sand, (**b**) medium-coarse sand, (**c**) clay, and (**d**) sandstone.

The relationship between the shear strength and the normal stress for the different soil–concrete interfaces was fitted (Figure 9) with the following equation:

$$\tau = \sigma \tan \varphi + c \tag{1}$$

where τ is the shear strength (kPa), σ is the normal stress (kPa), φ is the internal friction angle (°), and c is the cohesion (kPa).

Figure 10 shows the variations of the internal friction angle and cohesion with the moisture content. With the increase in the moisture content, the internal friction angle of the soil–concrete interface showed a decreasing trend. The friction angles for the silty sand and the medium-coarse sand interfaces were in the ranges of 29–32° and 35–38°, respectively, and the friction angle was not greatly affected by the moisture content. For the clay and sandstone, the friction angles decreased rapidly with the increase in the moisture content, in the ranges of 14–35° and 8–34°, respectively. Moreover, the friction angle was significantly affected by the moisture contents of the clay and sandstone. Furthermore, the cohesion of the different soil–concrete interfaces showed different variation trends with the moisture content. The cohesion of silty sand and medium-coarse sand decreased with the increase in the moisture content, although the magnitude of the change was only about 2 kPa. As the moisture content increased, the cohesion of the clayey soil and the sandstone first increased and then decreased, and the moisture content corresponding to the maximum cohesion was near the plastic limit of the soil. This was because the adsorption of water in the soil gradually increased the cohesion. When the moisture content reached the plastic limit, the water film thickness between the soil particles increased, and the effect of the water pressure became greater than the adsorption effect. Thus, the cohesion began to decrease.

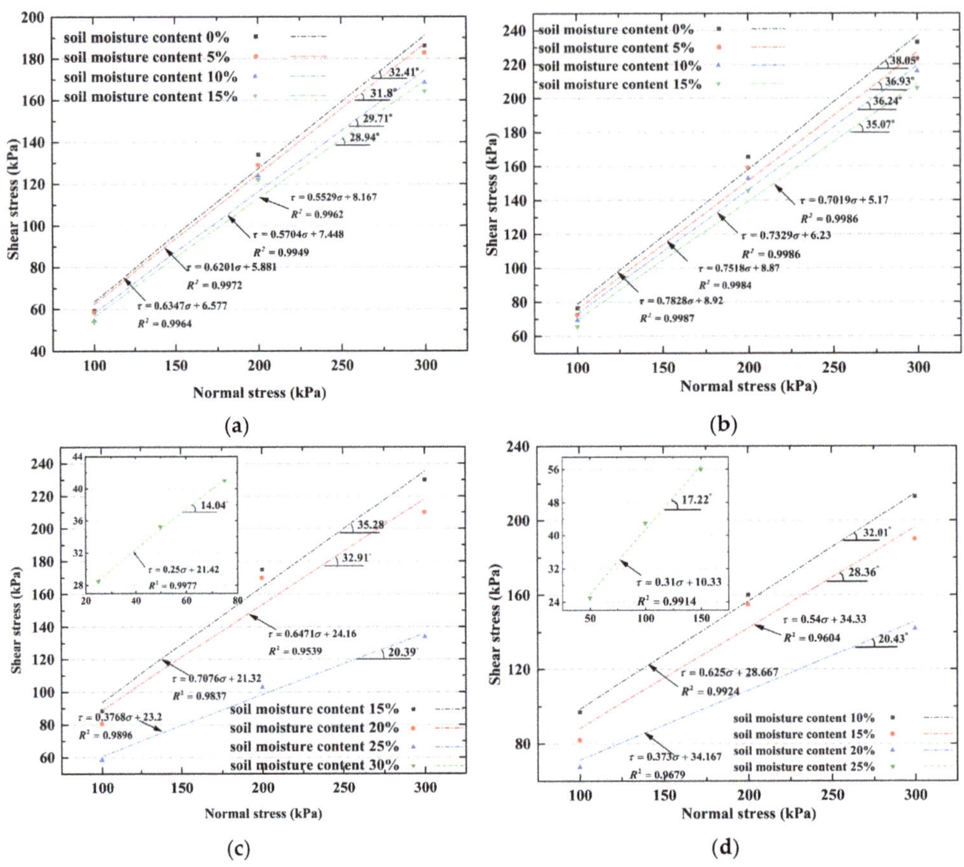

Figure 9. Fitted shear strength curves for different interfaces: (**a**) silty sand, (**b**) medium-coarse sand, (**c**) clay, and (**d**) sandstone.

3.2. Shear Test of Filled Soil–Concrete Interface

In order to study the influence of the filling materials on the shear characteristics of the soil–concrete interface, the soil–concrete interface was filled with 2 mm of medium-coarse sand or silty sand, and the silty sand–concrete interface was filled with 2 mm of silt. Figure 11 shows the variation of the shear stress with the shear displacement, and Figure 12 shows the variation of shear strength. Compared with the condition without filling materials, the shear stress of the concrete interface with a thin layer of sandy soil decreased slightly, indicating that the thin layer of sandy soil had the effect of reducing the shear strength of the interface. In terms of the strength parameters, the interfacial friction angle was 35.28° without filling material, which decreased to 33.39° and 33.54° after filling with thin layers of medium-coarse sand and silty sand, respectively. The cohesion without filling was 21.32 kPa, which changed to 16.14 and 16.803 kPa after filling with thin layers of medium-coarse sand and silty sand, respectively. The reduction in the friction angle and cohesion was mainly because the sliding friction on the soil–concrete interface was partially converted into rolling friction between the sand particles and the concrete, which reduced the friction coefficient. In addition, the sandy soil also played a certain role in reducing the adhesion of the clay to the concrete.

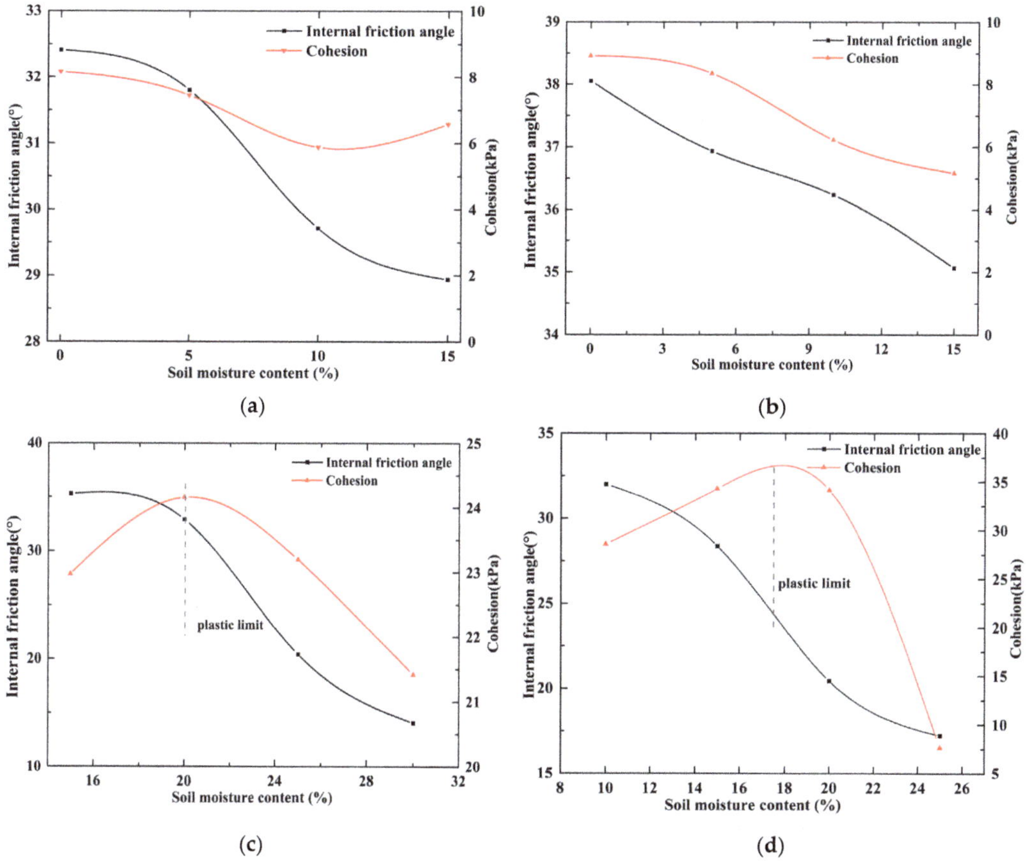

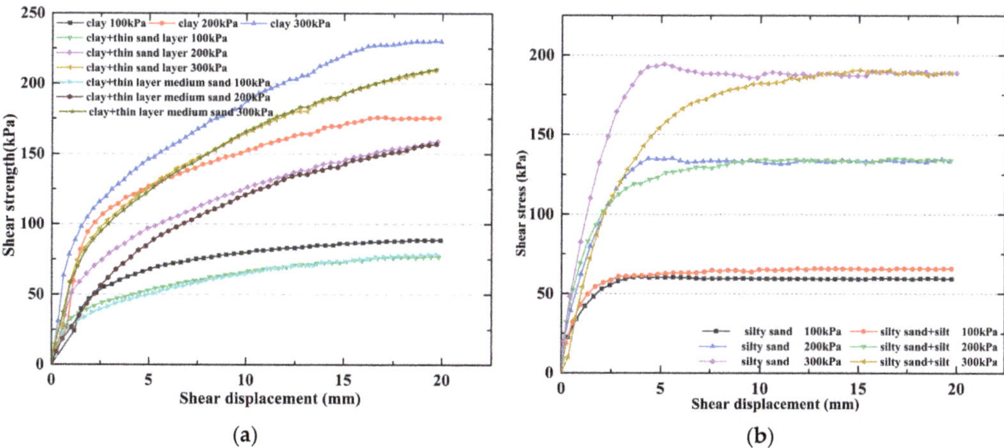

Figure 10. Variation of shear strength parameters of different interfaces with soil moisture content: (**a**) silty sand, (**b**) medium-coarse sand, (**c**) clay, and (**d**) sandstone.

Figure 11. Effect of filling material on shear characteristics of soil–concrete interface: (**a**) clay filled with sandy soil and (**b**) silty sand filled with silt.

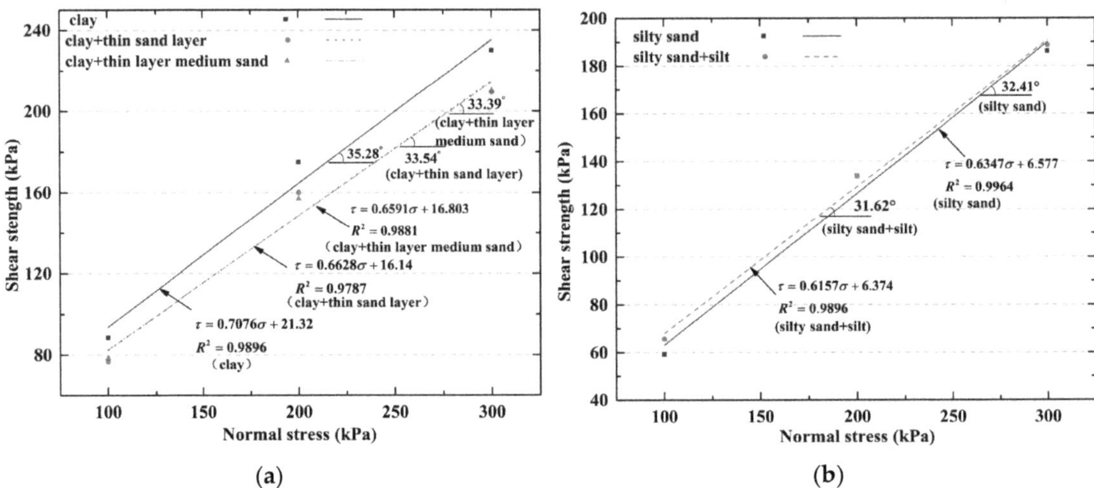

Figure 12. Variation of shear strength for soil–concrete interfaces with different filling materials: (**a**) shear strength of clay–concrete interface before and after filling with sandy soil and (**b**) shear strength of silty sand–concrete interface before and after filling with silt.

Silt with a 50% moisture content was filled in the silty sand–concrete interface, and the variation of the shear stress with the shear deformation was basically the same as that of the unfilled interface. The interfacial friction angles were 32.41° and 31.62° before and after silt filling, respectively, and the corresponding interfacial cohesion values were 6.37 and 6.577 kPa. Thus, the friction angle decreased, whereas the cohesion increased. However, the effect on the overall shear strength of the interface was minimal. This was mainly because the silt increased the interfacial adhesion between the soil and the concrete, resulting in an increase in the interfacial cohesion. Moreover, the silt had a certain lubricating effect, and therefore, reduced the interfacial friction angle.

4. Conclusions

Through large-scale direct shear tests of the interface between the concrete and the various soil types (i.e., medium-coarse sand, silty sand, silt, and sandstone), the influences of the normal stress, moisture content, and interface filling materials on the shear strength of the soil–concrete interface were analyzed, and the following conclusions were obtained:

(1) The shear stress of the soil–concrete interface increased initially and then stabilized with the increase in the shear displacement. Compared with the clayey soil, the sandy soil had a smaller shear displacement when the shear stress became stable. This was mainly because the sandy soil contained loose particles, which were more prone to redistribution than the flocculent structure of the clayey soil, and thus, the shear displacement of the sandy soil was smaller.

(2) With a high normal stress, the soil pore water was discharged and formed a lubricating layer in the soil–concrete interface, resulting in a decrease in the friction coefficient of the interface. The friction coefficient between the medium-coarse sand and the concrete was greater than that of the silty sand–concrete interface. This was mainly because the gaps between the large particles and between the soil particles and concrete resulted in an embedding effect, thereby improving the interfacial friction coefficient.

(3) The interfacial shear strength of the clayey soil, sandy soil, and concrete interfaces decreased with the increase in the soil moisture content. Compared with the clayey soil–concrete interface, the shear strength of the sandy soil–concrete interface was less sensitive to the moisture content. The friction angle and cohesion of the sandy soil–concrete interface decreased with the increase in the moisture content; yet, the

decrease magnitude was small. The friction angle of the clayey soil–concrete interface decreased rapidly with the increase in the soil moisture content, whereas the cohesion first increased and then decreased, and the peak cohesion was seen when the moisture content was near the plastic limit.

(4) For a given moisture content, filling the clay–concrete interface with a thin layer of sandy soil led to a reduction in the friction angle and the cohesion of the soil–concrete interface. Filling the silty sand–concrete interface with a thin layer of silt resulted in a decrease in the friction angle of the soil–concrete interface and an increase in the cohesion. However, the filling had little effect on the overall shear strength of the interface. In this study, the number of tests of the effects of interfacial fillings on the shear strength was limited. Further studies are needed to investigate the influences of the moisture content, filling material, and filling thickness on the shear parameters of different soil–concrete interfaces.

Author Contributions: Conceptualization, D.L. and C.S.; methodology, D.L.; validation, D.L., H.R. and B.L.; formal analysis, D.L.; investigation, C.S.; resources, D.L.; data curation, D.L.; writing—original draft preparation, D.L.; writing—review and editing, D.L.; visualization, H.R.; supervision, H.R.; project administration, D.L.; funding acquisition, C.S. All authors have read and agreed to the published version of the manuscript.

Funding: This work was supported by the National Natural Science Foundation of China (Grant No. 41,831,278 and Grant No. 51679071).

Institutional Review Board Statement: Not applicable.

Informed Consent Statement: Informed consent was obtained from all subjects involved in the study.

Data Availability Statement: Not applicable.

Acknowledgments: We would also like to thank Xutao Zeng for editing the manuscript.

Conflicts of Interest: The authors declare no conflict of interest.

References

1. Konkol, J.; Mikina, K. Some aspects of shear behavior of soft soil–concrete interfaces and its consequences in pile shaft friction modeling. *Materials* **2021**, *14*, 2578. [CrossRef] [PubMed]
2. Abdulghader, A.A.; Mohammad, T.R. Pile-soil interface characteristics in ice-poor frozen ground under varying exposure temperature. *Cold Reg. Sci. Technol.* **2021**, *191*, 103377. [CrossRef]
3. Johnson, K.; Karunasena, W.; Sivakugan, N.; Guazzo, A. Modeling pile-soil interaction using contact surfaces. In *Computational Mechanics—New Frontiers for the New Millennium*; Elsevier: Amsterdam, The Netherlands, 2001; pp. 375–380.
4. Yin, K.; Liu, J.; Vasilescu, A.R.; Di Filippo, E.; Othmani, K. A procedure to prepare sand–clay mixture samples for soil-structure interface direct shear tests. *Appl. Sci.* **2021**, *11*, 5337. [CrossRef]
5. Canakci, H.; Hamed, M.; Celik, F.; Sidik, W.; Eviz, F. Friction characteristics of organic soil with construction materials. *Soils Found.* **2016**, *56*, 965–972. [CrossRef]
6. Ilori, A.O.; Udoh, N.E.; Umenge, J.I. Determination of soil shear properties on a soil to concrete interface using a direct shear box apparatus. *Int. J. Geo-Eng.* **2017**, *8*, 17. [CrossRef]
7. Shakir, R.R.; Zhu, J.-G. Mechanical behavior of soil and concrete interface. In *Proceedings of SPIE—The International Society for Optical Engineering*; Elsevier B.V: Amsterdam, The Netherlands, 2008; Volume 7353, pp. 1543–1550. [CrossRef]
8. Muszyński, Z.; Wyjadłowski, M. Assessment of the shear strength of pile-to-soil interfaces based on pile surface topography using laser scanning. *Sensors* **2019**, *19*, 1012. [CrossRef] [PubMed]
9. Cen, W.J.; Wang, H.; Du, X.H.; Sun, Y.J. Experimental evaluation on cyclic shear behavior of geomembrane–concrete interfaces. *J. Test. Eval.* **2020**, *48*, 3561–3578. [CrossRef]
10. Yin, K.; Vasilescu, R.; Fauchille, A.-L.; Kotronis, P. Thermal effects on the mechanical behavior of Paris green clay-concrete interface. In *E3S Web of Conferences*; EDP Sciences: Les Ulis, France, 2020; p. 13006. [CrossRef]
11. Ravera, E.; Laloui, L. Failure mechanism of fine-grained soil-structure interface for energy piles. *Soils Found.* **2022**, *62*, 101152. [CrossRef]
12. Casagrande, B.; Saboy, F.; Tiban, S.; McCartney, J.S. Mechanical response of a thermal micro-pile installed in stratified sedimentary soil. In *E3S Web of Conferences*; EDP Sciences: Les Ulis, France, 2020; Volume 205, p. 05007. [CrossRef]

13. Wang, P.; Yin, Z.; Zhou, W.; Chen, W. Micro-mechanical analysis of soil–structure interface behavior under constant normal stiffness condition with DEM. *Acta Geotech.* **2022**, *17*, 2711–2733. [CrossRef]
14. Liu, T.; Chen, H.; Buckley, R.M.; Quinteros, V.S.; Jardine, R.J. Characterisation of sand-steel interface shearing behaviour for the interpretation of driven pile behaviour in sands. In *E3S Web of Conferences*; EDP Sciences: Les Ulis, France, 2019; Volume 92, p. 13001. [CrossRef]

applied
sciences

MDPI

Article

Impact Resistance and Flexural Performance Properties of Hybrid Fiber-Reinforced Cement Mortar Containing Steel and Carbon Fibers

Jong-Gun Park [1,*], Dong-Ju Seo [2] and Gwang-Hee Heo [3]

[1]   Public Safety Research Center (PSRC), Konyang University, 121, Daehak-ro, Nonsan-si 32992, Korea
[2]   School of Disaster and Safety Engineering, Konyang University, 121, Daehak-ro, Nonsan-si 32992, Korea
[3]   Department of International Civil and Plant Engineering, Konyang University, 121, Daehak-ro, Nonsan-si 32992, Korea
*   Correspondence: park2630@konyang.ac.kr; Tel.: +82-41-730-5644

Abstract: Fiber-reinforced cement mortar (FRCM) has been widely used since it has many advantages compared to plain mortar (PM), and various fibers are highly applicable as repair and reinforcement materials for concrete. In the present paper, an experimental study was planned to investigate the properties, such as flexural performance (flexural strength and toughness), compressive strength, and impact resistance of mono fiber-reinforced cement mortar (MFRCM) containing only steel fiber (SF) or carbon fiber (CF), as well as hybrid fiber-reinforced cement mortar (HyFRCM) containing different combinations of SF and CF. The fiber content was used in five levels (0.0, 0.25, 0.5, 0.75, and 1.0%) at a total volume fraction of 1.0% by volume. The results show that HyFRCM containing 0.75% SF and 0.25% CF improved compressive strength, flexural strength, and impact resistance compared to MFRCM and other HyFRCM, resulting in a synergistic effect of hybrid reinforced fibers. It is noted that, in the case of HyFRCM containing 0.5% SF and 0.5% CF, the flexural strength was slightly lower, but the highest flexural toughness was obtained, which led us to judge that the result shown in this investigation can be the optimal fiber combination to improve toughness and energy absorption capacity.

Keywords: fiber-reinforced cement mortar; steel fiber; carbon fiber; flexural toughness; impact resistance

Citation: Park, J.-G.; Seo, D.-J.; Heo, G.-H. Impact Resistance and Flexural Performance Properties of Hybrid Fiber-Reinforced Cement Mortar Containing Steel and Carbon Fibers. *Appl. Sci.* **2022**, *12*, 9439. https://doi.org/10.3390/app12199439

Academic Editors: Francisco B. Varona Moya and Mariella Diaferio

Received: 23 August 2022
Accepted: 16 September 2022
Published: 21 September 2022

Publisher's Note: MDPI stays neutral with regard to jurisdictional claims in published maps and institutional affiliations.

1. Introduction

Cement-based composites have been widely used worldwide as some of the most frequently employed construction materials for buildings and civil structures, owing to their excellent quality and durability in comparison with the cost [1,2]. Nevertheless, cement-based composites have well-known disadvantages, such as weak flexural and tensile strength, and low performance in terms of strain capacity, ductility, toughness, and fatigue. As a means to improve these drawbacks, fiber-reinforced cement mortar (FRCM) has been developed by inputting chopped fibers in cement composites irregularly and discontinuously dispersed [3–5]. In general, fibers are used to improve the shrinkage cracking, flexural performance (flexural strength and toughness) and impact resistance of cement composites [6–8]. In mono fiber-reinforced cement mortar (MFRCM), which contains one type of fiber, it is difficult to expect a synergistic effect of fiber reinforcement over a certain level within a limited range [9–12]. However, a hybrid fiber-reinforced cement mortar (HyFRCM) in which two or more different fibers are properly incorporated can achieve much improved mechanical properties and impact resistance because different fibers share their roles in it [13–16].

The most commonly applied fibers in the construction of concrete structures are classified into micro and macro fibers according to their size (length and diameter). Micro-fibers have a length of 5 to 10 mm and a diameter of 7 to 30 μm and, according to the material, they are classified into carbon fiber (CF), glass fiber (GF), and basalt fiber (BF) [17–20]. On the other hand, macro-fibers have a length of 25 to 60 mm and a diameter of 0.2 to 0.8 mm, in such types as steel fiber (SF) and polyvinyl alcohol fiber (PVAF) [21–23].

Figure 1 shows the action of the hybrid fibers of different sizes in relation to crack bridging occurring in the cement matrix. As seen in Figure 1, the micro-fibers are cross-linked at the stage of micro-cracks, and at the stage where these micro-cracks develop into macro-cracks, the macro-fibers cross-linked in the cracks, which is expected to improve flexural performance and impact resistance. In other words, short micro-fibers more effectively control micro-cracks, while relatively long macro-fibers more effectively control large cracks [24–26].

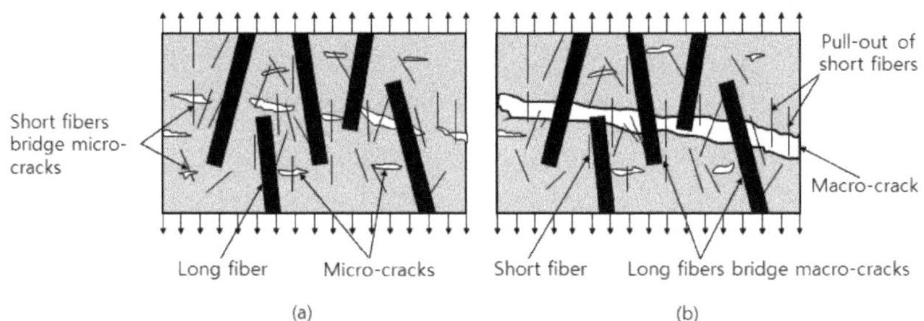

Figure 1. Bridging action of the hybrid with different size; (**a**) first phase of loading, and (**b**) second phase of loading.

Although MFRCM has been used only for a single fiber, various studies have been conducted recently on HyFRCM, which can maximize the effect that a single fiber cannot exert if two or more fibers with different material properties are used in an appropriate ratio. Even in the case of HyFRCM containing different combinations of SF and CF, it is considered to be very effective in terms of improving mechanical properties, such as impact resistance, flexural performance, and compressive strength, but the problem is that information on fiber combinations is too limited. In consideration of constructability and economic feasibility particularly, fibers with a fiber combination of 1.0% or less are widely applied to the construction of concrete structures, but research on this is necessary.

Therefore, it is intended in this study to evaluate such properties as flexural performance, compressive strength, and impact resistance of HyFRCM containing SF and CF with the goal of developing FRCM containing fibers with different material properties to obtain a synergistic effect that cannot be achieved with single fibers alone. In addition, their performance is compared and reviewed with plain mortar (PM). For this objective, the optimal fiber combination of HyFRCM incorporated with SF and CF was examined.

2. Materials and Methods

2.1. Materials

The cement used in this investigation was Ordinary Portland Cement (OPC) manufactured by Company S, with a specific gravity of 3.13 and a fineness of 3860 cm^2/g. For fine aggregate, Jumunjin standard sand produced in Jumunjin-eup, Gangneung-si, Gangwon-do was used in order to make a homogeneous mortar mixture. The specific gravity of the fine aggregate with a dry and saturated surface was 2.65. The admixture used is a polycarboxylic acid-based product manufactured by Company D in Korea, and a light yellow, high-performance water reducing agent with a specific gravity of 1.04 and pH 5.0 ± 1.5. In this investigation, two types of fibers (SF and CF) with different properties were used. Table 1 shows the physical properties of the SF used in this investigation. The SF has a diameter of 0.5 mm, a length of 30 mm, and a tensile strength of 1100 MPa.

Table 1. Physical properties of SF.

Length (mm)	Diameter (mm)	Aspect Ratio (L/D)	Density (g/cm^3)	Tensile Strength (MPa)	Elongation (%)	Elastic Modulus (GPa)
30	0.5	60	7.85	1100	>3.5	>210

In addition, the physical properties of CF used in this investigation are shown in Table 2. CF has a diameter of 7 μm and a length of 6 mm, and the tensile strength and modulus of elasticity are 4900 MPa and 230 GPa, respectively. It is manufactured by Japanese company T, which uses acrylic nitrile (polyacrylonni-trile, PAN) as a raw material. Figure 2 is a photographic representation of the SF and CF used in this investigation.

Table 2. Physical properties of CF.

Length (mm)	Diameter (μm)	Aspect Ratio (L/D)	Density (g/cm^3)	Tensile Strength (MPa)	Elongation (%)	Elastic Modulus (GPa)	Carbon Content (%)
6	7	857	1.8	4900	2.1	230	>92

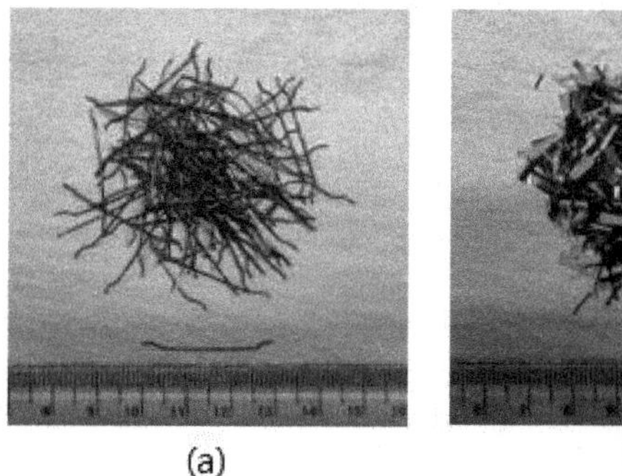

(a) (b)

Figure 2. Morphology of fibers; (**a**) SF, and (**b**) CF.

2.2. Mix Proportion and Mixing Procedure

The mixture designations and proportions used in this investigation are summarized in Table 3. As seen in Table 3, a mixture of six types of cement mortars was prepared. The water:cement ratio (W/C) was set to be 0.46, and the cement mortar was mixed in a ratio (mass ratio) of cement:fine aggregate:water = 1:2:0.46. Specimens were prepared by adding 0.5% to 1.5% of SP (super-plasticizer) based on the mass of cement in order to secure a certain fluidity while mixing. In the case of PM or SF, SP was not added separately. As for the mixing method, cement and fine aggregate were put first, and dry mixing was performed at a low speed for 30 s. Subsequently, 1/3 of the mixing water was added to the mixture and further mixed for 30 s. In order to secure the dispersibility of the fibers, fibers were added and mixed at a medium speed for 1.5 min. Then, the remaining water and SP were added and mixed for 30 s immediately. Following the stop for 30 s, the attached cement mortar was removed and, finally, the mixer was operated again and mixed at a high speed for 2 min. The total mixing time was about 5 min. Compaction of cement mortars was achieved by using a vibrating table, and the vibration times varied depending on the stiffness of the mixtures. After the mold was completed, all specimens were covered with a plastic sheet for 36 h to prevent moisture evaporation, and the molds were demolded and cured in water until the age of 28 days in a curing tank in which the temperature was maintained at 20 ± 2 °C.

Table 3. Mixture designation and used materials proportions.

Designation of Mixtures	Fiber Type	Fiber Mix Proportion by Volume (V_f)		Total Volume Fraction (%)	Mixture Proportions (Unit Weight, kg/m^3)			SP (C × %)
		(%)	(kg/m^3)		W	C	FA	
SP0 (plain)	0	0.0	0.0	0.0	297	645	1290	-
SC1	SF	1.0	78.5					
	CF	0.0	0.0					
SC2	SF	0.75	58.87	1.0	297	645	1290	0.0~1.5
	CF	0.25	13.50					
SC3	SF	0.5	39.25					
	CF	0.5	9.00					
SC4	SF	0.25	19.62					
	CF	0.75	4.50					
SC5	SF	0.0	0.0					
	CF	1.0	18.0					

W, water; C, cement; FA, fine aggregate; SP, super-plasticizer.

2.3. Test Methods

2.3.1. Flexural Performance Test

For the flexural performance test, beam specimens of $100 \times 100 \times 400$ mm^3 suggested by ASTM C1609/C1609M [27] and the Korean regulations of KS F 2566 [28] were fabricated, and tests of all specimens were performed at the age of 28 days. Figure 3 shows the set-up of the test equipment installed by the third-point loading method for the flexural performance test. It is the test equipment of a universal testing machine (UTM) with a capacity of 2500 kN.

Figure 3. View of specimens and set-up for flexural performance test.

Meanwhile, deflection and applied load were measured using two linear variable displacement transducers (LVDTs) set up in the center of both sides of the specimen. The loading rate was controlled at a constant speed of 0.2 mm/min until the specimen was destroyed at L/1500 per minute, and the flexural load was recorded by using a load cell with a capacity of 850 kN. The flexural strength (f_r) in the third-point loading method is calculated by Equation (1) at the maximum load [27,28].

$$f_r = \frac{PL}{bh^2} \tag{1}$$

where, P is the maximum load (N), L is the span length (mm), b is the width of the specimen (mm), and h is the height of the specimen (mm).

Equivalent flexural strength is defined as the average flexural strength at a given deflection (L/150) in the load-deflection curve obtained from the flexural performance test. The equivalent flexural strength in the third-point loading method is calculated as in Equation (2) [27,28].

$$f'_r = \frac{A_b}{\delta_{tb}} \times \frac{L}{bh^2} \tag{2}$$

where, f'_r is the equivalent flexural strength (MPa), δ_{tb} is the deflection (mm) of L/150, L is the span length (mm), b is the width of the specimen (mm), and h is the height (mm) of specimen, and A_b is the area (J, kN·mm) from the load-deflection curve to δ_{tb}.

2.3.2. Evaluation Methods of Flexural Performance

Figure 4 is a graph presented in ASTM C1609/C1609M [27] and the domestic regulation of KS F 2566 [28], showing a typical load-center point pure deflection curve to obtain the flexural performance of a beam specimen. As evidenced in Figure 4, the flexural toughness evaluates the energy absorption capacity by accumulating the sum of the area under the load-deflection curve until the deflection reaches L/600 and L/150. In this study, the respective load values P_{600} and P_{150} are read from the load-deflection curve for the deflection values corresponding to L/600 and L/150, and then f_{600} and f_{150} are obtained by substituting the values into the flexural strength Equation (1). Thus, the flexural toughness measured at 0.5 mm with a deflection of L/600 is T_{600}, and that measured at 2.0 mm with a deflection of L/150 is T_{150}.

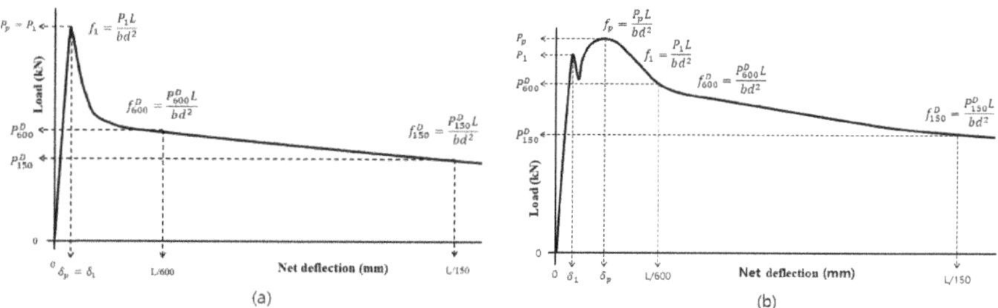

Figure 4. Definition of flexural toughness according to ASTM C1609/C1609M and KS F 2566; (**a**) when the maximum load and the first peak load match, and (**b**) when the maximum load and the first peak load are different.

2.3.3. Impact Resistance Test

The impact resistance test of the specimens was performed in compliance with the Korean regulations KS F 2221 [29]. Figure 5 indicates a schematic diagram of a cement mortar specimen for impact resistance test as well as the test equipment used. The size of all the specimens was 400 × 300 mm (width × length), and the thickness was 30 mm. A total of 18 specimens were prepared and measured at the age of 28 days. For a blow, the specimen was placed horizontally so as to be supported on the entire surface of the sand, and a 1.0 kg steel ball (41 mm in diameter) was allowed to naturally drop from a height of 900 mm above the center of the surface of the specimen. Then, the number of blows until the first crack appeared and the number of blows until the final fracture of the specimen were measured. Equation (3) was adopted to calculate the energy absorption at a specific height.

$$E_i = (N)\, mgh \tag{3}$$

where, E_i is the impact energy (N·m), N is the number of blows, m is the mass of drop hammer (N), g is the gravitational acceleration (9.80 m/s^2), and h is the height of drop hammer (m).

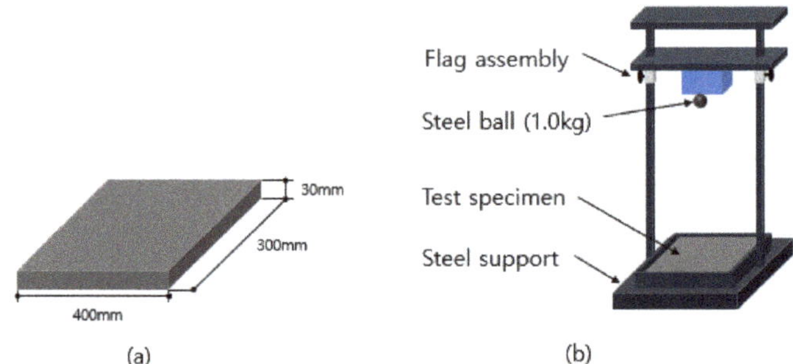

Figure 5. Impact resistance test; (**a**) schematic diagram for specimen fabrication, and (**b**) used apparatus for implementation of impact resistance test.

2.3.4. Compression Test

For the compression test, a mold was manufactured according to the test method of KS L ISO 679 [30], and the compressive strength was measured at the age of 28 days. The compressive strength of the specimens cut from $40 \times 40 \times 160$ mm^3 size was measured by using a universal testing machine (UTM, Korea, UT-100F, MTDI, Daejeon City, Republic of Korea) with a capacity of 100 kN, and the loading rate was applied at a constant speed under the conditions of 2400 N/s.

3. Results and Discussion

3.1. Properties of Mechanical

3.1.1. Flexural Strength Test Results

Figure 6 presents the average flexural strength test results of MFRCM and HyFRCM compared to PM. As is clear from Figure 6, the average flexural strength of PM at the age of 28 days was 3.09 MPa. The average flexural strength of MFRCM containing only CF or SF was 4.96 MPa and 5.85 MPa, respectively, which was about 60.5~89.3% compared to PM. According to the results, HyFRCM containing 0.75% SF and 0.25% CF had the highest average flexural strength of 7.68 MPa, revealing the highest flexural strength improved by about 148.5% compared to PM. It also turns out that the average flexural strength of HyFRCM containing 0.75% SF and 0.25% CF or 0.5% SF and 0.5% CF increased by about 31.3% and 28.2%, respectively, compared to the flexural strength of MFRCM containing only SF. This is thought to be because the incorporated fiber plays a cross-linking role in preventing cracks from growing and improving flexural strength through redistribution of stress. That is, it is considered that the performance is improved by controlling the rather macro-cracks of the relatively long steel fibers while the short carbon fibers control the micro-cracks [31]. Once SF and CF are used in an appropriate fiber combination, a synergistic effect could be obtained by increased flexural strength even in the low fiber volume fraction range. Since the optimal fiber combination to obtain the maximum flexural strength is found to be 0.75% SF and 0.25% CF in this experimental investigation, it is believed to be the most appropriate mixing ratio to improve flexural strength.

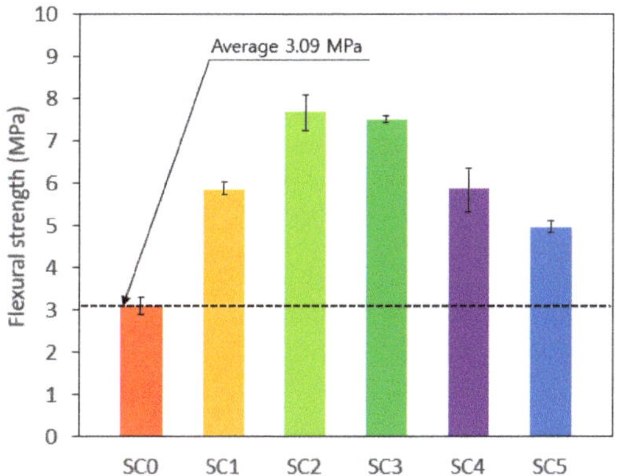

Figure 6. Flexural strength of MFRCM and HyFRCM versus PM.

3.1.2. Compressive Strength Test Results

Figure 7 shows the average compressive strength test results of MFRCM and HyFRCM compared to PM. As shown in Figure 7, the average compressive strength of PM at the age of 28 days was 45.9 MPa, and the average compressive strength of MFRCM containing only SF or CF was 45.3 MPa and 35.6 MPa, respectively. However, the average compressive strength of HyFRCM containing 0.75% SF and 0.25% CF was 46.9 MPa, the highest compressive strength that was ever obtained in this experimental investigation. The reason for the increase in compressive strength is that it is reinforced with micro-CF, which is relatively small compared to SF. In the case of MFRCM containing only CF, however, the average compressive strength was 35.6 MPa, a decrease by about 22.4% compared to PM. This seems to be because the compressive strength was lowered as the interfacial bonding force between the fibers and the matrix after curing became weak due to the non-hydrophilicity of the fiber surface [32]. If a large amount of CF was used, particularly, the dispersibility was lowered, and the workability and compressive strength were greatly affected due to the fiber balls phenomenon. Therefore, considering that the HyFRCM containing 0.75% SF and 0.25% CF was shown as the optimum fiber combination for obtaining the maximum compressive strength in this experimental investigation, it is considered to be the most appropriate mixing ratio to improve the compressive strength.

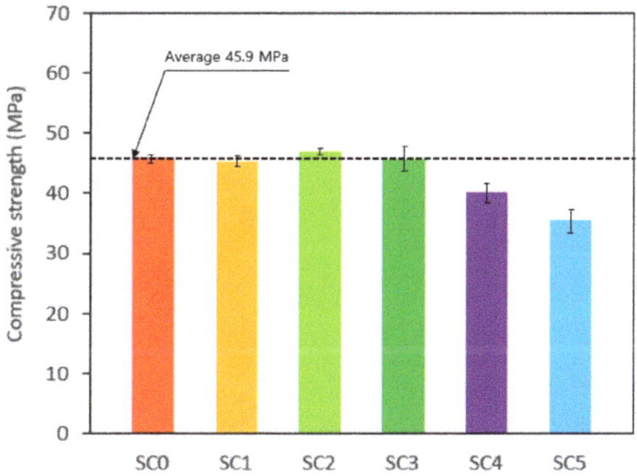

Figure 7. Compressive strength of MFRCM and HyFRCM versus PM.

3.1.3. Comparative Review of Equivalent Flexural Strength and Residual Flexural Strength

In Korea, the flexural toughness specification requires that the equivalent flexural strength be 68% or more of the maximum flexural strength until the deflection becomes 2 mm through the flexural performance test for the beam specimen, and the maximum flexural strength is required to satisfy the design flexural strength (4.5 MPa). The maximum flexural strength, equivalent flexural strength, the ratio of equivalent flexural strength, residual flexural strength, and the ratio of residual flexural strength for each specimen are summarized in Table 4, and the calculation results of equivalent flexural strength and residual flexural strength of each specimen are shown in a graph in Figure 8. As can be seen from Table 4 and Figure 8, the MFRCM containing only PM or CF was difficult to make a relative comparison due to its rapid brittle behavior after initial cracking, and it was destroyed before the deflection reached 0.5 mm for L/600 and 2.0 mm for L/150. However, MFRCM containing SF only and HyFRCM containing SF and CF failed at a deflection of 2.0 mm or more due to their ductile behavior. In the case of HyFRCM containing 0.5% SF and 0.5% CF especially, it exhibited the highest equivalent flexural strength, which could mean that the CF bears the stress, such as the SF when the initial crack occurs, and the SF bears the stress after the crack occurs, resulting in higher equivalent flexural strength by relieving the stress concentration borne by SF. On the other hand, the residual flexural strength of the MFRCM containing only PM or CF could not be obtained, while the HyFRCM containing 0.5% SF and 0.5% CF featured the highest residual flexural strength. It was made clear that the SF has a higher crack suppression effect and strain capacity compared to the CF when the specimen is broken. When MFRCM and HyFRCM were compared with PM, both maximum flexural strength and equivalent flexural strength tended to increase significantly, and the equivalent flexural strength of HyFRCM containing 0.5% SF and 0.5% CF increased by about 25.3 times. Therefore, as demonstrated in Tables 4 and 5, the application of equivalent flexural strength is judged to be a rather conservative evaluation, given that it has a predetermined residual flexural strength and a large amount of flexural toughness even under a deflection of 2.0 mm or more.

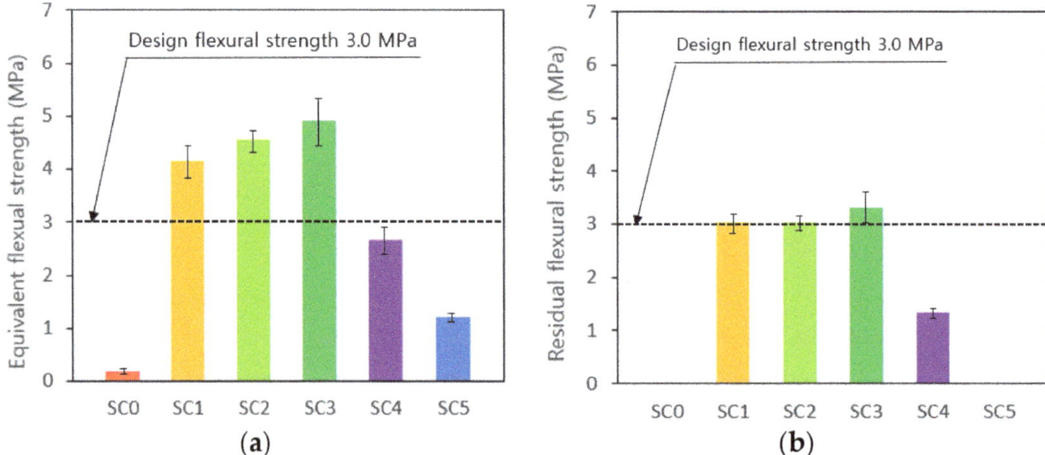

Figure 8. Variation of calculation results of (**a**) equivalent flexural strength, and (**b**) residual flexural strength.

Table 4. Calculation results of equivalent flexural strength and residual flexural strength.

Mixture ID	Fiber Mix Proportion by Volume (%)		Total Volume Fraction (%)	f_r (MPa) (1)	f'_r (MPa) (2)	f'_r/f_r (2)/(1)	R_r (MPa) (3)	R_r/f_r (3)/(1)
	SF	CF						
SC0-1				2.94	(0.21)	(0.07)	-	-
SC0-2	0	0	0	3.24	(0.15)	(0.05)	-	-
SC0-3				3.09	(0.18)	(0.06)	-	-
Average				3.09	(0.18)	(0.06)		
SC1-1				5.79	3.97	0.69	2.88	0.50
SC1-2	1	0		5.97	4.41	0.74	3.30	0.55
SC1-3				5.79	4.08	0.70	2.94	0.51
Average				5.85	4.15	0.71	3.04	0.52
SC2-1				8.16	4.76	0.58	3.12	0.38
SC2-2	0.75	0.25		7.20	4.40	0.61	2.88	0.40
SC2-3				7.68	4.51	0.59	3.09	0.40
Average				7.68	4.56	0.59	3.03	0.39
SC3-1			1.0	7.50	4.41	0.59	2.96	0.40
SC3-2	0.5	0.5		7.44	5.36	0.72	3.51	0.47
SC3-3				7.56	4.95	0.65	3.45	0.46
Average				7.50	4.91	0.65	3.31	0.44
SC4-1				6.42	2.93	0.46	1.44	0.22
SC4-2	0.25	0.75		5.34	2.33	0.44	1.32	0.25
SC4-3				5.88	2.76	0.47	1.26	0.21
Average				5.88	2.67	0.46	1.34	0.23
SC5-1				4.83	(1.28)	(0.26)	-	-
SC5-2	0	1		5.10	(1.19)	(0.23)	-	-
SC5-3				4.95	(1.23)	(0.25)		
Average				4.96	(1.23)	(0.25)		

f_r, maximum flexural strength; f'_r, equivalent flexural strength; f'_r/f_r, equivalent flexural strength ratio; R_r, residual flexural strength; R_r/f_r, residual flexural strength ratio; values in brackets indicate the value destroyed ones when the span did not reach $L/150$ (2 mm).

Table 5. Test results of flexural performance indices using ASTM C 1609 and KS F 2566.

Mixture ID	Fiber Mix Proportion by Volume (%)		Total Volume Fraction (%)	P_{600} (kN)	P_{150} (kN)	f_{600} (MPa)	f_{150} (MPa)	T_{600} (J)	T_{150} (J)
	SF	CF							
SC0-1				-	-	-	-	-	-
SC0-2	0	0	0	-	-	-	-	-	-
SC0-3				-	-	-	-	-	-
Average									
SC1-1				18.4	9.60	5.52	2.88	5.01	26.5
SC1-2	1	0		18.5	10.8	5.55	3.24	5.99	29.4
SC1-3				18.9	9.80	5.67	2.94	5.50	27.2
Average			1.0	18.6	10.1	5.58	3.02	5.50	27.7
SC2-1				25.3	10.5	7.59	3.15	5.05	31.7
SC2-2	0.75	0.25		21.9	9.91	6.57	2.97	4.95	29.3
SC2-3				22.9	10.2	6.87	3.06	4.99	30.1
Average				23.4	10.2	7.01	3.06	5.00	30.4

Table 5. *Cont.*

Mixture ID	Fiber Mix Proportion by Volume (%)		Total Volume Fraction (%)	P_{600} (kN)	P_{150} (kN)	f_{600} (MPa)	f_{150} (MPa)	T_{600} (J)	T_{150} (J)
	SF	CF							
SC3-1				24.3	9.79	7.29	2.93	7.71	29.4
SC3-2	0.5	0.5		23.6	11.7	7.08	3.51	8.50	35.7
SC3-3				24.0	11.5	7.20	3.45	8.11	33.0
	Average			24.0	11.0	7.19	3.30	8.11	32.7
SC4-1				18.2	4.80	5.46	1.44	4.75	19.5
SC4-2	0.25	0.75		12.4	4.30	3.72	1.29	4.15	15.5
SC4-3				15.9	4.20	4.77	1.26	4.60	18.4
	Average			15.5	4.43	4.65	1.33	4.50	17.8
SC5-1				15.2	-	4.56	-	4.59	-
SC5-2	0	1		15.3	-	4.59	-	4.44	-
SC5-3				15.2	-	4.56	-	4.50	-
	Average			15.2		4.57		4.51	

3.2. Evaluation of Flexural Performance

Flexural performance is an important parameter in evaluating the effect of fibers on the post-peak behavior of FRCM. To obtain the flexural performance properties of FRCM, various methods have been developed, and intensive studies have been carried out [33–35]. Therefore, in order to evaluate the flexural performance of the specimen in this study, tests were performed using the third-point loading method according to the test method specified in ASTM C 1609/C 1609/M and Korean regulation KS F 2566. Table 5 summarizes the flexural performance test data with parameters such as P_{600}, P_{150}, f_{600}, f_{150}, T_{600}, and T_{150}, and Figure 9 shows the flexural toughness measurement results of MFRCM and HyFRCM at L/600 and L/150. As shown in Table 5, in the case of MFRCM containing only CF, the flexural toughness could not be measured at T_{150}, and the average flexural toughness at T_{600} was the lowest 4.51 J. On the other hand, in the case of HyFRCM containing 0.5% SF and 0.5% CF, the average flexural toughness at T_{600} and T_{150} was the highest 8.11 J and 32.7 J, respectively, and the flexural toughness significantly improved the post-cracking, indicating that it had the highest load-carrying capacity. It is judged that the flexural toughness is improved as the SF controls the macro-cracks relatively, while the CF controls the micro-cracks.

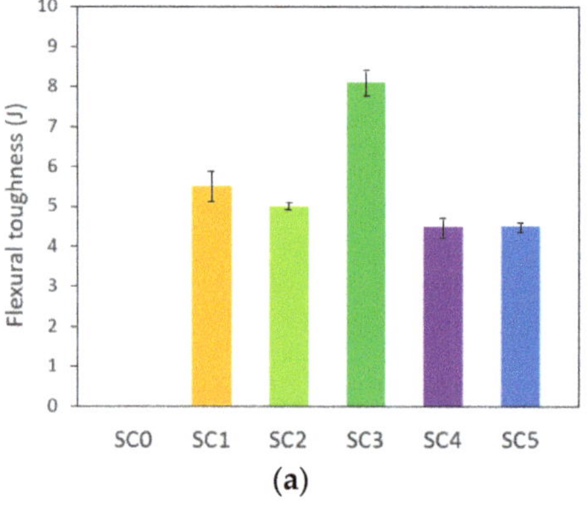

(a)

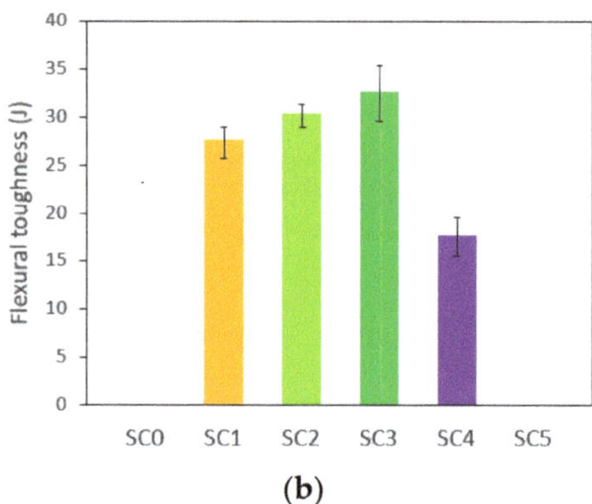

(b)

Figure 9. Flexural toughness of MFRCM and HyFRCM; (**a**) when the deflection is L/600 (0.5 mm), and (**b**) when the deflection is L/150 (2.0 mm).

3.3. Evaluation of Impact Resistance

In Table 6, the test results for the impact resistance of each specimen are summarized and Figure 10 also shows the number of blows for the first crack and the number of blows for final failure. As seen in Table 6 and Figure 10, it is obvious that the impact resistance is greatly improved compared to the PM specimen if SF and CF are incorporated. The improvement of the impact resistance can be mainly due to the randomly distributed fibers in the cement matrix. These fibers, each acting as a miniature energy-absorbing mechanism, support a certain percentage of the load during each impact event. Thus, MFRCM and HyFRCM specimens can bear more impact blows and improve the impact resistance to the first crack compared to PM specimens. After the first crack occurs and is followed by the others, the fibers across these cracks act not only as energy-absorbing mechanisms but also as load transfer mechanisms. The mechanisms may defend the cracked MFRCM and HyFRCM disks against the tendency to fall into different pieces, therefore, improving the failure impact resistance. However, SF showed better performance than CF, having smaller dimensions and the same volume fraction (%) due to higher elastic modulus and tensile stress. The impact resistance of the PM specimen turned out to be about 1 to 2 blows from the first crack to the final failure, revealing that the impact ductility in the specimen had little effect. On the other hand, although there was a slight difference depending on the amount of SF and CF mixture in the case of MFRCM and HyFRCM specimens, the average number of blows from the first crack to final failure reached 39 times in the case of HyFRCM specimen containing 0.75% SF and 0.25% CF, showing that the fiber and impact ductility acted on the specimen even after the first crack occurred, resulting in a slight increase of the number of blows. Therefore, it is judged that the HyFRCM specimen containing 0.75% SF and 0.25% CF is dispersed throughout the specimen due to the cross-linking action of the fibers, and the time from the first crack to the final failure is delayed.

Table 6. Results obtained from the impact resistance test.

Mixture ID	No. of Blows for First Crack (1)	No. of Blows for Final Failure (2)	Increased (2) − (1)	Absorbed Energy (First Crack) (N-m)	Absorbed Energy (Final Failure) (N-m)
SC0-1	1	3	2	8.82	26.46
SC0-2	2	3	1	17.64	26.46
SC0-3	2	4	2	17.64	35.28
Average	2	3	2	14.70	29.40
SC1-1	93	124	31	820.26	1093.68
SC1-2	96	129	33	846.72	1137.78
SC1-3	98	133	35	864.36	1173.06
Average	96	129	33	843.78	1134.84
SC2-1	101	138	37	890.82	1217.16
SC2-2	104	142	38	917.28	1252.44
SC2-3	104	145	41	917.28	1278.90
Average	103	142	39	908.46	1249.50
SC3-1	87	117	30	767.34	1031.94
SC3-2	81	113	32	714.42	996.66
SC3-3	72	105	33	635.04	926.10
Average	80	112	32	705.60	984.90
SC4-1	40	67	27	352.80	590.94
SC4-2	35	57	22	308.70	502.74
SC4-3	32	51	19	282.24	449.82
Average	36	58	23	314.58	514.50
SC5-1	13	18	5	114.66	158.76
SC5-2	12	17	5	105.84	149.94
SC5-3	12	16	4	105.84	141.12
Average	12	17	5	108.78	149.94

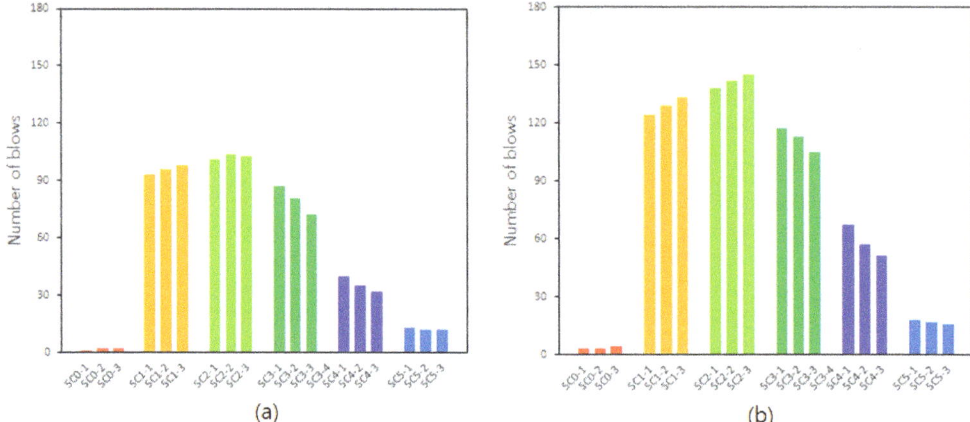

Figure 10. Number of blows for specimens; (**a**) first crack, and (**b**) final failure.

Meanwhile, Figure 11 shows the crack and fracture patterns of the fractured surface after the impact resistance test. As is clear from Figure 11, the specimen exhibited a brittle fracture pattern at the same time as the occurrence of abrupt micro-cracks in the vertical direction in the case of the PM specimen, whereas the MFRCM and HyFRCM specimens containing SF and CF showed cracking in the vertical or diagonal direction. It is analyzed here that the impact resistance increases as the fibers in the cement matrix exert a tensile force against the impact transmitted to the outside. As a consequence, it seems that after the first crack appearance, CF could not bridge on macro-cracks due to their short length. On the other hand, SF bridged on macro-cracks and arrested them.

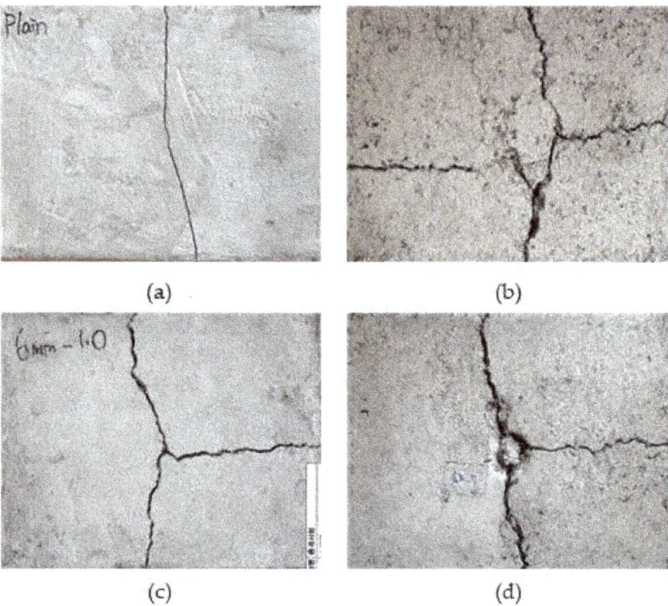

Figure 11. Failure surface of specimens under impact resistance; (**a**) PM, (**b**) MFRCM (SF), (**c**) MFRCM (CF), and (**d**) HyFRCM (SF and CF).

4. Conclusions

In this study, the properties, such as flexural performance, compressive strength, and impact resistance of MFRCM and HyFRCM, were investigated. From the obtained experimental study, the following conclusions can be drawn:

1. The compressive strength of MFRCM and HyFRCM did not improve significantly compared to PM, whereas the flexural strength showed a high improvement effect of about 60.5~148.5%.

2. The mechanical properties of HyFRCM with high SF content were relatively higher than those with low SF content, and HyFRCM containing 0.75% SF and 0.25% CF featured higher strength and impact resistance than other HyFRCM and MFRCM.

3. Although the flexural strength of HyFRCM containing 0.5% SF and 0.5% CF was slightly low, the maximum effect was exhibited in terms of flexural toughness, and the effect of improving flexural toughness by hybridization of SF and CF could be confirmed.

4. Compared to PM, both MFRCM and HyFRCM significantly improved flexural performance, such as maximum flexural strength, flexural toughness, and equivalent flexural strength. In the case of HyFRCM, the application of equivalent flexural strength is judged to be a rather conservative evaluation since it has a predetermined residual flexural strength and a large amount of flexural toughness even under a deflection of 2.0 mm or more.

5. The flexural toughness test results showed that the average flexural toughness of T600 and T150 in HyFRCM containing 0.5% SF and 0.5% CF was 8.1 J and 32.7 J, respectively. After cracking, flexural toughness was greatly improved, which was the most effective in terms of load-carrying capacity.

6. From the impact resistance test, the number of blows from the first crack to the final fracture of the PM specimen is found to be about 1 to 2, whereas, in the case of MFRCM and other HyFRCM specimens, there was a slight difference depending on the amount of SF and CF incorporated but, HyFRCM specimen incorporated with 0.75% SF and 0.25% CF showed the utmost impact resistance. It is thus believed that the effect of increasing the impact ductility by SF and CF is large.

Author Contributions: Conceptualization, J.-G.P. and G.-H.H.; methodology, J.-G.P. and D.-J.S.; validation, G.-H.H. and D.-J.S.; formal analysis, J.-G.P. and D.-J.S.; investigation, J.-G.P. and D.-J.S.; writing—original draft preparation, J.-G.P.; writing—review and editing, G.-H.H. and D.-J.S.; visualization, J.-G.P. and D.-J.S.; supervision, G.-H.H.; project administration, G.-H.H.; funding acquisition, G.-H.H. All authors have read and agreed to the published version of the manuscript.

Funding: This research was funded by Basic Science Research Program through the National Research Foundation of Korea (NRF) of the Ministry of Education, Republic of Korea (Grant no. NRF-2018R1A6A1A03025542).

Institutional Review Board Statement: Not applicable.

Informed Consent Statement: Not applicable.

Data Availability Statement: Not applicable.

Acknowledgments: This research was supported by Basic Science Research Program through the National Research Foundation of Korea (NRF) funded by the Ministry of Education (Grant no. NRF-2018R1A6A1A03025542).

Conflicts of Interest: The authors declare that there is no conflict of interest.

References

1. Malchiodi, B.; Marchetti, R.; Barbieri, L.; Pozzi, P. Recovery of cork manufacturing waste within mortar and polyurethane: Feasibility of use and physical, mechanical, thermal insulating properties of the final green composite construction materials. *Appl. Sci.* **2022**, *12*, 3844. [CrossRef]
2. Nili, M.; Afroughsabet, V. The long-term compressive strength and durability properties of silica fume fiber-reinforced concrete. *Mater. Sci. Eng. A* **2012**, *531*, 107–111. [CrossRef]
3. *ACI 544.2R-89*; Measurement of Properties of Fiber Reinforced Concrete (Reapproved 2009). ACI Committee 544: Farmington Hills, Mi, USA, 2009.
4. Brandt, A.M. Fibre reinforced cement-based (FRC) composites after over 40 years of development in building and civil engineering. *Compos. Struct.* **2008**, *86*, 3–9. [CrossRef]
5. Yoo, D.-Y.; Banthia, N. Mechanical properties of ultra-high-performance fiber-reinforced concrete: A review. *Cem. Concr. Compos.* **2016**, *73*, 267–280. [CrossRef]
6. Hussein, Z.M.; Khalil, W.L.; Hisham; Ahmed, K. Impact strength and shrinkage of sustainable fiber reinforced crushed brick aggregate concrete. *Mater. Proc.* **2021**, *42*, 3022–3027. [CrossRef]
7. Mastali, M.; Dalvand, A. Use of silica fume and recycled steel fibers in self-compacting concrete (SCC). *Constr. Build. Mater.* **2016**, *125*, 196–209. [CrossRef]
8. Won, J.-P.; Park, C.-G. Effects of specialty cellulose fibers on improvement of flexural performance and control of cracking of concrete. *J. Korea Concr. Inst.* **2000**, *4*, 89–98.
9. Heo, G.-H.; Park, J.-G.; Kim, C.-G. Evaluating the resistance performance of the VAEPC and the PAFRC composites against a low-velocity impact in varying temperature. *Adv. Civ. Eng.* **2020**, *2020*, 7901512. [CrossRef]
10. Kim, K.-C.; Yang, I.-H.; Joe, C. Effects of single and hybrid steel fiber lengths and fiber contents on the mechanical properties of high-strength fiber-reinforced concrete. *Adv. Civ. Eng.* **2018**, *2018*, 7826156. [CrossRef]
11. Nili, M.; Afroughsabet, V. The effects of silica fume and polypropylene fibers on the impact resistance and mechanical properties of concrete. *Constr. Build. Mater.* **2010**, *24*, 927–933. [CrossRef]
12. Yoo, D.-Y.; Yoon, Y.-S. A review on structural behavior, design, and application of ultra-high-performance fiber-reinforced concrete. *Int. J. Concr. Struct. Mater.* **2016**, *10*, 125–142. [CrossRef]
13. Banthia, N.; Soleimani, S.M. Flexural response of hybrid fiber-reinforced cementitious composites. *ACI Mater. J.* **2005**, *102*, 382–389.
14. Kim, D.-J.; Park, S.-H.; Ryu, G.-S.; Koh, K.-T. Comparative flexural behavior of hybrid ultra high performance fiber reinforced concrete with different macro fibers. *Constr. Build. Mater.* **2011**, *25*, 4144–4155. [CrossRef]
15. Lee, J.-H.; Park, C.-G. Effect of polyvinyl alcohol fiber volume fraction on pullout behavior of structural synthetic fiber in hybrid fiber reinforced cement composites. *J. Korea Concr. Inst.* **2011**, *23*, 461–469. [CrossRef]
16. Rashiddadash, P.; Ramezanianpour, A.A.; Mahdikhani, M. Experimental investigation on flexural toughness of hybrid fiber reinforced concrete (HFRC) containing metakaolin and pumice. *Constr. Build. Mater.* **2014**, *51*, 313–320. [CrossRef]
17. Ivorra, S.; Garcés, P.; Catalá, G.; Andión, L.G.; Zornoza, E. Effect of silica fume particle size on mechanical properties of short carbon fiber reinforced concrete. *Mater. Des.* **2010**, *31*, 1553–1558. [CrossRef]
18. Tassew, S.T.; Lubell, A.S. Mechanical properties of glass fiber reinforced ceramic concrete. *Constr. Build. Mater.* **2014**, *51*, 215–224. [CrossRef]
19. Monaldo, E.; Nerilli, F.; Vairo, G. Basalt-based fiber-reinforced materials and structural applications in civil engineering. *Compos. Struct.* **2019**, *214*, 246–263. [CrossRef]
20. Ralegaonkar, R.; Gavali, H.; Aswath, P.; Abolmaali, S. Application of chopped basalt fibers in reinforced mortar: A reviewBasalt-based fiber-reinforced materials and structural applications in civil engineering. *Constr. Build. Mater.* **2018**, *164*, 589–602. [CrossRef]
21. Atahan, H.N.; Pekmezci, B.Y.; Tuncel, E.Y. Behavior of PVA fiber-reinforced cementitious mposites under static and impact flexural effects. *J. Mater. Civ. Eng.* **2013**, *25*, 1438–1445. [CrossRef]
22. Huang, J.; Zhou, Y.; Yang, X.; Dong, Y.; Jin, M.; Liu, J. A multi-scale study of enhancing mechanical property in ultra-high performance concrete by steel-fiber@nano-silica. *Constr. Build. Mater.* **2022**, *342*, 128069. [CrossRef]
23. Pakravan, H.R.; Ozbakkaloglu, T. Synthetic fibers for cementitious composites: A critical and in-depth review of recent advances. *Constr. Build. Mater.* **2019**, *207*, 491–518. [CrossRef]
24. Markovic, I. High-Performance Hybrid-Fibre Concrete–Development and Utilisation. Ph.D. Thesis, Technische University, Delft, The Netherlands, 2006.
25. Pakravan, H.R.; Latifi, M.; Jamshidi, M. Hybrid short fiber reinforcement system in concrete: A review. *Constr. Build. Mater.* **2017**, *142*, 280–294. [CrossRef]
26. Vandewalle, L. Postcracking behaviour of hybrid steel fiber reinforced concrete. In Proceedings of the 6th International Conference on Fracture Mechanics of Concrete and Concrete Structures, Catania, Italy, 17–22 June 2007.
27. *ASTM 1609/C 1609M*; Standard Test Method for Flexural Performance of Fibre-Reinforced Concrete (Using Beam with Third-Point Loading). American Society for Testing Materials (ASTM): West Conshohocken, PA, USA, 2008.
28. *KS F 2566*; Standard Test Method for Flexural Performance of Fiber Reinforced Concrete. Korean Agency for Technology and Standards (KATS): Eumseong County, Korea, 2014.

29. *KS F 2221*; Test Method of Impact for Building Boards. Korean Agency for Technology and Standards (KATS): Eumseong County, Korea, 2009.
30. *KS L ISO 679*; Methods of Testing Cements-Determination of Strength. Korean Agency for Technology and Standards (KATS): Eumseong County, Korea, 2016.
31. Betterman, L.R.; Ouyang, C.; Shah, S.P. Fiber-matrix interaction in microfiber-reinforced mortar. *Adv. Cem. Base Mater.* **1995**, *2*, 53–61. [CrossRef]
32. Heo, G.-H.; Park, J.-G.; Song, K.-C.; Park, J.-H.; Jun, H.-M. Mechanical properties of SiO_2-coated carbon fiber-reinforced mortar composites with different fiber length and fiber volume fractions. *Adv. Civ. Eng.* **2020**, *2020*, 8881273. [CrossRef]
33. *ASTM C 1018*; Standard Test Method for Flexural Toughness and First-Crack Strength of Fibre-Reinforced Concrete (Using Beam with Third-Point Loading). American Society for Testing Materials (ASTM): West Conshohocken, PA, USA, 1997.
34. *JSCE-SF4*; Methods of Tests for Flexural Strength and Flexural Toughness of Steel Fiber Reinforced Concrete. Japan Society of Civil Engineers (JSCE): Tokyo, Japan, 1984.
35. Banthia, N.; Trottier, J.F. Test methods for flexural toughness characterization of fiber reinforced concrete: Some concerns and a proposition. *ACI Mater. J.* **1995**, *92*, 48–57.

applied sciences

Article

Earthquake Retrofitting of "Soft-Story" RC Frame Structures with RC Infills

George C. Manos *, Konstantinos Katakalos, Vassilios Soulis and Lazaros Melidis

Laboratory of Strength of Materials and Structures, Aristotle University, 54006 Thessaloniki, Greece
* Correspondence: gcmanos@civil.auth.gr

Abstract: Multi-story, old reinforced concrete (RC) structures with a "soft-story" on the ground floor, sustain considerable damage to the soft story during earthquakes due to the presence of masonry infills in the upper stories. Aspects of such masonry infill–RC frame interaction are briefly discussed and a particular retrofitting scheme for the soft story is studied. It consists of RC infills, added within the bays of the ground floor frames and combined with RC jacketing of the surrounding frame, aiming to avert such soft-story deficiency. The impact of such a retrofit is studied through the measured response of 1/3 scaled single-story, one-bay frames subjected to cyclic seismic-type horizontal loads. It is shown that this retrofit results in a considerable beneficial increase in stiffness, strength, and plastic energy consumption. The importance of the presence of effective steel ties connecting this RC infill with the surrounding frame is also demonstrated. In order to achieve these desired beneficial effects to such vulnerable buildings, additional design objectives are established with the aim of avoiding premature failure of the RC infill panel and/or fracture of the steel ties and to protect the surrounding RC frame from undesired local damage. A numerical methodology, which is validated by using the obtained experimental results, is shown to be capable of predicting reasonably well these important response mechanisms and can therefore be utilized for design purposes.

Keywords: upgrading old RC frame structures; soft-story retrofit; RC infills; masonry infills

Citation: Manos, G.C.; Katakalos, K.; Soulis, V.; Melidis, L. Earthquake Retrofitting of "Soft-Story" RC Frame Structures with RC Infills. *Appl. Sci.* **2022**, *12*, 11597. https://doi.org/10.3390/app122211597

Academic Editor: Andrea L. Rizzo

Received: 21 October 2022
Accepted: 10 November 2022
Published: 15 November 2022

Publisher's Note: MDPI stays neutral with regard to jurisdictional claims in published maps and institutional affiliations.

1. Introduction

The seismic vulnerability of multi-story reinforced concrete (RC)-framed structures built according to relatively old seismic code provisions increases significantly when their ground floor bays are left open to function as parking spaces, whereas the bays of the upper stories are infilled with unreinforced masonry infill panels (UMI). It was demonstrated by extensive past research that the dynamic behaviour of such structures, with a relatively flexible ground floor ("soft story") and stiff upper stories due to the presence of UMI, results in increased earthquake demands on the ground floor columns and shear walls, which are not designed to withstand such force levels. This is due to the interaction of the UMI with the surrounding RC frames, which contributes to a substantial increase of the upper story stiffness compared to the story stiffness of the ground floor. This, in turn, leads to structural damage (Figures 1–3), unless the RC structural elements at the ground floor are properly designed [1–7]. Current revised seismic codes include provisions against such unfavourable seismic responses [4–7]. However, there are many existing old RC structures with such a soft-story seismic deficiency. This undesired stiffness irregularity in elevation, due to the presence of UMI, could also be detected in plans in which UMI are placed irregularly, thus introducing a significant torsional response. It is well known that, apart from the masonry infills, the geometry of an RC building in plan and the location of the vertical structural elements (shear walls and columns) are among the primary causes of torsional response, thus contributing to a further increase in the vulnerability of such buildings [4,8,9]. Current seismic design includes provisions against such unfavourable torsional seismic responses [5–7].

Figure 1. Damaged columns at the soft story.

Figure 2. Temporary wooden story. Damaged shear wall at the soft story.

Figure 3. Damaged columns at the soft-story and temporary steel shoring.

The formation of short columns by constructing UMI within the bays of RC frames also leads to severe structural damage of such RC columns (Figure 4). Retrofit schemes should try to introduce specific countermeasures to effectively address these causes of severe structural damage. During post-earthquake activity, further damage accumulation or even collapse can be prevented by temporary shoring schemes, as depicted in Figures 1–3, where either wooden or steel structural members are employed. Such effective temporary countermeasures, being part of a prompt post-earthquake preparedness plan, could save damaged buildings from total collapse. For buildings with a soft story the objectives of a permanent retrofitting scheme usually are an extension of the temporary countermeasures, e.g., to provide the ground floor story with an increased stiffness, shear capacity, and possibly ductility. In the provisions of relevant guidelines [4], the designer is provided with a number of distinct choices for a retrofitting technique denoted as reinforced concrete infill, whereby reinforced concrete infill panels (RC-IP) are cast in place, filling selected bays of an RC-framed structure. It is also important to identify the irregular placing of UMI within a story or UMI which could trigger the short column failure mechanism and try to introduce effective countermeasures. Valente and Milani [10] by an extensive experimental and numerical study investigate the effectiveness of a number of alternative retrofitting strategies to prevent the failure of existing underdesigned reinforced concrete frames potentially vulnerable to horizontal earthquake loads. They investigated retrofitting schemes with FRP composites or RC jacketing of the columns, eccentric steel bracing, or

RC infill walls spread along the height of a building in order to compare the effectiveness of such retrofitting schemes.

Figure 4. Damaged short columns from the interaction of unreinforced masonry infills (UMI), built within the bay of an RC frame structure, with the adjacent RC columns.

From a variety of retrofitting schemes, this manuscript studies a particular retrofitting scheme that focuses in confining, if possible, the structural intervention only at the ground floor of a soft-story vulnerable building. This retrofit consists of constructing RC infills at the soft-story level together with RC jackets for the members of the surrounding RC frames. In this way, this intervention aims to counteract the described detrimental effect resulting from the increased stiffness of the upper stories due to the presence of the masonry infills. Such a retrofit can be selectively extended, if required, to the upper stories. In this framework, the objectives of the present study are:

(a)   to briefly describe the influence of the presence of masonry infill within an RC frame, which leads to the formation of such a soft story, by presenting a summary of the effects of UMI–RC frame interaction. The importance of the peripheral mortar joint at the interface between the UMI and the surrounding RC frame is underlined, and the necessity of including this interface response mechanism in the vulnerability assessment is highlighted (Section 2);

(b)   to study the impact of the studied particular retrofit consisting of constructing an RC infill within the bay of an existing RC frame. This is done by an experimental study employing 1/3 scaled single-story one-bay RC frames. It is shown that such a retrofit, can be beneficial because it results in considerable increase of the stiffness, strength, and plastic energy consumption. The importance of the presence of properly designed steel ties connecting the RC infill with the surrounding frame is also shown (Section 3); and

(c)   to show the capabilities of a numerical methodology aimed at predicting the RC infill interaction effects, including the nonlinear response of the RC infill, the nonlinear response of the RC frame combined with the nonlinear response of the steel ties, and additional nonlinear response mechanisms at the interface. This numerical approximation, which is validated by utilizing the measured response, is shown to be capable of predicting reasonably well these important response mechanisms. It can therefore have the potential to be utilized for design purposes (Section 4).

2. The Unreinforced Masonry Infill (UMI)—RC Frame Interaction

The in-plane UMI–RC frame interaction has been widely studied in the past by many researchers through analytical treatments [11–16], focused experiments [17–23], and/or numerical modeling [24–36]. The in-plane seismic response of UMI frames problem has also been investigated by tests employing scaled models [37–39]. The influence of UMI is studied by experimentally comparing the response of identical RC frame samples with or without such infills [17–23]. These comparisons document the influence of the inclusion of UMI on the horizontal stiffness, strength, and various nonlinear mechanisms (damage forms). Observed damage during testing develops at either the UMI or/and at the surrounding RC frame members replicating similar damage patterns in prototype structures. A common feature of most of these studies is the in-plane nature of the applied horizontal loads. Thus, the measured stiffness, strength, and damage forms are related mainly to the in-plane response. The usual UMI damage is either the horizontal shear sliding or the diagonal tension/compression within the masonry volume. In addition, compression crushing at the corners of the infill near the RC column to beam joints of the surrounding frame also appears, which can be accompanied by the crushing of the RC column to beam joints themselves. The prevailing form of damage depends on the diagonal or compressive strength of the masonry infill and the cross-section detailing of the RC members. These forms of damage are documented from in situ and laboratory observations. Initially, Stanford–Smith [13,15] proposed an ingenious analytical approach based on the diagonal strut analogy in order to approximate the influence of UMI within an RC frame. Following this basis, many researchers proposed various approximations that numerically simulate with some success these basic behavioural characteristics. Most of these numerical simulations make use of the finite element method, discretizing both the infill and the frame, employing a combination of nonlinear mechanisms for the masonry and the RC structural members [24–36]. A number of researchers base the numerical simulation effort on focused experiments, thus increasing the realism of such numerical predictions [40–46]. The degree of realism depends on how well the numerical predictions simulate the changes of stiffness and strength caused by the inclusion of UMI, as well as on the realism in replicating the forms of damage to the UMI and/or the RC frame. Very few of these numerical approximations recognize the important role played by the interface between the masonry infill and the surrounding frame. As demonstrated [23,44], when the in-plane horizontal load is applied, it results in deformations of both the RC members and the masonry infill that lead to a partial separation of the masonry from the RC frame shown in an amplified form in Figure 5a,b. The shape and distribution of this separation depends on the deformation and strength properties of the RC members, of the UMI, and of the peripheral mortar joint that lies between the UMI and the surrounding frame. The state of stress and the subsequent modes of failure of the masonry and/or of the RC members depend on these separation regions between the RC members and the infill. Therefore, the realism of the numerical predictions obviously depends on the realism to successfully simulate the nonlinear stress-deformation mechanism of all the participating media, e.g., the RC members the masonry infill (UMI) and the peripheral mortar joint at the interface.

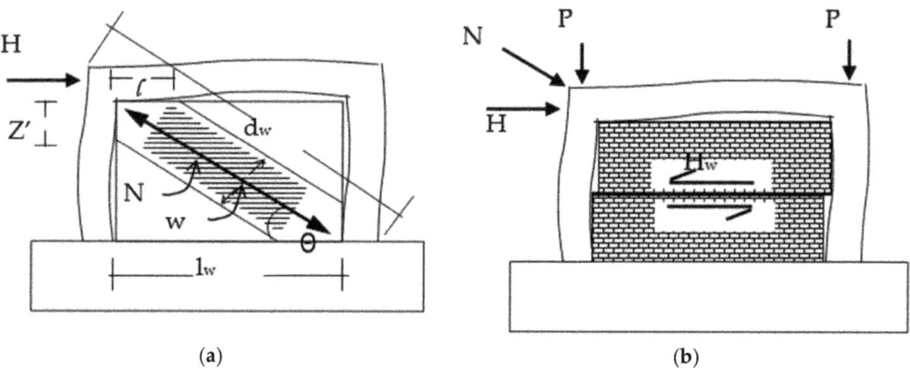

(a)   (b)

Figure 5. (**a**) UMI as a diagonal strut. (**b**) Shear-sliding failure of UMI.

The important aspects of such an interaction have been identified [23], and a comprehensive numerical simulation methodology was proposed that includes all these aspects [23,41,44]. More specifically, this comprehensive numerical simulation includes the following non-linear mechanisms: (a) The development of plastic hinges at either the top and bottom of the RC frame columns or the right and left ends of the RC frame beams. (b) The development of damage within the volume of the masonry infill. (c) The separation of the masonry infill from the surrounding RC frame structural members when a tensile stress field normal to these boundaries develops at the contact surface. (d) Relative sliding between the masonry infill and the frame when the shear stress field parallel to the contact boundaries exceeds certain limit values. (e) Non-linear compressive behaviour of mortar joint at the contact surface depending on the mechanical properties of this peripheral mortar joint. In this way, the interaction between the masonry infill and the surrounding RC frame includes all these nonlinear mechanisms, resulting in a realistic prediction of the actual contact regions, thus resulting in a realistic prediction of the subsequent stress fields of both the RC frame structural members and the masonry infill which governs their potential damage. It is of importance in this approximation that all the mechanical properties of the linear and non-linear range for the UMI, for the peripheral mortar joint, for the RC frame materials and for their contact surfaces are specified through simple testing. This methodology has been described in detail in a past publication [44]. To validate the proposed numerical simulation process, one-bay, one-story RC frame samples were initially tested without either UMI or RC-IP ("bare" frame with the code name F1BN) [23]. Next, additional samples were formed with identical one-bay, one-story RC frames hosting within their bays unreinforced masonry infills constructed with clay bricks (see Table 1 and Figures 6–8).

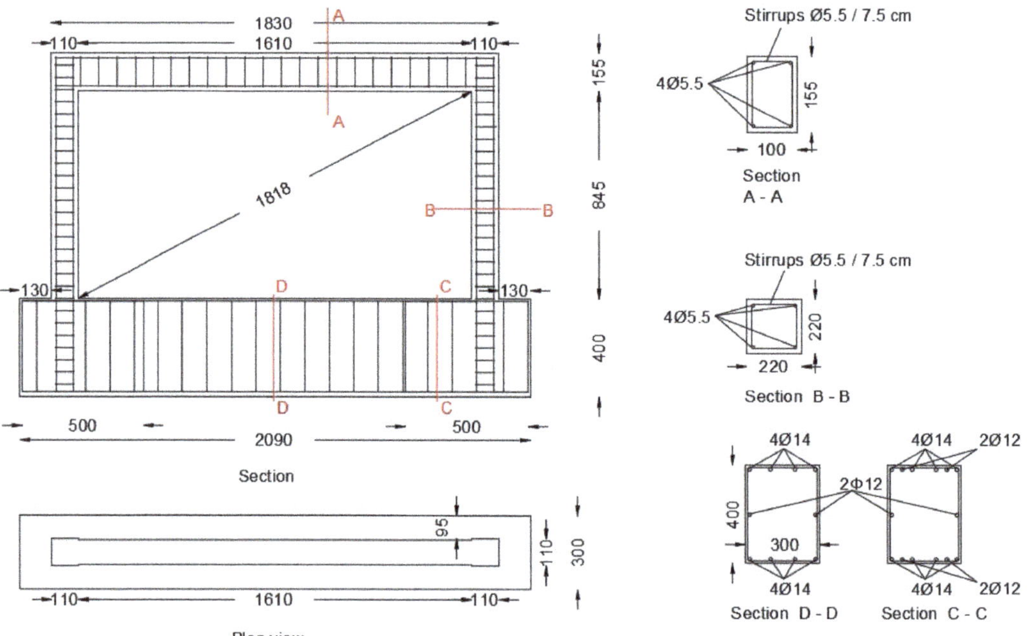

Figure 6. Structural details of the RC frame models hosting UMI tested at Aristotle University [23,44].

One such specimen with the code name F2N is formed with well-built clay masonry of 60 mm nominal thickness by using clay bricks of nominal dimensions 60 mm × 100 mm × 190 mm (4.8 MPa compressive strength) and relatively weak mortar (see Table 1, compressive/tensile strength equal to 1.13/0.11 MPa, E = 100 MPa). Another masonry-infilled specimen (F3NP) is built in the same way as F2N but having each of its two facades covered with a relatively thin layer (15 mm thickness) of strong cement mortar (17 MPa compressive strength) hosting within its thickness a net of thin steel wires, as shown in Figure 7a–c. The clay masonry had a compressive strength of 2.5 MPa and a diagonal tensile strength of 0.15 MPa. In the experimental sequence, the mechanical characteristics of the peripheral

10-mm-thick mortar joint were obtained from tests and were almost the same as those of the mortar joints of the masonry infill (Table 1).

Table 1. Basic mechanical properties of all the building materials [23].

Concrete Compressive Strength (MPa)	Steel Reinforcement Yield/Ultimum Stress (MPa)	Clay Brick Compressive Strength (MPa)	Mortar Compressive/Tensile Strength (MPa)	Masonry Compressive Strength (MPa)	Masonry Diagonal Tension Strength (MPa)	Compressive Strength of Mortar Applied to the Façade (MPa)
(1)	(2)	(3)	(4)	(5)	(6)	(7)
23.3	311/425	4.8	1.13 / 0.12	2.5	0.15	17.0

(a)   (b)   (c)

Figure 7. RC frame + UMI specimen F3NP [23] (**a**) before applying the mortar facades. (**b**) attaching the wire net. (**c**) applying the cement mortar hosting the wire net to produce (Last row of Table 2).

Table 2. Summary of the obtained peak response of the bare and the infilled specimens.

Code Name	Initial hor. Stiffness (kN/mm)	H_{max} Maximum * Hor. Load (kN)	Normalized ** Inter-Story Drift at H_{max}	Cumulative Plastic Energy Till H_{max} (kN/mm)	Initial Stiffness Ratio ***	Maximum Hor. Load Ratio ***	Cumulative Energy Ratio ***
(1)	(2)	(3)	(4)	(5)	(6)	(7)	(8)
Bare F1BN	8.75	15.20	1.04%	472/10 mm	1	1	1
	Decrease Hor. Load 12.50		2.65%	1927/25 mm			
F2N infilled	22.12	35.30	0.80%	924/8 mm	2.53	2.32	1.96
	Further testing stopped						
F3NP infilled	77.84	78.10	0.80%	2743/8 mm	8.90	5.14	5.81
	Decrease Hor. Load 45.00		2.55%	8076/25 mm			

* Average value of measured minimum and maximum load. ** Measured hor. displacement at maximum load normalized by the frame height. *** Ratio of the infilled specimen response by the response of the corresponding bare specimen.

The response of this peripheral mortar joint and the separation of the masonry infill from the surrounding frame were monitored throughout testing. Prior to applying the horizontal cyclic load, each specimen was subjected to a vertical load in such a way that each one of its columns was loaded with an axial load of 50 KN that was kept constant throughout the horizontal loading sequence (Figure 8a,b). Thus, the experimental setup used approximates the stress field on such a subassembly generated by in-plane seismic actions. The same loading arrangement was also used for the specimens of the RC frames infilled with RC panels (Section 3). More information on the masonry infills can be found in a past publication [44]. A brief summary of the measured overall response of the tested three specimens is listed in Table 2; the observed damage patterns are depicted in Figure 8c,d. Apart from the plastic hinges at the column and beam ends, the damage of specimen

F2N as characterized as widespread diagonal cracking of the masonry infill, whereas for specimen F3NP, due to the addition of strong reinforced mortar facades together with a stiff peripheral mortar joint, the damage was concentrated at the corners of the masonry infill and the RC column to beam joints. The following summarizes the main observations.

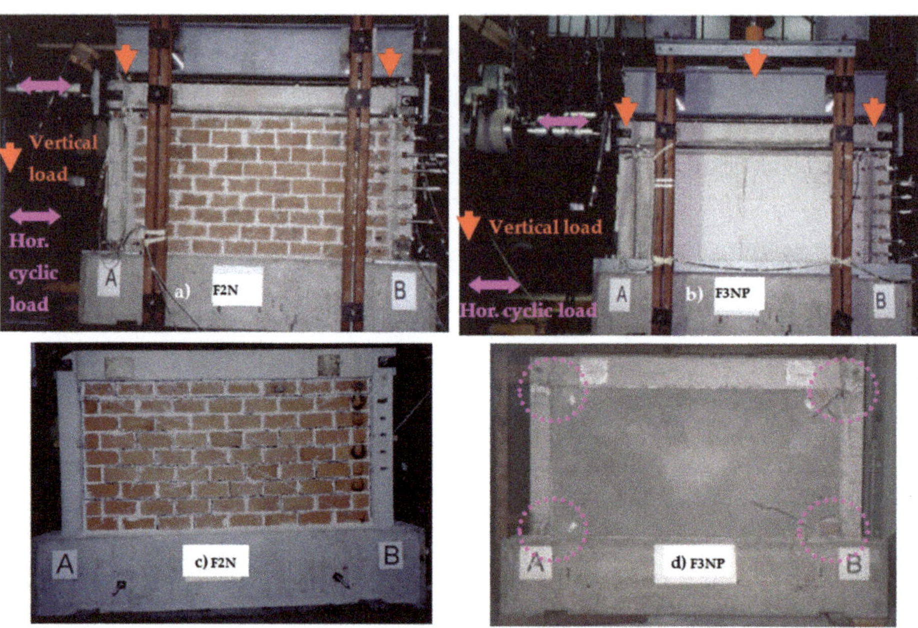

Figure 8. (**a**) The loading arrangement for F2N-UMI specimen (second row of Table 2) (**b**) The loading arrangement for F3NP- UMI specimen after having applied the cement mortar with the wire net (last row of Table 2). (**c**) Observed damage for F2N-UMI specimen (**d**) Observed damage fir F3NP specimen [23,44].

(a)   The presence of UMI results in a significant increase of the horizontal in-plane stiffness and bearing capacity. The presence of reinforced mortar attached on the facades of the UMI results in a further increase in the horizontal stiffness and load-bearing capacity of the subassembly of a UMI and RC frame (columns 2, 3, 8, and 9 of Table 2).

(b)   The deformability and strength of the peripheral mortar joint interface between the UMI and the surrounding frame can influence the results of the infill–frame interaction. A very stiff peripheral mortar joint results in the narrowing of the compressive contact region that may lead to the crushing of either the corners of the infill panel or/and the damage of the RC frame at the same regions. These undesirable consequences are more likely to occur when a nonflexible interface is combined with an UMI of large stiffness, as is the case when strong mortar facades are used as retrofitting without combining them with measures to protect the corners of either the infill or the surrounding frame from crushing (Figure 8d).

(c)   The premature failure of the UMI (Figure 8c) results in a sudden decrease of all the beneficial effects in the horizontal stiffness and load-bearing capacity. Moreover, the in-plane damage of the masonry infill may also facilitate its out-of-plane collapse because of the simultaneous out-of-plane seismic actions.

These important aspects of the interaction between unreinforced masonry infill and the surrounding frame, including the important effect of the peripheral mortar joint, were demonstrated by appropriate numerical simulations that include all these possible nonlinear mechanisms, using the full set of test results for validation [44].

From the previous discussion, the detrimental effects of the presence of UMI are: (a) the undesired forms of damage of underdesigned soft-story systems (Figures 1–3); and (b) the unintentional formation of short columns (Figure 4). The response from both these undesired influences resulting from the presence of UMI could be well predicted, and the structural system could be thus protected by specific detailing during design or by retrofitting countermeasures, among which is the one studied in Section 3. Basic parameters are the mechanical properties of the masonry infill combined with those of the peripheral mortar joint at the interface between the masonry infill and the surrounding frame. A realistic numerical approximation of the seismic response of one-story, one-bay RC frame is the basis for approximating the seismic response of the multistory RC building including UMI [47,48]. The influence of door or window openings should also be included [49–52]. When the detrimental effects of UMI are prohibited, the interaction of UMI with surrounding framed structures can be beneficial because they represent a first line of defense, due to their added stiffness, strength, and energy consumption, against horizontal earthquake loads. The premature failure of UMIs during their in-plane (IP) seismic response represents a considerable repair cost. For strong earthquake excitations, the partial or total out-of-plane (OOP) collapse of UMI panels (Figure 9) is a serious hazard that is being currently investigated [53–55]. Such undesired OOP response interacts with the IP response interaction, being influenced by the boundary conditions and the progress of the damage for the UMI and the surrounding frame.

Figure 9. Out-of-plane collapse of unreinforced masonry infills (UMI) panles from RC frame structures due to earthquake excitation.

3. Cyclic Behaviour of a RC One-Bay, One-Story Frame Retrofitted with an RC Infill Panel

In this section, the impact of the studied particular retrofit consisting of constructing an RC infill within the bay of an existing RC frame is investigated [56–66] as a countermeasure of the presence of the soft story. This is done through an experimental study employing 1/3 scaled single-story, one-bay RC frames, described in Section 3.1., in order to identify significant nonlinear mechanisms from the interaction of the RC infills with the surrounding frame and to quantify the result of such a retrofit on stiffness, strength, ductility, and energy consumption. This retrofitting scheme is usually combined with the jacketing of the columns or beams of the surrounding frame. Choi et al. [57] conducted a series of experiments employing one-bay, single-story scaled RC frames having RC-IP, which were connected by using various structural details to the surrounding frame. They concluded that the addition of an RC infill wall significantly reduced the drift and improved the stiffness and the ultimate strength. Moreover, they observed that the failure mode of the RC infill walls was dominated by a shear compression effect. Anil and Altin [58] and Altin, Anil, and Kara [59] conducted a series of experiments with 1/3 scaled one-bay RC frames, of either one or two stories, including RC infills in various forms that replicated door and window openings. They concluded that such infills increased substantially the stiffness, strength, and energy dissipation of these frames with RC infills when compared with the corresponding bare frames. They also performed analytical studies and predicted quite successfully the strength of these RC infilled frames. Manos [60] and Manos et al. [61,62] investigated the behaviour of RC infilled frames located at the ground floor of buildings subjected to cyclic seismic-type loads. These RC infills were either not connected or connected with the surrounding frame being retrofitted with RC jackets. This manuscript is a summary of the findings of this investigation. Biskinis et al. [63] proposed calculation models for the yield moment, the secant-to-yield-point-stiffness, the cyclic ultimate chord rotation, and the shear resistance (due to diagonal tension, shear sliding, squat-wall effects, etc.) of the composite wall produced by infilling the space between two RC columns with RC panels without encapsulating them. They validated this methodology by comparing their predictions with experimental measurements. Moreti et al. [64] and Papatheocharis et al. [65] employed the diagonal strut model, broadly used as a design tool in the case of masonry-infilled RC frames, for the design of reinforced concrete (RC)-infilled frames. They conducted an extensive literature review and an experimental investigation to support this study, concluding that for the best performance of an RC-infilled frame with non-ductile members, the columns should be strengthened with RC jackets concurrently with the casting of the RC infill. Furthermore, the RC infill should be connected to the frame through dowels, preferably only along the horizontal interfaces with the frame, in order to avoid early failure in the columns. They also concluded that the failure of RC-infilled frames with non-ductile reinforcement detailing is likely to occur in a frame component and not in the infill. Chrysostomou et al. [66] employed the pseudo-dynamic method to test a full-scale 3-D RC four-story building, having the central bay of two opposite three-bay frames retrofitted with RC infills. They used different connection details between the RC infill and the surrounding frame as well as reinforcement percentages for the two infilled frames. They concluded that this is a viable method for retrofitting and can be used to strengthen existing ductility and strength-deficient structures. The current research was initiated by the relevant provisions of the design guidelines (OASP 2012 [67], which cover a wide range of a number of distinct design choices for a retrofitting technique denoted as reinforced concrete infill panel (RC-IP). These choices are: (a) an RC-IP without any connections to the surrounding frame; (b) an RC-IP with light connections to the surrounding frame; and (c) casting an RC-IP with structural details connecting it with the surrounding frame. This represents the construction of a new shear wall extending to the whole building height. The construction of (a) and (b) are limited to the soft story of multi-story buildings with open ground floor parking space where such a retrofitting scheme can be easily applied. The construction of a new shear wall extending to the whole building height is a very effective retrofit in increasing the stiffness, strength, and ductility for an underdesigned RC multistory building [10]. However, such a construction can be a complex, costly, and time-consuming operation for an existing building because it involves a large number of structural interventions and alterations to the functioning spaces at many levels; it also requires strengthening of the foundation for the new shear walls, which represents an additional complex and costly operation. The retrofitting scheme which is studied here is aimed as a structural intervention at the soft story in the following two alternatives:

1. The RC-IP is not structurally connected to the surrounding R/C structural elements (columns or beams). Alternatively, a limited connection between the RC-IP and the upper/lower horizontal frame interface is constructed.

2. The RC-IP is constructed together with steel ties connecting the RC-IP with the surrounding RC structural elements strengthened by jacketing within the bay of a frame (columns or beam). The thickness of this RC-IP is usually smaller than the width of the beams and columns that form the infilled bay of the frame.

For this purpose, the one third (1/3) one-bay, single-story scaled RC specimen is used with all information of its geometry, mechanical characteristics of all the materials and structural detailing defined by testing, as shown in Figure 7, Figure 8 and Table 1. This specimen is step-wise transformed to: (a) a retrofitted frame with RC jackets (Figure 10); (b) a 50-mm-thick unreinforced concrete infill panel (UC-IP) connected to the same retrofitted frame with steel anchors. (Figure 11); and (c) the same as before but having a lightly reinforced RC-IP (Figure 11). Prior to applying the horizontal cyclic load, each specimen was subjected to a vertical load in a way that each one of its columns was loaded with an axial load of 50 KN that was kept constant throughout the horizontal loading sequence. Together with this vertical load, an in-plane horizontal cyclic load was applied at the top bay of the frame of continuously increasing amplitude (Figures 10 and 11). This experimental setup approximates in a realistic way the stress field of such a prototype subassembly subjected to in-plane seismic actions. The measured response is detailed in Section 3.1, and it subsequently utilized to compare it with corresponding numerical predictions, thus validating a specific numerical analysis process (Section 4). There are five specimens in total. Two specimens are RC frames without RC infills (Bare 1 and Bare jacket1,2) and three RC frames are the previous RC frames hosting either UC or RC infills (UC-IP 1, UC-IP 2, RC-IP 3), which are described in detail in this Section 3.1.

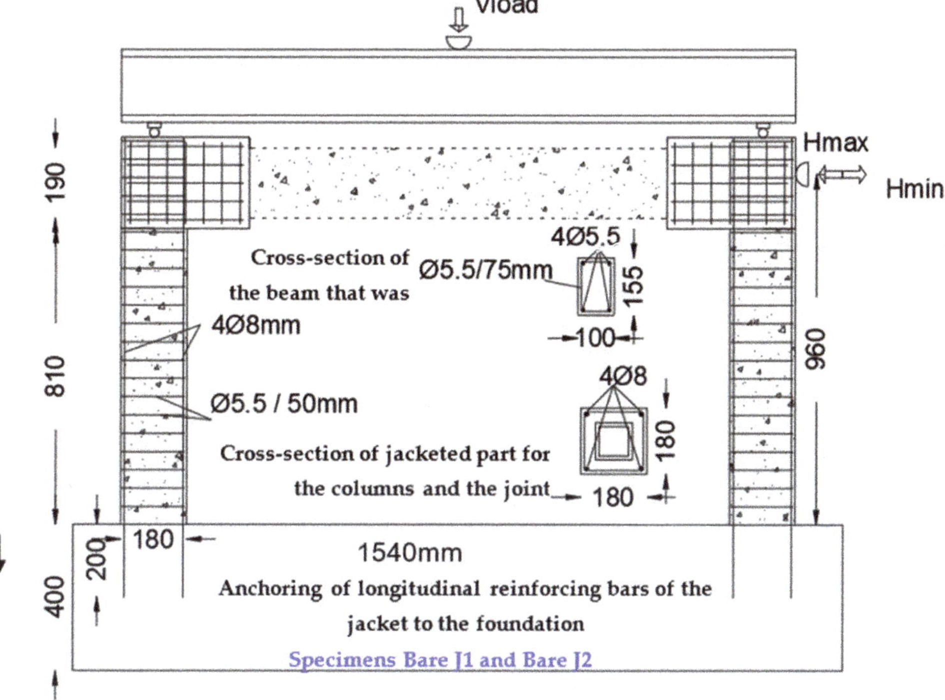

Figure 10. Specimens Bare J1 (Bare Jacket1) and Bare J2 (Bare Jacket2).

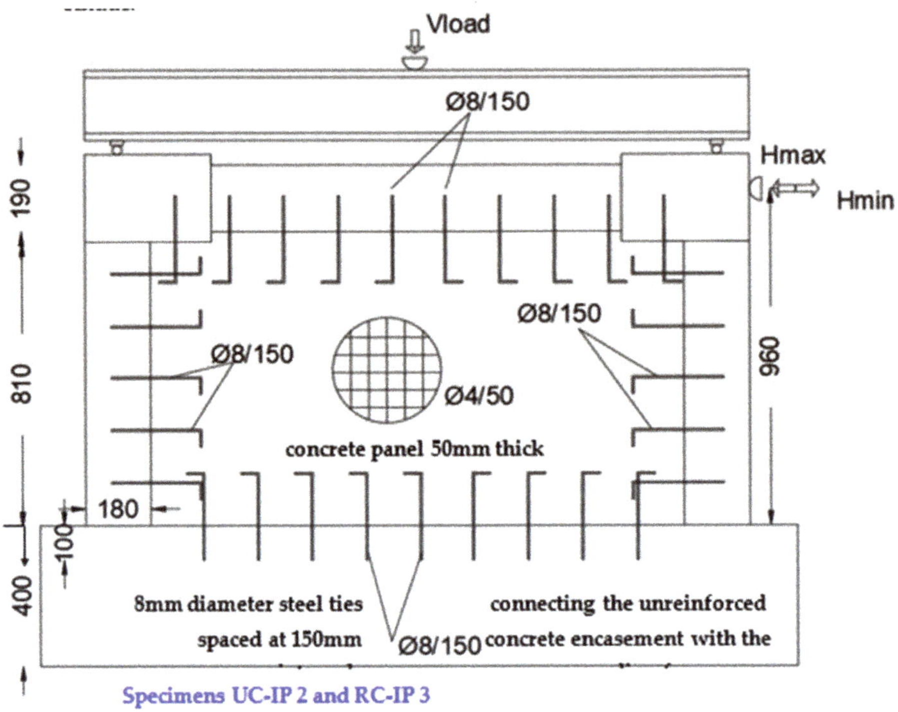

Figure 11. Specimens UC-IP 2 and RC-IP 3.

3.1. Unit T-Beams with Open Hoop CFRP Strips Employing Specific Mechanical Anchoring Devices

- **Bare 1.** This is an RC frame specimen without any jacketing of its columns and top beam with the code name Bare 1. The same is true of the one depicted in Figure 6 for the masonry infill study. The basic mechanical properties of the materials employed in their construction are listed in Table 1. The obtained cyclic response is depicted in Figure 12a.

- **Unreinforced Concrete–Infill Panel 1.** This specimen was formed by filling the bay of specimen Bare 1 with an unreinforced concrete panel having a thickness of 50 mm with concrete compressive strength equal to 22 MPa. This concrete panel was simply cast within the bay without the use of any ties between this concrete panel and the structural members of the surrounding frame. The code name for this specimen is UC-IP 1. Its cyclic response is depicted in Figure 12c.

- **Bare Jacket1.** This specimen (Figure 10) was formed by removing the fractured unreinforced concrete panel of specimen UC-IP 1 and by retrofitting the columns and part of the beam of the RC frame near the beam-to-column joints with concrete jackets. The cross-section of the jacket was 180 mm by 180 mm, having at each of its four corners longitudinal steel reinforcing bars of 8 mm diameter and 570 MPa yield stress (Figure 10). These jackets were also reinforced with closed hoop steel stirrups of 5.5 mm diameter spaced at 50-mm intervals. These jackets were cast with high-strength concrete having a compressive stress of 40 MPa. The code for this specimen is Bare jacket1. The obtained cyclic response for this retrofitted frame is depicted in Figure 12b.

- **Unreinforced Concrete-Infill Panel 2.** This specimen was formed by filling the bay of specimen Bare J1 with an unreinforced concrete panel having a thickness of 50 mm with the concrete compressive strength equal to 22 MPa. This time the unreinforced concrete panel was cast within the bay by using 8-mm-diameter steel ties, with a yield stress equal to 570 MPa, spaced at 150 mm intervals connecting this panel and the structural members of the surrounding frame, as shown in Figure 11. These ties were anchored at the concrete columns and beam, before casting the unreinforced concrete infill, by drilling holes and using a special resin to ensure bonding. The code name for this specimen is UC-IP 2. The obtained cyclic response for this sample is depicted in Figure 12d.

- *Bare Jacket2.* This specimen resulted from removing the fracture unreinforced concrete panel of the previous specimen, UC-IP 2. It is almost the same as the previous retrofitted frame specimen Bare jacket1, having been subjected to a loading sequence as part of the previous specimen UC-IP 2. The code name for this specimen is Bare jacket2. Its cyclic response is depicted in Figure 12b, together with the corresponding response of Bare jacket1.

- *Reinforced Concrete–Infill Panel 3.* This specimen (Figure 11) was formed by filling the bay of specimen Bare jacket2 by using the same steel ties as before for specimen UC-IP 2. The reinforcement of the concrete panel was a net of steel reinforcing bars of 4.5 mm diameter (500 Mpa nominal yield stress) spaced at 85-mm intervals in both horizontal and vertical directions. The code name for this specimen is RC-IP 3. Its cyclic response is depicted in Figure 12e.

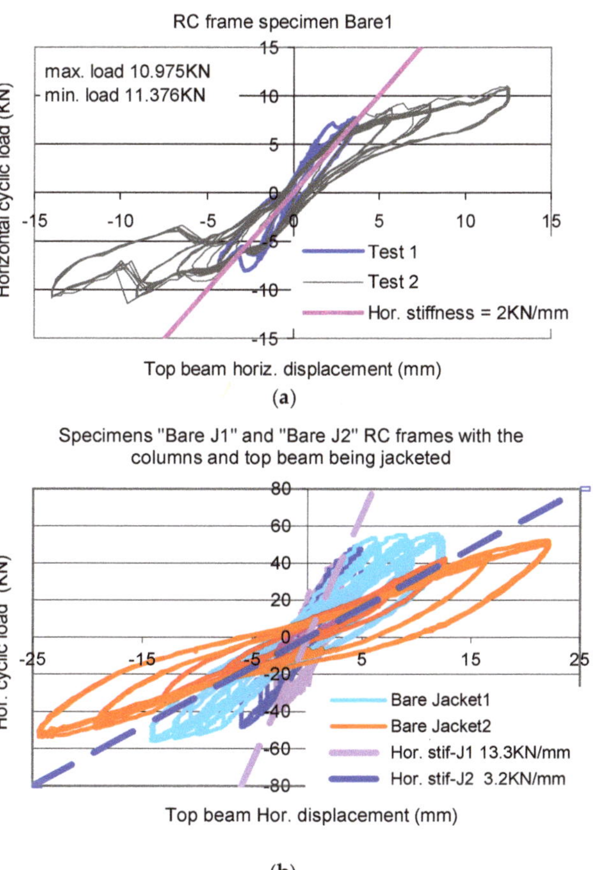

Figure 12. *Cont.*

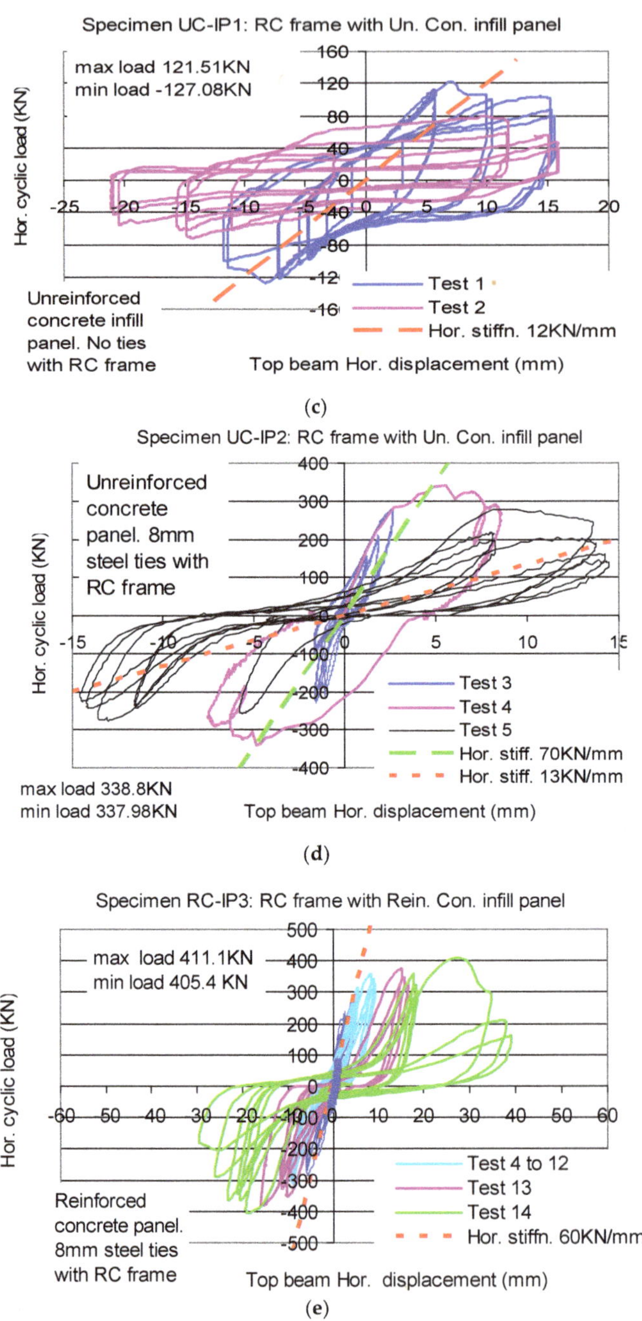

Figure 12. (**a**) Horizontal displacement versus horizontal load cyclic response of specimen Bare-1. (**b**) Horizontal displacement versus hor. load for specimens Bare Jacket1 and Bare Jacket2. (**c**) Horizontal displacement versus horizontal load cyclic response of specimen UC-IP 1. (**d**) Horizontal displacement versus horizontal load cyclic response of specimen UC-IP 2. (**e**) Horizontal displacement versus horizontal load cyclic response of specimen RC-IP 3.

All specimens were subjected to a number of imposed horizontal displacement cycles with a frequency of 0.1 Hz and with a continuously increasing amplitude within sequential loading groups, each group having three cycles of constant horizontal displacement amplitude.

Figure 13a–c depicts the observed damage of specimens UC-IP 1, UC-IP 2, and UC-IP 3, respectively. Summary results of the observed response are listed in Table 3. Column 2 in this table lists the observed initial horizontal stiffness, column 3 the measured maximum horizontal load, and column 4 the plastic energy accumulated from the beginning of loading until the test in which the maximum horizontal load was measured. Columns 5 and 6 in this table list the ratio of the measured maximum response of the tested specimen by the response of the corresponding Bare specimen.

Table 3. Summary of the obtained peak response of the Bare and the Infilled specimens.

Code Name	Initial Hor. Stiffness (kN/mm)	H_{max} Maximum * Hor. Load (kN)	Normalized ** Inter-Story Drift at H_{max}	Cumulative Plastic Energy Till H_{max} (kN/mm)	Initial Stiffness Ratio ***	Maximum Hor. Load Ratio ***	Cumulative Energy Ratio ***
(1)	(2)	(3)	(4)	(5)	(6)	(7)	(8)
Bare 1	2	11.12	1.48%	1241.8	6.0	11.18	11.3
UC-IP 1	12	124.30	0.7%	14,048.6			
Bare jacket1	13.3	55.81	1.48%	7574.3	5.26	6.06	1.98
UC-IP 2	70	338.4	1.48%	15,028.3			
Bare jacket2	3.2	52.99	2.4%	5913.4	18.75	7.7	10.75
RC-IP 3	60	408.25	2.86%	63,597.9			

* Average value of measured minimum and maximum load. ** Measured hor. displacement at maximum load normalized by the frame height. *** Ratio of the Infilled specimen response by the response of the corresponding Bare specimen.

- As expected, the construction of the concrete infill panel, with or without metal ties with the surrounding frame increases considerably the initial horizontal stiffness. The jacketing of the frame columns and part of the top beam also increase, to a lesser degree, this initial horizontal stiffness (Figure 12a,e and column 2 of Table 3).
- The jacketing of the frame columns and part of the top beam increased almost five times the horizontal load-bearing capacity of the initial Bare frame. The Bare frame both before and after jacketing responded in a ductile manner reaching normalized interstory drift values in the range of 1.5% without a considerable decrease in the maximum horizontal load value. The inclusion of an infill panel increases even further, more than six times the horizontal load-bearing capacity of the jacketed frame (Figure 12a–e and column 3 of Table 3).
- The jacketing of the frame columns and part of the top beam increased by more than five times the cumulative plastic energy until the maximum horizontal load was reached when compared with the corresponding cumulative plastic energy of the initial Bare frame. The inclusion of an infill panel increased even further this cumulative plastic energy (Figure 12a–e and column 5 of Table 3). This increase is quite spectacular for specimen RC-IP 3 and is due to the combination of the following contributing mechanisms First, the inclusion of the steel ties connecting the infill panel with the surrounding frame alters the interaction between frame and panel in such a way that the concentration of compressive stresses at narrow zones near the corners of the frame is avoided and neither the concrete panel nor the frame is crushed prematurely at these regions. However, when the infill panel is unreinforced (UC-IP2), it cannot sustain the large forces that develop due to its large stiffness, and it fails in a way depicted in Figure 13b. Instead, the capacity of the RC infill panel to large forces

is substantially increased by the inclusion of the steel reinforcing net. Secondly, the development of cracking within the panel is controlled by the net reinforcement in such a way that it does not lead to a sudden decrease to its bearing capacity, even for much larger normalized interstory drift values reaching 2.5% (column 4 Table 3). This contributing mechanism mobilizes even further the steel ties connecting the panel with the frame, which represents an additional plastic energy accumulation medium. This combined effect for specimen RC-IP 3 explains the eleven times increase in cumulative plastic energy when compared to the jacketed frame (column 8 of Table 3), whereas the corresponding increase for the UR-IP 2 specimen is only twofold.

- The fact that the RC infill panel sustains large horizontal forces and interacts successfully with the surrounding frame can also be seen by the level of shear strains which develop within the infill panel when the maximum horizontal load capacity is reached (Figure 14c).
- All the above is expected to be valid for prototype structures. Constructing such an RC infill in a prototype frame bay needs proper design. The numerical simulation in Section 4 presents such a methodology.

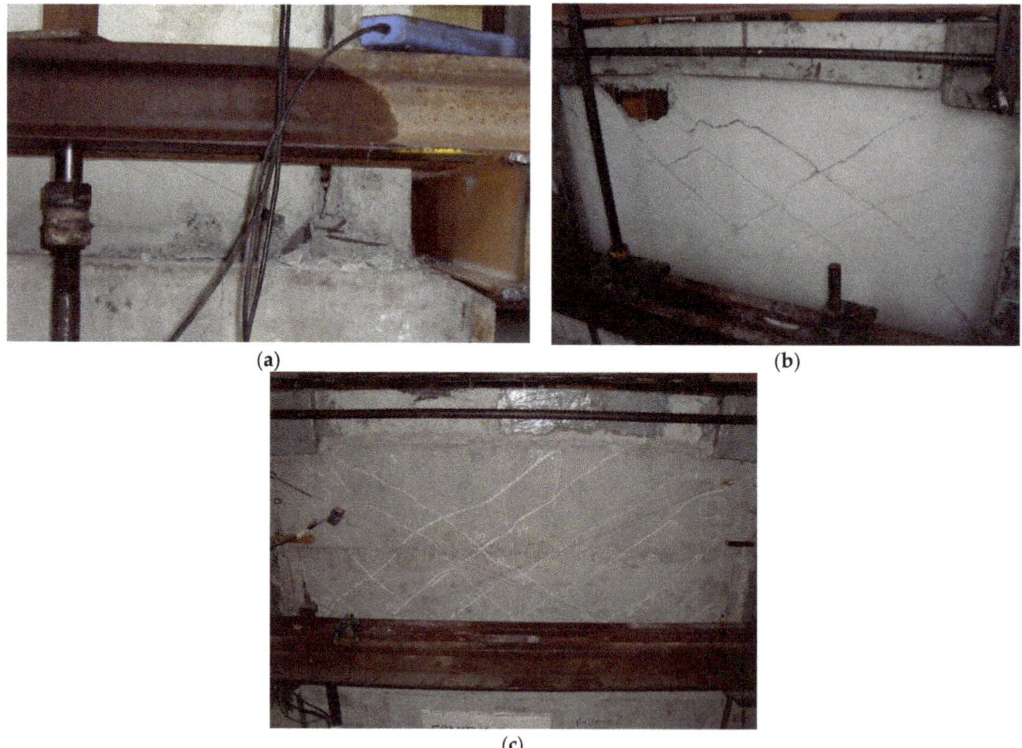

(a)

(b)

(c)

Figure 13. Damage of UC-IP 1, UC-IP 2, and RC-IP 3. (**a**) Crushing of the UC infill panel and the RC frame at the corner. Specimen UC-IP 1. (**b**) Large cracking of the UC infill panel. Specimen UC-IP 2 (**c**) Control of cracking of the reinforced infilled panel. Specimen RC-IP 3.

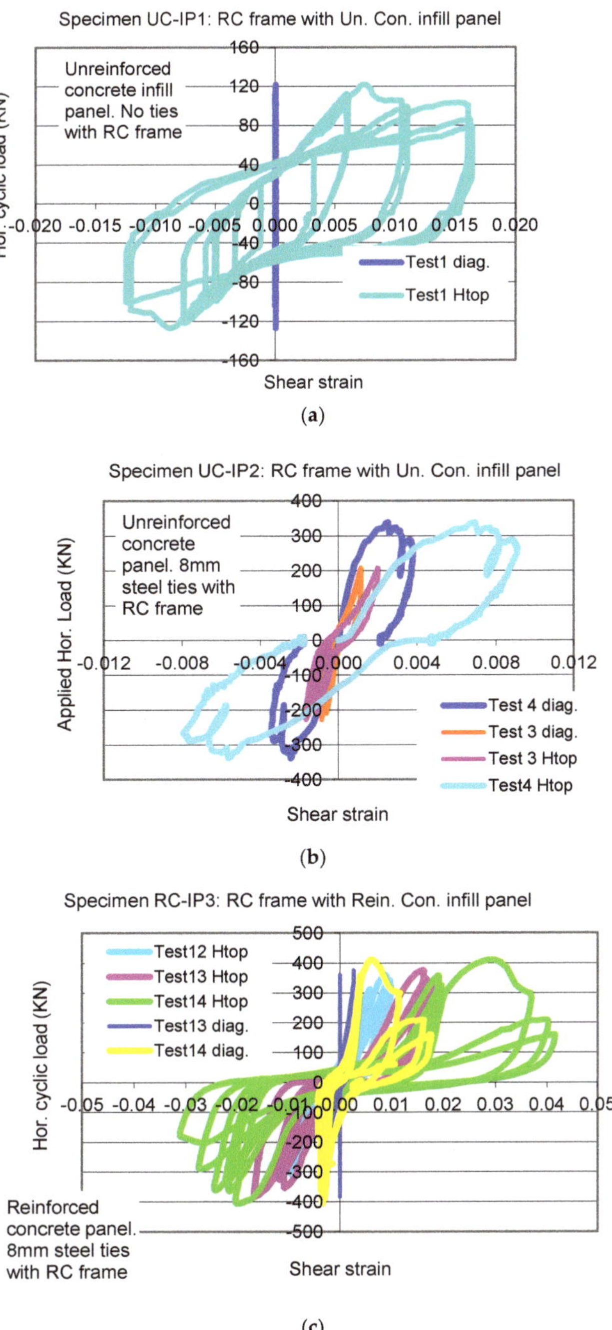

Figure 14. (**a**) Shear strain response of UC-IP 1. (**b**) Shear strain response of UC-IP 2. (**c**) Shear strain response of UC-IP 3.

In order to monitor the behaviour of the UC or RC infill panel itself, a number of displacement transducers were attached at both facades of this panel during testing measuring the variation of the length of its two main diagonals throughout the loading sequence. By combining these measurements, the average shear strain response of the concrete infill panel was obtained, and it is plotted in Figure 14a–c against the applied load (diag). For comparison purposes, an equivalent shear strain for the whole infilled frame is also plotted based on the measured horizontal displacement of the top beam of each tested specimen, which is identical to the drift value for this single-story frame (H-top). From Figure 14a, it can be deduced that for specimen UC-IP 1 the maximum applied load (127 kN) results in equivalent shear strain (drift) of approximately 1.5%. This is mainly due to the formation of the plastic hinges at the columns and beam and the separation of the infill panel from the surrounding frame, due to the absence of any ties, whereas the shear strain of the infill panel itself is almost zero. As already discussed, during this test the concrete of the infill panel was crashed locally at its four corners without the development of any diagonal tension cracking within the panel itself. This explains the zero shear strain in Figure 14a deduced from the instrumentation readings of the displacement transducers located within the infill panel away from the corners. It was underlined that this limit state is rather undesirable because this localized damage limits the horizontal load-bearing capacity of such a retrofitting scheme. This local damage at the corners of the RC infill may also inflict damage on the joints of the RC frame.

Figure 14c demonstrates that the presence of net reinforcement in the RC-IP3 leads this panel to sustain larger shear strain values within itself and for larger horizontal load (408 kN) than the case in which the infill panel was unreinforced (Figure 14b, UC-IP2, applied horizontal load 338 kN).

The displacement response in terms of sliding and separation between the infill panels and the two columns (E-East, and W-West) of the surrounding RC frame close to the region of the column-to-beam joints is also monitored. This response versus the applied load is depicted in Figure 15a–c for specimens UC-IP1, UC-IP2, and RC-IP3, respectively. The absence of any steel ties results in large sliding and separation displacement response at this interface (Figure 15a). The presence of steel ties (Figure 11) combined with a reinforced concrete infill panel (RC-IP3, Figure 15c) results in small separation displacement response, while allowing for considerable sliding displacements for relatively large horizontal load values (408 kN). In this way, the steel ties contribute to the accumulation of plastic energy through dowel action, which is an additional beneficial effect toward seismic resistance. The effect of the presence of the steel is also seen for the unreinforced concrete specimen but to a lesser extent (UC-IP2, Figure 15b). The absence of met reinforcement leads to the premature failure of the unreinforced panel (Figure 13b).

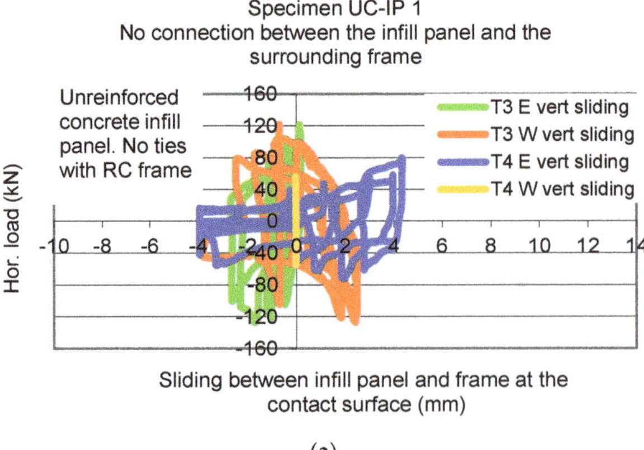

(**a**)

Figure 15. *Cont.*

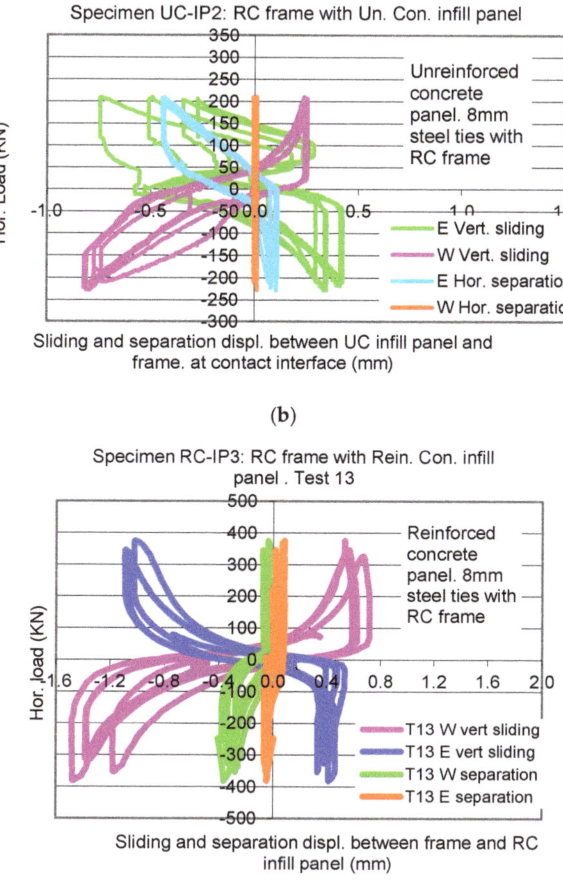

(b)

(c)

Figure 15. (**a**) Horizontal load cyclic response versus sliding at the contact surface of the concrete infill panel and surrounding frame for specimen UC-IP 1. (**b**) Horizontal load cyclic response versus sliding or separation at the contact surface of the concrete infill panel and surrounding frame for specimen UC-IP 2. (**c**). Horizontal load cyclic response versus sliding or separation at the contact surface of the concrete infill panel and surrounding frame for specimen UC-IP 3.

In Figure 16, the top horizontal load versus the equivalent shear strain of the infilled frame is shown for all three tested specimens with concrete infill panels in the form of envelope curves. As can be seen in this figure (see also Table 3), the steel ties result in a substantial increase of the in-plane horizontal load-bearing capacity. However, the lack of reinforcement of the infill concrete panel in this case results in its premature failure limiting this specimen in sustaining this load-bearing capacity for relatively low equivalent shear strain values (0.8%); this limitation is resolved by providing the infilled concrete panel with partial reinforcement. In this case (RC-IP 3), the maximum horizontal load-bearing capacity attains an even higher value (408 kN) than for UC-IP 2 (338 kN); moreover, this horizontal load-bearing capacity is sustained for equivalent strain (drift) values of nearly 4%. The specific limitations of the UC-IP 1, in terms of both load-bearing capacity and deformability, have been noted.

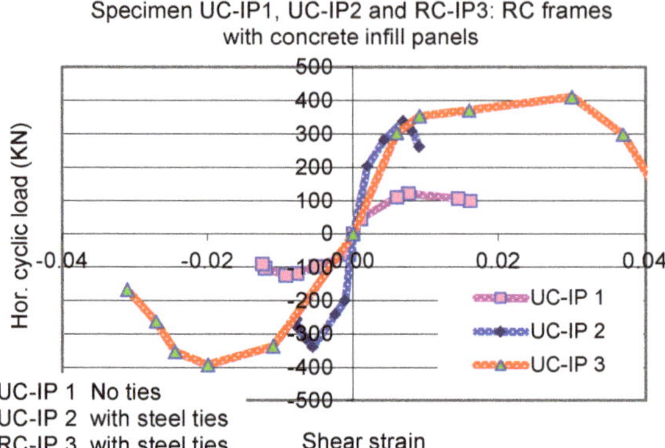

Figure 16. Horizontal load cyclic response versus sliding or separation at the contact surface of the concrete infill panel and surrounding frame for specimens UC-IP 1, UC-IP 2, and RC-IP 3.

4. Numerical Simulation of the RC Frame—Infill Concrete Panel Interaction

Summary results of a numerical simulation are presented, predicting the behaviour experimentally studied in Section 3 infilled RC frames. In this numerical simulation, the same methodology presented in detail by Manos and Soulis [44] is followed employing commercial software [68,69]. The following important mechanisms were included in this simulation [44,60–62].

(a) The possibility of either the two columns or the beam of the RC frame to form plastic hinges at their ends. To this end, the surrounding frame is simulated with linear frame elements based on the relevant cross-sections detailing specific locations for plastic hinges with nonlinear properties (bending moment against rotation) obtained from the cross-sectional reinforcing details and the mechanical properties of the concrete and the longitudinal reinforcement (Figure 6, Figure 10 and Table 1). This was done by employing an in-house-developed software based on the work by Mahin and Bertero [70] and the RC detailing [71]. Next, use was made of a number of commercial software packages that could simulate numerically the non-linear mechanism of the frame depicted in Figure 17 for specimen Bare jacket1 [60–62]. These commercial packages were employed in a combined way during the various stages of the numerical investigation [68,69].

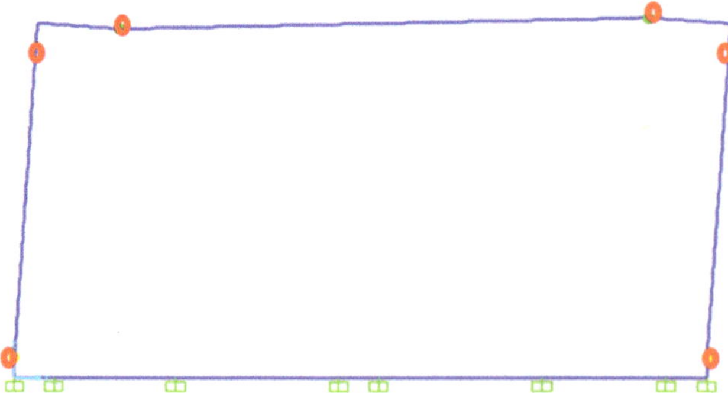

Figure 17. Numerical simulation of the surrounding RC frame with the location of the plastic hinges.

In these numerical analyses, in order to trace the nonlinear solution path, a combined incremental/Newton–Raphson equilibrium iterative procedure included in these commercial packages has been utilized. The solution process follows a step-by-step incremental static material nonlinear analysis adopting initially a constant displacement increment. A line search technique was adopted together with a number of convergence parameters in an effort to improve the convergence rate and to achieve equilibrium from one step to the next. The resulting envelope curve in terms of a numerically predicted horizontal load against the top beam displacement is compared with the corresponding experimental cyclic response in Figure 18 demonstrating reasonable agreement.

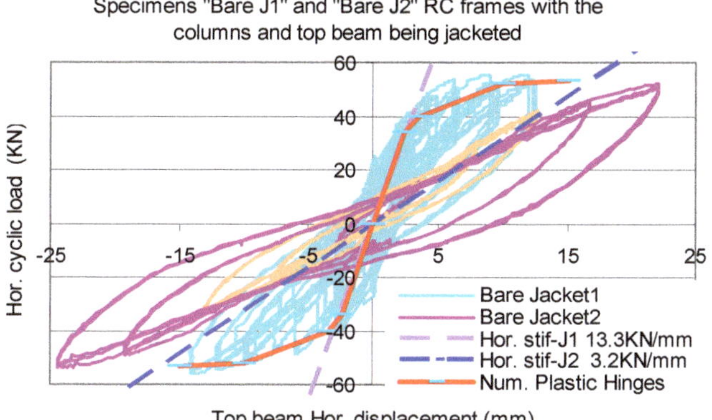

Figure 18. Horizontal displacement versus horizontal load cyclic response of specimens Bare Jacket1,2.

(b)   The same numerical simulation for the RC frame described before is also used here when forming the numerical model of the RC frame together with the UC-RC infill panels. The concrete infills were simulated with shell elements in all cases, together with a failure envelope for numerically predicting the formation of nonlinear limit states within the concrete infill (modified Von-Mises, Figure 19b) which was available in the used commercial software [68]. To quantify the parameters of this failure envelope, a number of square concrete panels with dimensions 730 mm × 730 mm and a thickness of 50 mm were constructed with the same concrete mix and at the same time with the concrete panels of the specimens presented in Section 3. These panels were next subjected to diagonal compression-tension (Figure 19a). Two of these diagonal compression panels were unreinforced (similar to UC-IP 1 and UC-IP 2) and the other two were reinforced with the same reinforcement as RC-IP 3 (see Figure 11). Figure 19a depicts the typical diagonal cracking, which developed along the main vertical diagonal during these tests [72–75]. This testing arrangement was numerically simulated by adopting the same failure envelope (Figure 19b) but with different sets of values—one set for the unreinforced panels and another set for the reinforced square panels. The values shown in Figure 19b for uniaxial compression or tension limits, equal to fc = 22 MPa and ft = 0.22 Mpa, correspond to the unreinforced panel. The corresponding values adopted for the reinforced panel are equal to fc = 26 Mpa and ft = 0.26 Mpa. These values were found from back analysis by utilizing these diagonal compression-tension tests (Figure 19a) and were employed for numerically simulating the behaviour of the RC infilled frames presented in Section 3.

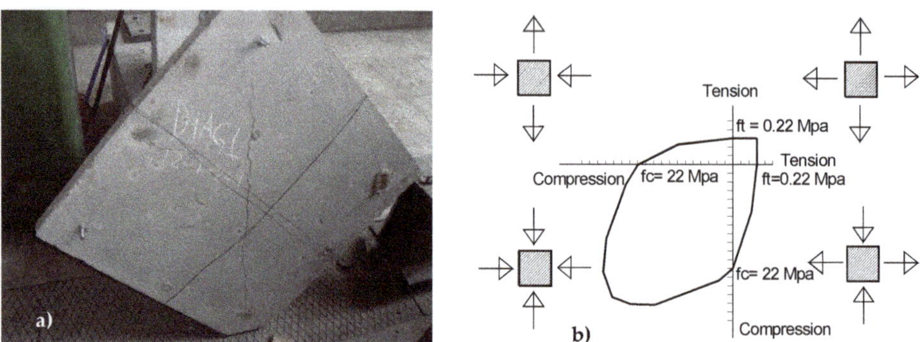

Figure 19. (**a**) Observed damage. (**b**) Numerically simulated diagonal tension cracking of square concrete panel. (**b**) Assumed combined compression–tension failure envelope included in the numerical simulation of the square concrete panel subjected to diagonal compression as well as of RC the infilled panels.

As can be seen in Figure 20a, good agreement is reached between the numerical and measured diagonal compression–tension response through this nonlinear numerical simulation for these square panels tested in diagonal compression-tension. Based on this good agreement, the same numerical simulation is followed for the infilled RC frame specimens investigated in this study. This numerical simulation reproduced as a limit state the plastified region depicted in Figure 20b (where the "x" signs denote the areas reaching the adopted tensile limit state), which is along the same main vertical diagonal region depicted in Figure 19a.

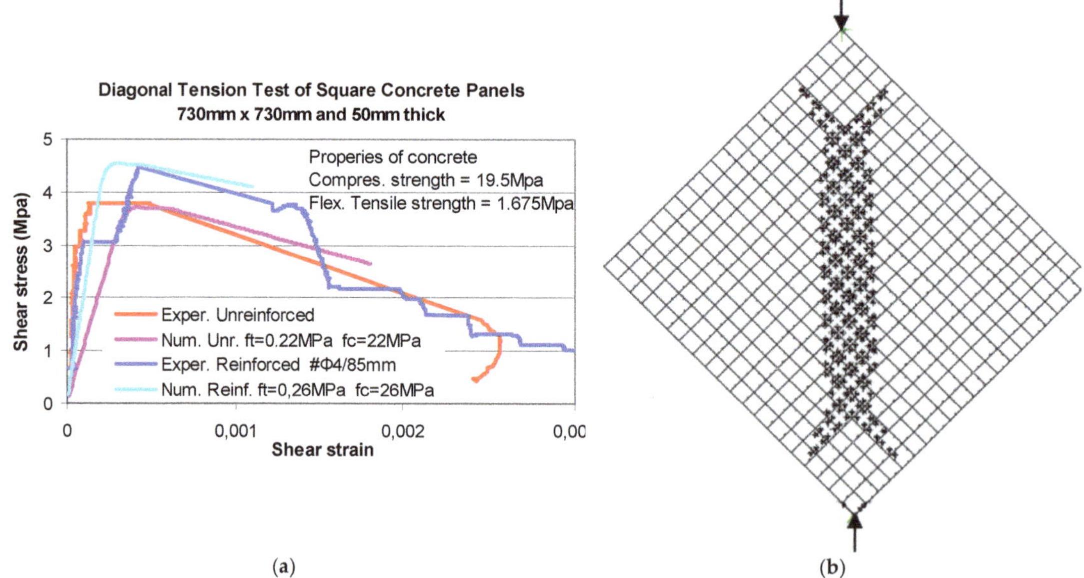

(**a**) (**b**)

Figure 20. Diagonal compression–tension. (**a**) Comparison between observed and numerically simulated behaviour of the tested square concrete panels. (**b**) Predicted plastified regions from the numerical simulation.

(c)   A first group of nonlinear link elements, which could not sustain any tension, was used to numerically simulate the possibility of the concrete infill to be separated from the surrounding frame. These elements could transfer shear at the interface when no separation was detected [44]. An additional second group of nonlinear link elements simulated the steel ties at their exact location, and they could transfer axial tensile and shear forces at the interface between the shell finite elements representing the infill panel and the linear finite elements simulating the surrounding frame. The nonlinear behaviour in axial tension of these link elements was derived from their diameter (8 mm) and yield stress (570 Mpa). These link elements were also provided with nonlinear properties to simulate the dowel action (shear force transfer) at this interface. A number of provisions with relevant empirical formulas are included in [67] that describe the nonlinear shear transfer mechanism between two concrete parts connected with a steel tie. In addition, an extensive experimental sequence was carried out at Aristotle University in an effort to quantify this shear force transfer for steel ties of diameters varying from 8 mm, identical to the ones depicted in Figure 11, up to 14 mm [60–62]. The used experimental setup is depicted in Figure 21 wherein a portion of a jacketed column together with a portion of the RC infill connected with these steel ties are subjected to combined loads indicated by the red arrows. All steel ties used in this experimental sequence as well, as for the specimens reported in Section 3, were embedded within the concrete volume in a way which prohibited any undesirable pullout.

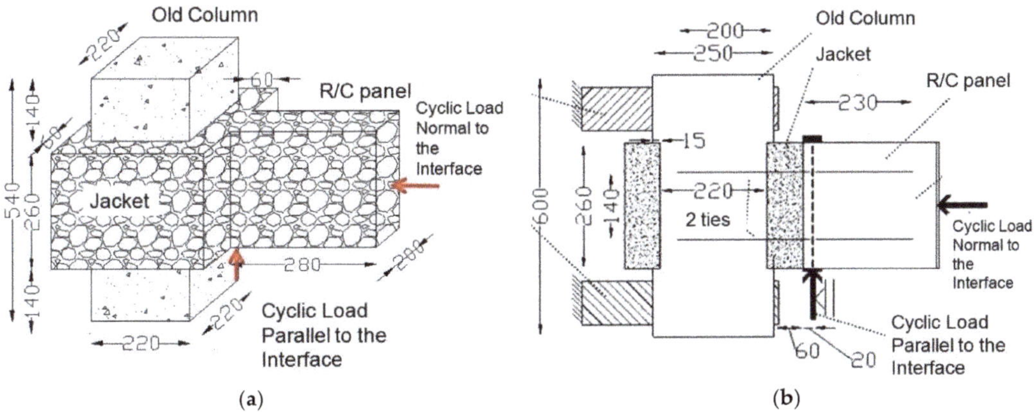

Figure 21. (**a**) Experimental setup to study the force transfer mechanism between portions of the RC panel and the jacketed RC column with or without steel ties. (**b**) Cross-section [60–62].

Twenty-three specimens were built at the same time with the same geometry and the same quality concrete. Three of these specimens were built without any steel ties whereas the rest were provided with steel ties of the same steel grade having diameters varying of 8 mm, 10 mm, 12 mm, and 14 mm. Figure 21b depicts a cross-section of such specimens wherein the upper and lower parts of the "old" column are rigidly supported by the strong reaction frame. Initially, the load normal to the interface was applied as indicated in this figure, keeping its amplitude constant during one test. Then, a cyclic load with a direction parallel to the interface was applied with an eccentricity of 20 mm from this interface. The amplitude of this load was gradually increased until a limit state condition was reached. By comparing the nonlinear behaviour of the specimens with steel ties to that of the control specimen (without the steel ties), the non-linear shear force versus the relative sliding displacement at the interface due to the presence of the steel ties could be determined. This was used in the current numerical methodology to define the nonlinear constitutive behaviour of the second group of links in their tangential direction (parallel to the interface).

The obtained numerical response is presented in Figures 22–27. The numerically predicted response of horizontal displacement versus load in the form of an envelope curve is compared against the corresponding cyclic measured response (Figures 22, 24 and 26). The results also include numerical predictions of regions that the numerical solution indicates damage (Figures 23a, 25a and 27a where the "x" signs denote the areas reaching the adopted tensile limit state). In Figures 23b, 25b and 27b is the corresponding observed damage.

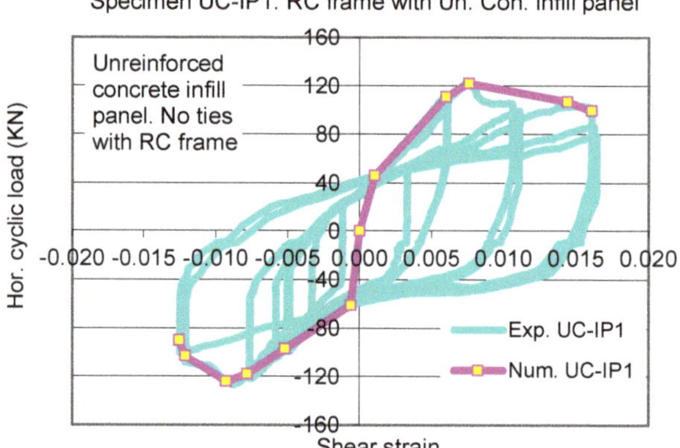

Figure 22. Horizontal displacement versus horizontal load cyclic response of specimen UC-IP 1. Comparison between observed and numerically predicted behaviour.

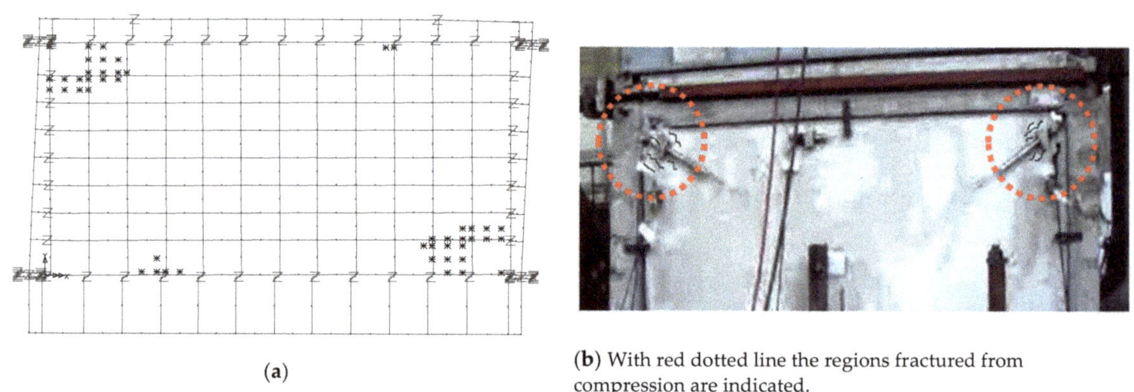

(**a**)

(**b**) With red dotted line the regions fractured from compression are indicated.

Figure 23. (**a**) The numerically predicted compression limit state for specimen UC-IP 1. (**b**) The observed damage for specimen UC-IP 1.

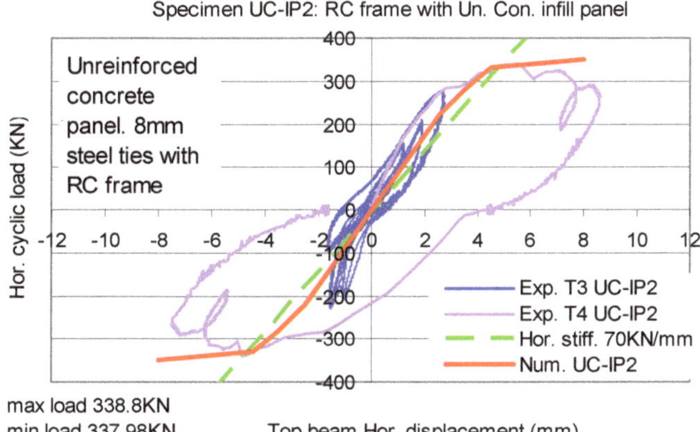

Figure 24. Horizontal displacement versus horizontal load cyclic response of specimen UC-IP 2. Comparison between observed and numerically predicted behaviour.

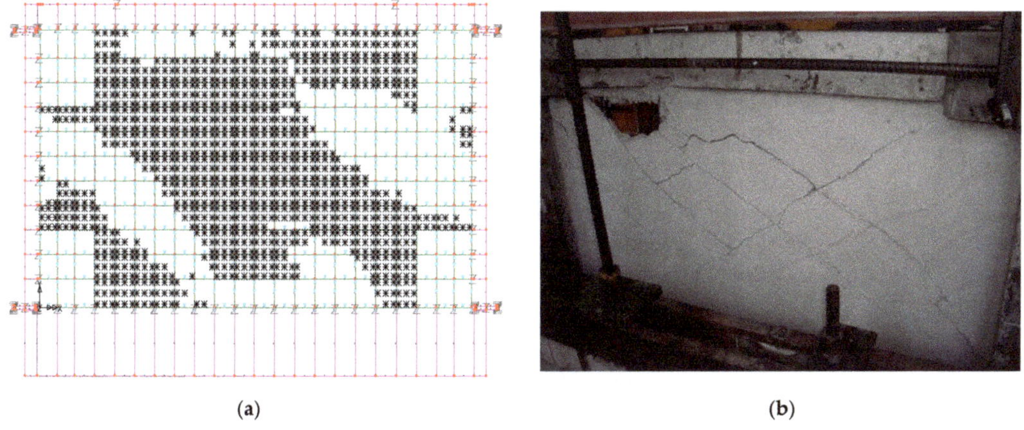

(**a**)   (**b**)

Figure 25. (**a**) The numerically predicted diagonal tension limit state for specimen UC-IP 2. (**b**) The observed damage for specimen UC-IP 2.

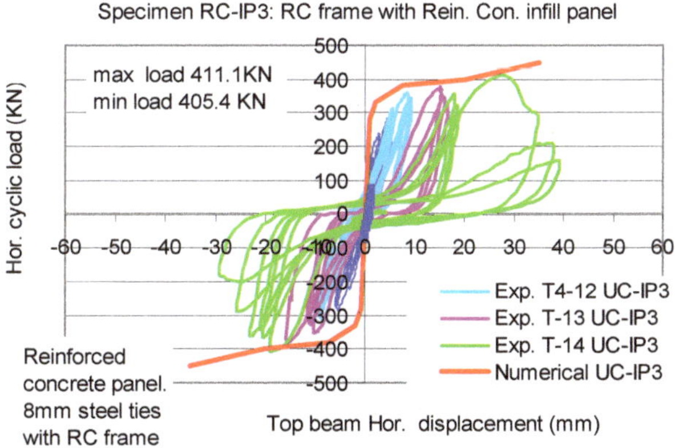

Figure 26. Horizontal displacement versus horizontal load cyclic response of specimen RC-IP 3. Comparison between observed and numerically predicted behaviour.

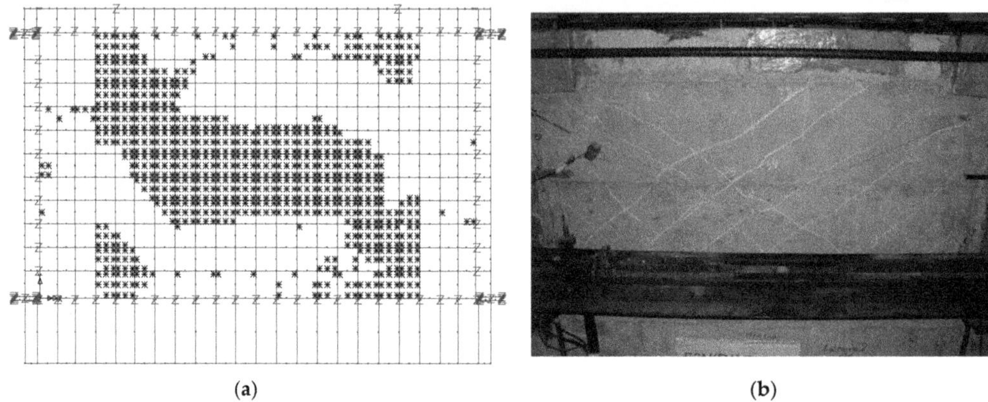

Figure 27. (**a**) The numerically predicted diagonal tension limit state for specimen RC-IP 3. (**b**) The observed damage for specimen RC-IP 3.

Numerical Simulation of UC-IP 1: The previously described numerical methodology was first applied for specimen UC-IP 1. The numerical response in terms of envelope curve of the horizontal load versus the horizontal displacement at the top beam is depicted in Figure 22, being compared with the corresponding measured experimental cyclic response. The predicted peak load horizontal value is equal to 135 kN and compares reasonably well with the peak measured value equal to 127 kN. The predicted shear strain versus horizontal load response distribution compares reasonably well with the corresponding measured response. The numerically predicted compressive limit state at the corners of the infill panel (Figure 23a) also compares well with the crushing of the corners, which was observed for this specimen during testing (Figure 23b).

Numerical Simulation of UC-IP 2: The numerical response for specimen UC-IP 2, in terms of envelope curve of the horizontal load versus the horizontal displacement at the top beam, is depicted in Figure 24. This numerical response is compared with the corresponding experimentally measured cyclic response. The predicted peak horizontal load value is equal to 350 kN and compares reasonably well with the corresponding measured value equal to 338 kN. The predicted shear strain versus horizontal load response distribution compares reasonably well with the corresponding measured response. The observed sudden drop of the bearing capacity for specimen UC-IP 2 is not predicted by this numerical simulation. Figure 25a shows the diagonal tensile limit state numerically predicted at a wide region of this unreinforced infilled panel. This numerical limit state compares reasonably well with the quite wide diagonal cracks spread at most parts of the unreinforced concrete panel, which was observed for this specimen during testing (Figure 25b).

Numerical Simulation of RC-IP 3: The numerical response in terms of envelope curve of the horizontal load against the horizontal displacement of the top beam is depicted in Figure 26. In the same figure, the measured cyclic response is also plotted. The numerically predicted peak horizontal load value is equal to 450 kN and compares reasonably well with the peak measured value of 408 kN. The predicted shear strain versus horizontal load response distribution compares reasonably well with the corresponding measured response. The observed sudden drop of bearing capacity for specimen RC-IP 3 is not predicted by this numerical simulation, due to the adopted constitutive law for the RC panel. In Figure 27a, the numerically predicted diagonal tensile limit state is shown; this numerical limit state compares reasonably well with the diagonal cracks observed for this specimen during testing (Figure 27b).

The preceding comparison demonstrated that the outline simulation could predict quite well the in-plane bearing capacity of the RC frames infilled with concrete (UC or RC) panels and the corresponding limit states. The numerical simulation of the limit state of the RC panel needs further improvement. All the limit state numerical approximations are based on nonlinear constitutive laws of either the concrete infill panel itself or the steel ties that were obtained from specific experimental tests.

5. Conclusions

1. The construction of RC infill panels, connected with metal ties to the surrounding jacketed frame, considerably increases the horizontal in-plane stiffness, bearing capacity, and energy dissipation of the resulting RC frame and RC panel when compared with those of the initial bare frame before retrofit. This study highlighted the important role of the steel ties for such retrofit, which has not been investigated by other researchers when studying the performance of other retrofit schemes. It is relatively easy to apply the retrofit studied here to multiple bays of a soft story at the ground floor level, thus counteracting this soft-story deficiency and decreasing to a degree the seismic vulnerability of such old RC buildings.

2. Properly designed RC infills, RC frame jackets, and the connecting steel ties could prohibit undesired stress concentration and local damage at the corners of the RC infill and frame and also protect them from unstable out-of-plane response of the RC infill panels. Such effective retrofitting of multiple bays of the frames at ground floor, consisting of RC infills, RC jackets, and their steel ties, for each individual frame-bay should aim to upgrade bearing capacity and ductility and to prohibit premature damage of either the RC panel or the connecting steel ties.

3. The presented numerical simulation is proposed as a tool for designing such an effective retrofit. This numerical simulation includes important nonlinear mechanisms that could develop at the RC structural members, the RC infill at the steel ties. For all these nonlinear mechanisms the nonlinear material mechanical properties are required in order to predict with an acceptable degree of approximation. The validity of simulating each one of these nonlinear mechanisms was performed in a step-by-step process utilizing in each step experimental measurements obtained from testing. Its usefulness lies in its ability to be applied to prototype multi-story buildings. Directly applying the proposed methodology for each individual bay in a complete numerical model for the whole multistory building is too complex and costly and needs very large computational effort.

4. For multistory, frames Manos and Soulis [47] proposed an alternative approximation. The seismic demands, in terms of displacement or force for each individual infilled frame, can be found by utilizing a less complex equivalent 3D simulation of such a building, substituting each masonry or RC infill with an equivalent nonlinear diagonal truss element. The nonlinear properties of each equivalent diagonal truss are defined by simulating the in-plane behaviour of each one-story, one-bay frame, which is part and forms the whole 3D building. For each such subassembly two different models are formed having the same RC frame simulation (Figure 17). The first model is formed following the methodology presented in Section 4, which includes the infill panel and steel ties in all their details. The second model includes instead an equivalent nonlinear diagonal truss element [47]. By comparing the horizontal displacement versus force response of the two models, an effort is made to obtain reasonable agreement between their behaviour. This is done by using back analysis and altering the nonlinear properties of the equivalent diagonal truss aiming to reach reasonable agreement. Next, an equivalent 3D model is finally formed having all its infills replaced with such equivalent diagonal truss elements with nonlinear properties defined as described. The seismic demands are found from a "push over" analysis of this equivalent 3D model, in terms of horizontal displacement or force for each individual single-story infilled frame. The final step is to compare these demands to the available capacities and possible limit states of all parts (RC panel, steel ties, and RC members) for each infilled one-story, one-bay frame again utilizing its model, which was formed according to the proposed complex methodology.

5. Multi-story frame old RC buildings in earthquake-active regions that have soft stories at their ground floor and UMI at all upper stories are quite vulnerable and prone to serious damage to their RC columns of this level. A retrofit aiming to counter this particular deficiency is the addition of RC infill panels at the bays of the ground floor RC frames together with RC jackets of the adjacent RC columns. This particular retrofit, when shown to be sufficient, is comparatively less cumbersome and costly compared to other structural interventions which extend to all floors. This is because these bays are left without infills as the ground floor serves as a parking space. When seismic retrofit is required for all floors of a vulnerable building, it becomes quite difficult and expensive and sometimes incompatible with basic functions of the facades and/or the interior, although it is more effective. It also requires the dislocation of the inhabitants for long periods. Seismic retrofit of existing multi-story RC building poses many practical difficulties and therefore requires additional research, despite the progress made so far, in order to validate ingenious and effective solutions. These difficulties differ from country to country as they are linked to a variety of past design and construction practices as well as the legal framework which governs the multiple ownership, which is usually the case in these buildings.

Author Contributions: G.C.M. was the principal investigator responsible for the whole project concept, for the design/execution of all the test sequences, the supervision of all the tests, the recording and analysis of all the experimental results, the writing of the manuscript and the observations and the conclusions drawn. K.K. participated in the experimental sequence and the study of the measured response. V.S. contributed the numerical investigation. L.M. contributed in plotting of the measurements and in the preparedness of this manuscript. All authors have read and agreed to the published version of the manuscript.

Funding: This research had the partial support of the Hellenic Earthquake Preparedness and Planning Organization (OASP).

Data Availability Statement: Any additional information can be requested by the authors.

Conflicts of Interest: The authors declare no conflict of interest.

References

1. Manos, G. Consequences on the urban environment in Greece related to the recent intense earthquake activity. *Int. J. Civ. Eng. Archit.* **2011**, *5*, 1065–1090.
2. Manos, G.C. The 30th of October Samos-Greece Earthquake. Issues relevant to the protection of structural damage caused by strong earthquake ground motions. *J. Archit. Eng.* **2020**, *5*, 3–17. Available online: https://aej.spbgasu.ru/index.php/AE (accessed on 1 January 2020). [CrossRef]
3. Manos, G.C.; Papanaoum, E. Earthquake behaviour of a R/C building constructed in 1933 before and after its repair. In Proceedings of the Structural Studies Repairs and Maintenance of Heritage Architecture STREMAH XI, Tallin, Estonia, 22–24 June 2009; WIT Transactions on the Built Environment. WIT Press: Southampton, UK, 2009; Volume 109, pp. 465–475, ISBN 978-1-84564-196-2. Available online: www.witpress.com (accessed on 20 October 2022).
4. Manos, G.C.; Papanaoum, E. Assessment of the earthquake behaviour of Hotel Ermionio in Kozani, Greece constructed in 1933 before and after its recent retrofit. In *Earthquake Engineering Retrofitting of Heritage Structures, Design and Evaluation of Strengthening Techniques*; Syngellakis, S., Ed.; Wessex Institute of Technology: Southampton, UK, 2013; pp. 25–40, ISBN 978-1-84564-754-4. eISBN 978-1-84564-755-1.
5. Organization of Earthquake Planning and Protection of Greece (OASP). *Guidelines for Level—An Earthquake Performance Checking of Buildings of Public Occupancy*; OASP: Athens, Greece, 2001.
6. Provisions of Greek Seismic Code. *Organization of Earthquake Planning and Protection of Greece (OASP), Dec. 1999*; Revisions of Seismic Zonation Introduced in 2003, Government Gazette, Δ17α /115/9/ΦN275, No. 1154; OASP: Athens, Greece, 2003.
7. EN 1998-1/2005-05-12; Eurocode 8: Design of Structures for Earthquake Resistance—Part1: General Rules, Seismic Actions and Rules for Buildings. European Committee for Standardization: Brussels, Belgium, 2004.
8. Ruggieri, S.; Chatzidaki, A.; Vamvatsikos, D. Reduced-order models for the seismic assessment of plan-irregular low-rise frame buildings. *Earthq. Eng. Struct. Dyn.* **2022**, *51*, 3327–3346. [CrossRef]
9. Ruggieri, S.; Francesco Porco, F.; Uva, G. A practical approach for estimating the floor deformability in existing RC buildings: Evaluation of the effects in the structural response and seismic fragility. *Bull. Earthq. Eng.* **2020**, *18*, 2083–2113. [CrossRef]

10. Valente, M.; Milani, G. Alternative retrofitting strategies to prevent the failure of an under-designed reinforced concrete frame. *Eng. Fail. Anal.* **2018**, *89*, 271–285. [CrossRef]

11. Holmes, M. Steel Frames with Brickwork and Concrete Infilling. *Proc. Inst. Civ. Eng.* **1961**, *19*, 473–478. [CrossRef]

12. Holmes, M. Combined Loading on Infilled Frames. *Proc. Inst. Civ. Eng.* **1963**, *25*, 31–38. [CrossRef]

13. Smith, B.S. Behaviour of Square Infilled Frames. *J. Struct. Div.* **1966**, *92*, 381–404. [CrossRef]

14. Mallick, D.V.; Severn, R.T. The behaviour of infilled frames under static loading. *Proc. Inst. Civ. Eng.* **1967**, *38*, 639–656. [CrossRef]

15. Smith, B.S.; Carter, C. A Method of Analysis for Infilled Frames. *Proc. Inst. Civ. Eng.* **1969**, *44*, 31–48.

16. Mainstone, R.J. *Supplementary Note on the Stiffnesses and Strengths of Infilled Frames*; Current Paper CP 13/74; Building Research Station: Watford, UK, 1974.

17. Klingner, R.E.; Bertero, V.V. *Infilled Frames in Earthquake—Resistant Construction*; EERC, Report No. 76-32; University of California: Berkeley, CA, USA, 1976.

18. Zarnic, R.; Tomazevic, M. The Behaviour of Masonry Infilled Reinforced Concrete Frames Subjected to Cyclic Lateral Loading. In Proceedings of the 8WCEE, San Francisco, CA, USA, 21–28 July 1984.

19. Styliniades, K. Experimental Investigation of the Behaviour of Single-Story Infilled R/C Frames under Cyclic Quasi-Static Horizontal Loading (Parametric Analysis). Ph.D. Thesis, Department of Civil Engineering, Aristotle University of Thessaloniki, Thessaloniki, Greece, 1985.

20. Pook, L.L.; Dawe, J.L. Effects of interface conditions between a masonry shear panel and surrounding steel frame. In Proceedings of the 4th Canadian Masonry Symposium, Fredericton, NB, Canada, 2–4 June 1986; University of New Brunswick Press: Fredericton, NB, Canada, 1986; pp. 910–921.

21. Buonopane, S.G.; White, R.N. Pseudo-dynamic testing of masonry infilled reinforced concrete frame. *J. Struct. Eng.* **1999**, *125*, 578–589. [CrossRef]

22. Da Porto, F.; Grendene, M.; Mosele, F.; Modena, C. In-plane cyclic behaviour of load bearing masonry walls. In Proceedings of the 7th International Masonry Conference, London, UK, 30 October–1 November 2006.

23. Thauampteh, J. Experimental Investigation of the Behaviour of Single-Story R/C Frames Infills, Virgin and Repaired, under Cyclic Horizontal Loading. Ph.D. Thesis, Department of Civil Engineering, Aristotle University of Thessaloniki, Thessaloniki, Greece, 2009. (In Greek)

24. Dhanasekar, D.; Page, A.W. The influence of brick masonry infill properties on the behaviour of infilled frames. *Proc. Inst. Civ. Eng.* **1986**, *81*, 593–605.

25. Lourenco, P.B. Computational Strategies for Masonry Structures. Ph.D. Thesis, Delft University of Technology, Delft, The Netherlands, 1996.

26. Lourenco, P.; Rots, J.G. *On the Use of Micro-Models for the Analysis of Masonry Shear-Walls*; Pande, G.N., Middleton, J., Eds.; Computer Methods in Structural Masonry-2; Books & Journals International: Swansea, UK, 1993; pp. 14–25, ISBN 187414902X, 9781874149026.

27. Mehrabi, A.B.; Shing, P. Finite element modelling of masonry-infilled RC frames. *J. Struct. Eng.* **1997**, *123*, 604–613. [CrossRef]

28. Karapitta, L.; Mouzakis, H.; Carydis, P. Explicit finite element analysis for the in-plane cyclic behaviour of unreinforced masonry structures. *Earthq. Eng. Struct. Dyn.* **2011**, *40*, 175–193. [CrossRef]

29. Arnau, O.; Sandoval, C.; Murià-Vila, D. Determination and validation of input parameters for detailed micro- modelling of partially grouted reinforced masonry walls. In Proceedings of the Tenth Pacific Conference on Earthquake Engineering, Sydney, Australia, 6–8 November 2015; Volume 101.

30. Sandoval, C.; Arnau, O. Experimental characterization and detailed micro-modeling of multi-perforated clay brick masonry structural response. *Mater. Struct.* **2017**, *50*, 34. [CrossRef]

31. Bolhassani, M.; Hamid, A.A.; Lau, A.C.; Moon, F. Simplified micro modeling of partially grouted masonry assemblages. *Constr. Build. Mater.* **2015**, *83*, 159–173. [CrossRef]

32. Malomo, D.; DeJong, M.J.; Penna, A. A Homogenized Distinct Macro-Block (HDM) Model for Simulating the In-Plane Cyclic Response of URM Walls. In Proceedings of the 13th North American Masonry Conference, Salt Lake City, UT, USA, 16–19 June 2019; Volume 1, pp. 1042–1054.

33. Ghosh, A.K.; Amde, A.M. Finite Element Analysis of Infilled Frames. *J. Struct. Eng.* **2002**, *128*, 881–889. [CrossRef]

34. Asteris, P.G.; Antoniou, S.T.; Sophianopoulos, D.S.; Chrysostomou, C.Z. Mathematical Macromodeling of Infilled Frames: State of the Art. *J. Struct. Eng.* **2011**, *137*, 1508–1517. [CrossRef]

35. Asteris, P.G.; Cotsovos, D.M.; Chrysostomou, C.Z.; Mohebkhah, A.; Al-Chaar, G.K. Mathematical micromodeling of infilled frames: State of the art. *Eng. Struct.* **2013**, *56*, 1905–1921. [CrossRef]

36. Dias-Oliveira, J.; Rodrigues, H.; Asteris, P.G.; Varum, H. On the Seismic Behaviour of Masonry Infilled Frame Structures. *Buildings* **2022**, *12*, 1146. [CrossRef]

37. Manos, G.C.; Yasin, B.; Thawambteh, J. The Dynamic Response of Multi-story R.C. Frame Structures with Masonry Infills. A Laboratory tested 7-Story R.C. Planar Model and an In-situ 5-story Building at the European Test Site at Volvi, Greece. In Proceedings of the 4th Internernational Symposium on Computer Methods in Structural Masonry, Florence, Italy, 3–5 September 1997.

38. Manos, G.C.; Yasin, B.; Thaumpta, J. The Simulated Earthquake Response of two 7-story R.C. Planar Model Structures—A Shear Wall and a Frame with Masonry Infills. In Proceedings of the 1998 11th European Earthquake Engineering Conference, Paris, France, 6–11 September 1998.
39. Manos, G.C.; Pitilakis, K.D.; Sextos, A.G.; Kourtides, V.; Soulis, V.; Thauampteh, J. Field experiments for monitoring the dynamic soil-structure-foundation response of model structures at a Test Site. *J. Struct. Eng. Am. Soc. Civ. Eng.* **2015**, *141*, D4014012. [CrossRef]
40. Chiou, Y.C.; Tzeng, J.-C.; Liou, Y.W. Experimental and analytical study of Masonry Infilled Frames. *J. Struct. Eng.* **1999**, *125*, 1109–1116. [CrossRef]
41. Soulis, V. Investigation of the Numerical Simulation of Masonry Infilled R/C Frame Structures under Seismic Type Loading. Ph.D. Thesis, Department of Civil Engineering, Aristotle University of Thessaloniki, Thessaloniki, Greece, 2009. (In Greek)
42. Da Porto, F.; Guidi, G.; Garbin, G.; Modena, C. In-plane behaviour of clay masonry walls: Experimental testing and finite element modelling. *J. Struct. Eng.* **2010**, *136*, 1379–1392. [CrossRef]
43. Domede, N.; Sellier, A. Experimental and numerical analysis of behaviour of old brick masonries. *Adv. Mater. Res.* **2010**, *133–134*, 307–312. [CrossRef]
44. Manos, G.C.; Soulis, V.J.; Thauampteh, J. The Behaviour of Masonry Assemblages and Masonry-infilled R/C Frames Subjected to Combined Vertical and Cyclic Horizontal Seismic-type Loading. *J. Adv. Eng. Softw.* **2011**, *45*, 213–231. [CrossRef]
45. Del Vecchio, C.; Molitierno, C.; Di Ludovico, M.; Verderame, G.M.; Prota Manfredi, G. Experimental response and numerical modelling of two-story infilled RC fame. In Proceedings of the CompDyn2021, Athens, Greece, 27–30 June 2021; pp. 131–140.
46. Lavado, L.; Gallardo, J.; Honma, C. Experimental and numerical analysis of behaviour in compression and shear of handmade clay brick masonry. In Proceedings of the 17th World Conference in Earthquake Engineering, Sendai, Japan, 13–18 September 2020. Paper 2i-0141.
47. Manos, G.C.; Soulis, V.J.; Thauampteh, J. A Nonlinear Numerical Model and its Utilization in Simulating the In-Plane Behaviour of Multi-Story R/C frames with Masonry Infills. *Open Constr. Build. Technol. J.* **2012**, *6* (Suppl. S1-M16), 254–277. [CrossRef]
48. Manos, G.C.; Soulis, V. Simulation of the in-plane seismic behaviour of masonry infills within Multi-story Reinforced Concrete Framed Structures. In Proceedings of the 9th International Masonry Conference, Guimarães, Portugal, 7–9 July 2014.
49. Rastegarian, S.; Sharifi, A. An Investigation on the Correlation of Inter-story Drift and Performance Objectives in Conventional RC Frames. *Emerg. Sci. J.* **2018**, *2*, 140–147. [CrossRef]
50. Umar, M.; Shah, S.A.A.; Shahzada, K.; Naqash, M.T.; Ali, W. Assessment of Seismic Capacity for Reinforced Concrete Frames with Perforated Unreinforced Brick Masonry Infill Wall. *Civ. Eng. J.* **2020**, *6*, 2397–2415. [CrossRef]
51. Rahem, A.; Djarir, Y.; Noureddine, L.; Tayeb, B. Effect of Masonry Infill Walls with Openings on Nonlinear Response of Steel Frames. *Civ. Eng. J.* **2021**, *7*, 278–291, E-ISSN 2476-3055. [CrossRef]
52. Romano, F.; Alam, M.S.; Zucconi, M.; Faggela, M.; Barbosa, A.; Ferracuti, B. Seismic loss analysis of Code-designed infilled RC buildings accounting for model class uncertainty. In Proceedings of the CompDyn2021, Athens, Greece, 27–30 June 2021; pp. 102–110.
53. Carydis, P.G.; Mouzakis, H.P.; Taflambas, J.M.; Vougioukas, E.A. Response of infilled frames with brick walls to earthquake motions. In Proceedings of the 10th World Conference Earthquake Engineering, Madrid, Spain, 19–24 July 1992; pp. 2829–2834.
54. Cavaleri, L.; Zizzo, M.; Asteris, P.G. Residual out-of-plane capacity of infills damaged by in-plane cyclic loads. *Eng. Struct.* **2020**, *209*, 109957. [CrossRef]
55. Di Domenico, M.; Paolo Ricci, P.; Verderame, G.M. Effect of In-plane/Out-of-plane interaction in infill walls on the floor spectra of Reinforced Concrete Buildings. In Proceedings of the CompDyn2021, Athens, Greece, 27–30 June 2021; pp. 84–101.
56. Sugiyama, T.; Matsuzaki, Y.; Nakano, K. Design for Structural Performances of R/C Frame with Cast in Place Non-Structural R/C Walls. Paper No. 1277. In Proceedings of the 13th World Conference on Earthquake Engineering, Vancouver, BC, Canada, 1–6 August 2004.
57. Choi, C.S.; Lee, H.Y. Rehabilitation of Reinforce Concrete Frames with Reinforced Concrete Infills. *Eng. Mater.* **2006**, *324–325*, 635–638. [CrossRef]
58. Anil, O.; Altin, S. Rnq1: An experimental study on reinforced concrete partially infilled frames. *Eng. Struct.* **2007**, *29*, 449–460. [CrossRef]
59. Altin, S.; Anil, O.; Kara, M.E. Strengthening of RC nonductile frames with RC infills: An experimental study. *Cem. Concr. Compos.* **2008**, *30*, 612–621. [CrossRef]
60. Manos, G. Investigation of the behaviour of RC infilled frames at the ground floor of buildings subjected to cyclic seismic-type loads—Study of the connections of the RC infills with the surrounding frame retrofitted with RC jackets. Internal Report submitted to the Greek Organization of Earthquake Planning and Protection. 2012. (In Greek)
61. Manos, G.C.; Soulis, V.; Katakalos, K.; Koidis, G. Numerical and Experimental study of seismic retrofitting for one-bay single-story reinforced concrete (R/C) frames with an encased R/C panel. *Comput. Methods Exp. Meas. XVI J.* **2013**, *55*, 39–411.
62. Manos, G.C.; Soulis, V.; Katakalos, K.; Koidis, G. Study of the in-plane behaviour of ground floor reinforced concrete (R/C) frames retrofitted with jacketing and an encased R/C Panel in order to withstand seismic forces. In Proceedings of the 4th ECOMASS Thematic Conference on Computational Methods in Structural Dynamics and Earthquake Engineering—COMPDYN 2013, Kos Island, Greece, 12–14 June 2013.

63. Biskinis, D.; Fardis, M.N.; Psaros-Andriopoulos, A. Strength, stiffness and cyclic deformation capacity of RC frames converted into walls by infilling with RC. *Bull. Earthq. Eng.* **2016**, *14*, 769–803. [CrossRef]
64. Moretti, M.L.; Papatheocharis, T.; Perdikaris, P.T. Design of Reinforced Concrete Infilled Frames. *J. Struct. Eng.* **2014**, *140*, 04014062. [CrossRef]
65. Papatheocharis, T.; Perdikaris, P.T.; Moretti, M.L. Response of RC Frames Strengthened by RC Infill Walls: Experimental Study. *J. Struct. Eng.* **2019**, *145*, 04019129. [CrossRef]
66. Chrysostomou, C.Z.; Poljansek, M.; Kyriakides, N.; Taucer, F.; Molina, F.J. Pseudo-Dynamic Tests on a Full-Scale Four-Story Reinforced Concrete Frame Seismically Retrofitted with Reinforced Concrete Infilling. *Struct. Eng. Int.* **2018**, *23*, 59–166. [CrossRef]
67. Organization of Earthquake Planning and Protection of Greece (OASP). *Guidelines for Retrofitting in Reinforced Concrete Buildings*; Government Gazzette, $\Delta17\alpha/04/5/\Phi$N 429-1, No 42; OASP: Athens, Greece, 20 January 2012.
68. LUSAS 13.3. *Finite Element System*; FEA Ltd.: Kingston, UK, 2000.
69. SAP. *Structural Analysis Program*; Computer and Structure Inc.: Walnut Creek, CA, USA, 2000.
70. RCCola, R.C.; Mahin, S.; Bertero, V. *A Computer Program for Reinforced Concrete Column Analysis*; User's Manual and Documentation; Department of Civil Engineering, University of California: Berkeley, CA, USA, 1977.
71. *EN 1992-1-1*; Eurocode 2: Design of Concrete Structures—Part 1.1: General rules and rules for buildings. CEN: Brussels, Belgium, 2004.
72. *ASTM E519-15*; Standard Test Method for Diagonal Tension (Shear) in Masonry Assemblages. American Society for Testing Material: West Conshohocken, PA, USA, 2015.
73. Focardi, F.; Manzini, E. Cyclic and Monotonic Diagonal Tension Tests on Various Shape Reinforced and Non-Reinforced Brick Panels. In Proceedings of the 8th European Conference on Earthquake Engineering, Lisbon, Portugal, 7–12 September 1986; Volume 4.
74. Malm, R. Predicting Shear Type Crack Initiation and Growth in Concrete with Non-Linear Finite Element Method. Ph.D. Thesis, Department of Civil and Architectural Engineering, Royal Institute of Technology (KTH), Stockholm, Sweden, 2009.
75. Mercan, B.; Schultz, A.E.; Stolarski, H.K. Finite element modeling of pre-stressed concrete spandrel beams. *Eng. Struct.* **2010**, *32*, 2804–2813. [CrossRef]

applied sciences

Article

Accounting for Resilience in the Selection of R Factors for a RC Unsymmetrical Building

S. Prasanth [1], Goutam Ghosh [1], Praveen Kumar Gupta [2], Claudia Casapulla [3,*] and Linda Giresini [4]

[1] Department of Civil Engineering, Motilal Nehru National Institute of Technology Allahabad, Prayagraj 211004, India
[2] Department of Civil Engineering, GLA University, Mathura 281406, India
[3] Department of Structures for Engineering and Architecture, University of Napoli Federico II, 80134 Napoli, Italy
[4] Department of Structural and Geotechnical Engineering, Sapienza University of Rome, 00184 Rome, Italy
* Correspondence: casacla@unina.it

Abstract: Several design codes consider the non-linear response of a building by using one of the most important seismic parameters, called the response reduction factor (R). The lack of a detailed description of the R factor selection creates the need for a deeper study. This paper emphasises a methodology for the selection of a proper R factor based on resilience aspects. Unsymmetrical/irregular buildings have become the most common in recent times due to aesthetic purposes. However, because of the complexity due to the torsional effect, the selection of the R factor is even more difficult for this type of building. Therefore, a high-rise G+10-storey L-shaped building is herein considered. The building has re-entrant corners based on the structural/plan arrangement. Different R factors were used in the building design, considering buildings subjected to both unidirectional and bidirectional seismic loading scenarios. The building response with respect to various R factors (R equal to 3, 4, 5 and 6) in terms of its performance level, functionality, damage ratio and resilience was assessed at two design levels, i.e., design basic earthquake (DBE) and maximum considered earthquake (MCE). The study concludes that, considering the above criteria along with the resilience aspect, a maximum R factor up to 4 can be recommended for unidirectional loading, whereas for bidirectional loading, the maximum recommended R factor is 3.

Keywords: seismic resilience; building functionality; ductility demand; response reduction factor; performance level

Citation: Prasanth, S.; Ghosh, G.; Gupta, P.K.; Casapulla, C.; Giresini, L. Accounting for Resilience in the Selection of R Factors for a RC Unsymmetrical Building. *Appl. Sci.* **2023**, *13*, 1316. https://doi.org/10.3390/app13031316

Academic Editors: Mariella Diaferio and Francisco B. Varona Moya

Received: 21 December 2022
Revised: 12 January 2023
Accepted: 15 January 2023
Published: 18 January 2023

1. Introduction

The response reduction factor (R) is crucial in the seismic design of any structure. The current research only covered a small portion of criteria, besides the level of seismic zone factors, that should be considered when choosing R factors for the design. The factor 'R' is referred to as the response modification coefficient [1], behaviour factor [2] and response reduction factor [3], indifferently. Most constructions make use of R factors to lessen the seismic loads and bring the structure closer to the inelastic range. To allow the structure to dissipate energy, a greater degree of deformation is therefore necessary.

To ascertain the impact of the R factor on the seismic structural performance, much research has been carried out. According to Indian Standards [4], the reported R factor is remarkably higher than the actual scenario. Regarding the earthquake series, an updated R factor value was proposed [5], and the analysis findings demonstrated that the adjusted R is lower than the desired R. The behaviour factor (q), which is the European factor equivalent to R in Indian Standards, for steel special moment resisting frames (SMRF) proposed by Eurocode 8 for low-rise buildings, was found to be inadequate to ascertain the effects of the storey height and of the column to beam capacity ratios [6]. Tamboli and Amin [7] performed a non-linear pushover analysis to determine how the bracing system configuration may affect the R factor, finding that the latter increased by placing bracing/shear walls in different bays. Nishanth et al. [8] focused on analysing the actual R values using pushover analysis and taking into account the influence of geometrical non-linearity, storey height, etc.; they found that the R value suggested by the Indian code (IS: 1893–2016) was on the higher side. A non-linear pushover analysis was then performed by Chaulagain et al. [9] showing that the load path, the ductility factor and the beam–column strength ratios have an impact on the R factor.

In addition, the effect of vertical connections in braced frames on R in accordance with the seismic demand and capacity of the frames was assessed by Mohsenian et al. [10]. Patel and Amin [11] used non-linear static pushover analysis to investigate how soil flexibility affects R, the time period and the overall structural performance. A non-linear time history analysis (NLTHA) for a three-storey medium seismicity healthcare facility was conducted by Pérez Jiménez and Morillas [12] with varying significance/importance factor (I) values of 1.0, 1.2, 1.4 and 1.5, which are different from R. The result shows that the building receives less damage when a crucial constituent is higher. Hussein et al. [13] investigated the effect of non-uniformity on R with respect to span and height, and the results indicate that non-uniformity has a considerable negative impact on the R value with respect to height. Attia and Irheem [14] looked at the impact of modifying the building boundary condition on the R factor. Keykhosravi and Aghayari [15] performed a study to evaluate the R factor for unbraced and steel-braced RC framed structures, and Kappos [16] conducted research to pinpoint several factors, including ductility and the overstrength factor, which affect the building response. Patel and Shah [17] investigated the characteristics required to define R for a framed RC structure, while the significance R factor for the structural behaviour of RC members was assessed by Galasso et al. [18]. According to Abdi et al. [19], the building non-linear behaviour for increasing ground motion levels is influenced by the R factor. Prasanth and Ghosh [20] utilised the cracking coefficient to take into consideration strength degradation in terms of stiffness. An imbalance in the boundary conditions causes a shift in the seismic design acceleration spectrum, which has an impact on how the building works. Using a range of design acceleration spectra, Sekar and Ghosh [21] evaluated the resilience of an existing high-rise concrete building. Decision makers for pre-disaster events may utilise the computational platform with a hybrid model proposed by Marasco et al. [22] to anticipate damage and resilience on a wide scale without addressing recovery.

In order to evaluate the effects of four different ground motions on the functionality and performance of structures, a study was conducted by Hashemi et al. [23] on a five-storey structure with limited ductility, located in a low seismicity area and with a soft-storey mechanism. Cimellaro et al. [24] suggested a quantitative method for assessing the earthquake resilience in a healthcare system. This method highlights the importance of repair downtime for structural loss recovery because it integrates social, environmental and structural losses, while the direct and indirect losses were quantified and caused by socio-structural deterioration [25].

The necessity for dependable infrastructures was highlighted by Hudson et al. [26], along with resilience-based design principles and suggestions for creating infrastructures with adequate resilience. According to Gallagher and Cruickshank [27], the resilience-based strategy was applied even during harsh weather conditions. Grigorian and Kamizi [28] proposed a hybrid rocking–stepping core, with energy-dissipating-grade beams and replaceable energy-dissipating moment connections as three alternatives for durable or long-lasting earthquake-efficient moment frames. Dukes et al. [29] developed a fragility/vulnerability model for bridges utilising logistic regression and the Monte Carlo simulation as a design tool to complement the performance-based strategy for improving their seismic resilience. To calculate the risk of collapse of typical structures based on various design codes, several research [30–32] was conducted. The majority of these investigations, however, made use of two-dimensional (2D) architectural models that were subjected only to a single direction of seismic forces.

Bidirectional loading, on the other hand, has little effect on beams that are normally oriented along the building's primary planes. Much research has revealed the significance of evaluating the effects of bidirectional seismic excitation on columns [33–38]. Gwalani et al. [39] evaluated the seismic response of the structure under uni- and bidirectional seismic excitations using three-dimensional models of a low- and a mid-rise frame structure. The result shows that the collapse capacity was reduced in the case of bidirectional loading due to the combined action. Moreover, in a study carried out by Hussain and Dutta [40], an asymmetrical structure was subjected to various ground motions with consideration of bidirectional effect, where the result shows that this effect decreases in the inelastic range of the structure due to the increase in the lateral period. The above research shows that the bidirectional effect was studied for the seismic performance of the structure and the structural members, but the effect on the building resilience with various R factors was not analysed.

Additionally, the behaviour of the unsymmetrical buildings under seismic events was of major concern when compared with symmetrical buildings because of torsional effects [41,42]. A review on torsional effects during seismic events on buildings was conducted by Anagnostopoulos et al. [43], who show new methods and techniques to overcome torsional effects on irregular buildings. The bidirectional effect on low-rise concrete buildings with varying plan orientation angles was studied by Cimellaro et al. [44], while a framework to make low complex structural models was proposed by Ruggieri et al. [45] and the results were related with the regular three-dimensional reduced model. Hence, unlike symmetrical buildings, the selection of appropriate R factors for unsymmetrical buildings plays an important role in seismic design.

From the previous studies, it was noted that much research was focused on evaluating and using realistic R values in design, but the basis of the R factor selection criteria for robust design has not been discussed yet. Hence, this study emphasises how crucial the resilience parameter is when choosing R variables because the resilience-based strategy aids effective recovery planning following seismic disasters. As code provisions in many countries were silent in these aspects, this study proposes a framework based on resilience with considerations for various factors such as building performance level and ductility demand for the selection of R factors. In particular, a high-rise unsymmetrical building is herein considered, and the maximum R factor is recommended based on the above factors in favour of safety, since a higher R factor leads to a higher loss of resilience.

2. Building Description and Seismicity Conditions

A high-rise, G+10-storey, L-shaped, unsymmetrical, reinforced, concrete building was considered in this study. The total plan area of the building is 324 m^2. The plan and elevation of the unsymmetrical building are shown in Figure 1, while its 3D view is shown in Figure 2. The total height of the building is 44 m (Figure 1b) with an inter-storey height of 4 m. The collapse phenomenon is the combination of plastic and dynamic behaviour of the structural members. Though distributed plasticity provides slightly better results, in terms of modelling difficulties the default plastic hinges are assigned as per ASCE 41–17. The hinges were assigned in beams and columns at the starting and end portion of the member. The beam elements were assigned with M3 hinges, whereas for columns it was P-M2-M3 hinges. In the study, the lateral load-resisting system with bracings, URM infills, etc., was not included. The slab members were provided along with the rigid diaphragm action which was considered at each floor level that inhibited the actual behaviour of the building.

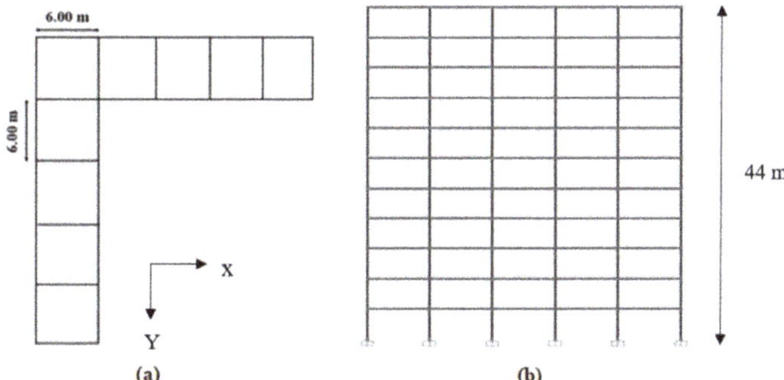

Figure 1. Geometric orientation of unsymmetrical building: (**a**) plan view; (**b**) elevation.

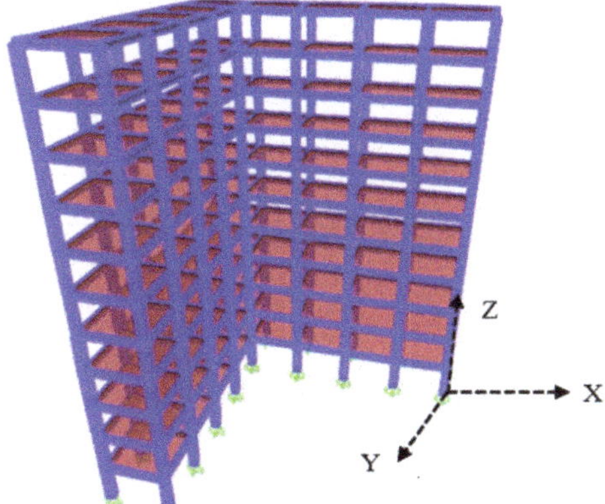

Figure 2. Three-dimensional view of L-shaped unsymmetrical building.

The building was designed in accordance with IS: 456–2000. The self-weight of concrete members such as beams, columns, slabs, and walls is included in the dead loads. Consideration was given to the wall load on the outside margin and the 150 mm slab thickness. The wall is 230 mm thick, and a floor finish of 1.5 kN/m^2 was taken into account. A live load of 3 kN/m^2 and a live roof load of 0.75 kN/m^2 were also considered. For beams and columns, the concrete grades M25 and M30, respectively, were taken into consideration, and steel reinforcement with a 500 MPa yield strength was employed, while 5% damping with medium soil (Type II) was assumed according to IS 1893:2016 (Part I). Five various ground motions, namely El Centro, Bam, Kobe, San Fernando and Tabas, were considered for the present study (Table 1), with the corresponding spectrum-compatible ground motion time histories sketched in Figures 3–7. Due to the asymmetry in the building orientation, unlike symmetrical building cases, the seismic loading was independently applied in both longitudinal (Ux) and transverse (Uy) directions. The behaviour of the building in both of the directions was therefore observed.

Table 1. Ground motion details [46].

Record	Event	Year	Magnitude 'Mw'	Station	PGA (g)
1	El Centro, California	1940	7.10	Imperial valley	0.319
2	Bam, Iran	2003	6.60	BAM	0.969
3	Kobe, Japan	1995	6.90	KJMA	0.821
4	San Fernando, California	1971	6.61	Pacoima dam	1.171
5	Tabas, Iran	1978	7.35	TABAS	0.861

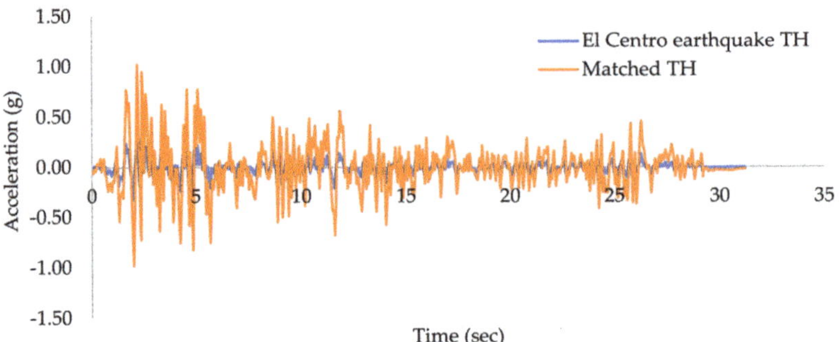

Figure 3. El Centro earthquake time history.

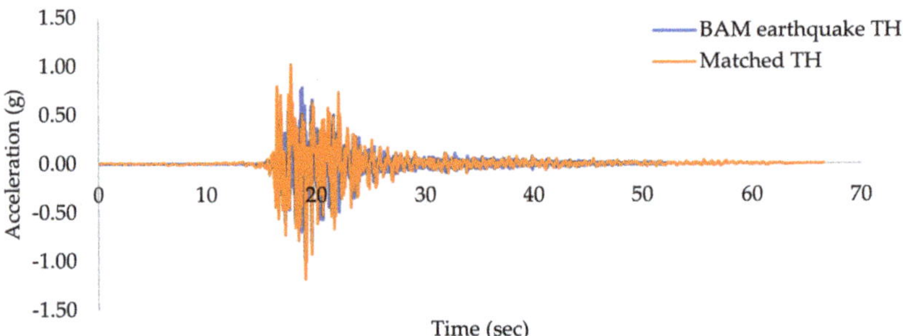

Figure 4. Bam earthquake time history.

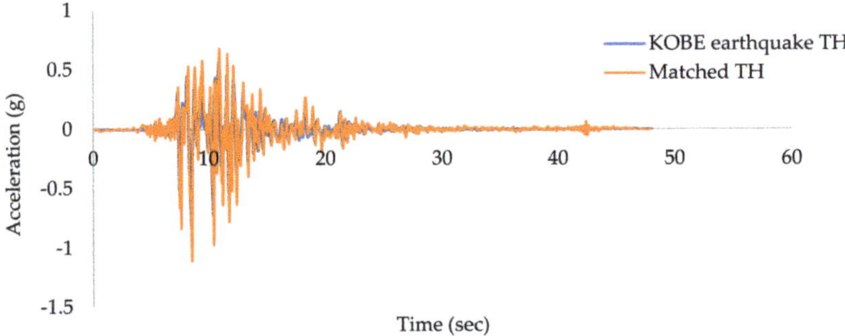

Figure 5. Kobe earthquake time history.

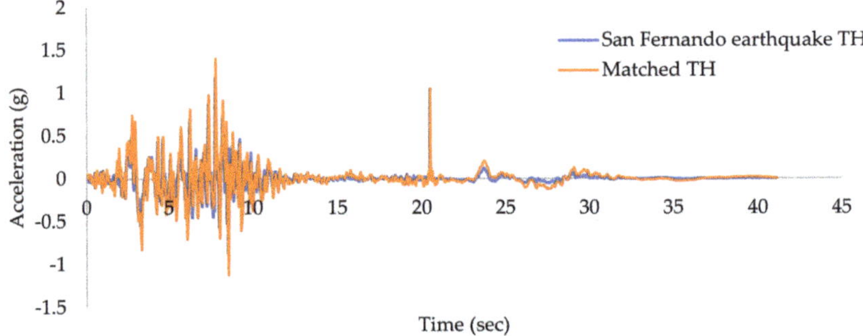

Figure 6. San Fernando earthquake time history.

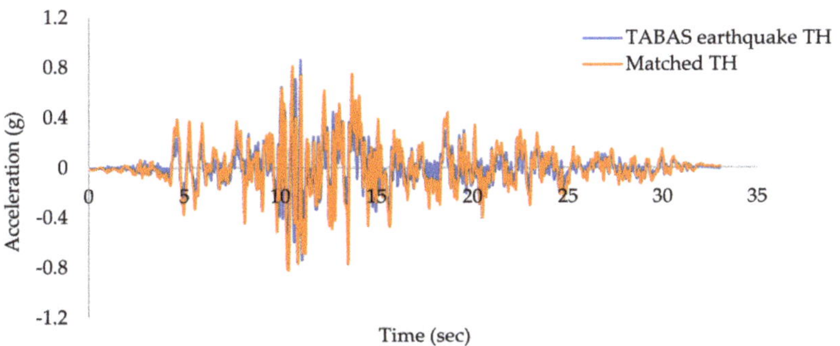

Figure 7. Tabas earthquake time history.

The building was designed with different cases of R (R equal to 3, 4, 5 and 6), considering 1.5 as the importance factor (I) and 0.36 as the zone factor (Z). It is worth highlighting that the response reduction factor is $R = V_e/V_d$, where V_e is the elastic base shear and V_d is the design base shear. The seismic behaviour of the building was analysed at each design level. The scale factor used is shown in Table 2 at design levels such as DBE and MCE levels.

Table 2. Scale factors at each design level.

S. No.	Design Level	Scale Factor (g)
1	DBE	2.6487
2	MCE	5.2974

3. Design Concept and Structural Details of the Building

The design concept of the building was implemented using SAP2000 V22 [47], as per Indian Standards with respect to R = 3, 4, 5 and 6. The maximum performance levels at DBE and MCE design levels were maintained at the IO-LS level and CP-C level, respectively, based on which maximum R factor was recommended. To find the capacity curve of the building, the non-linear static pushover analysis using FEMA-440 [48] was performed, which was later followed by a non-linear time history analysis (NLTHA). For each case of the unsymmetrical building, the structural details are shown in Table 3. The graphical representation of change in building recovery with respect to R factors is shown in Figure 8.

Table 3. Structural details of the unsymmetrical building, with Case I (R = 3), Case II (R = 4), Case III (R = 5), Case IV (R = 6).

Case No.	Structural Members		Cross Section		Area of Longitudinal Reinforcement 'A_{st}' (mm^2)	
			Width (mm)	Depth (mm)	Top	Bottom
I (R = 3)		Beam	400	600	1296	1296
	Column	C1 (up to 20 m)	750	750	20–25 Ø	
		C2	650	650	20–20 Ø	
II (R = 4)		Beam	380	550	1296	1296
	Column	C1 (up to 20 m)	700	700	20–25 Ø	
		C2	600	600	20–20 Ø	
III (R = 5)		Beam	300	500	603	603
	Column	C1 (up to 20 m)	600	600	12–20 Ø	
		C2	500	500	12–16 Ø	
IV (R = 6)		Beam	300	450	603	603
	Column	C1 (up to 20 m)	560	560	12–25 Ø	
		C2	420	420	12–20 Ø	

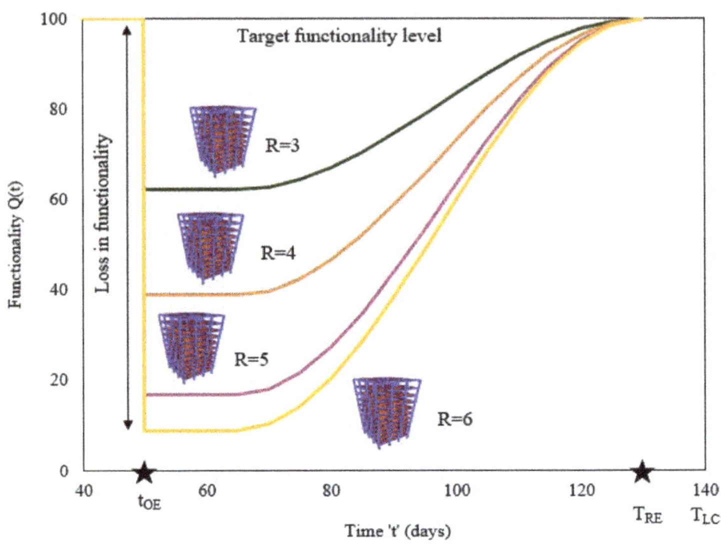

Figure 8. Graphical representation on functionality curves with respect to R factor. The time of occurrence of a seismic event is designated as t_{OE}, its recovery time is designated as T_{RE}.

3.1. Non-Linear Time History Analysis (NLTHA)

Using SAP2000, a NLTHA was carried out [49–52]. The investigation made use of the compatible ground motion time history (Figures 3–7). Since the building plan orientation is asymmetrical, each building instance was subjected to independent unidirectional (Ux direction) and bidirectional (Ux and Uy) loads. The structure must be constructed for the simultaneous impacts of a full design earthquake loading in one horizontal direction plus 30% of a design earthquake load along a second horizontal direction, according to IS: 1893–2016. The matching El Centro time history data were chosen for the bidirectional loading condition since the El Centro ground motion was more prominent in the previous research.

At each design level, the maximum displacement was determined based on the loading conditions. The maximum displacement was recorded at the control node, which was assumed to be the top roof node, implying that local demands are controlled by global demands according to traditional design practice (using R-factors). For each building scenario, the maximum roof displacement among Ux and Uy directions under unidirectional and bidirectional loading circumstances was considered (Tables 4 and 5).

Table 4. Maximum roof displacement under unidirectional loading.

S. No.	Design Level	Maximum Roof Displacement 'Δ_u' (mm)			
		Case I	Case II	Case III	Case IV
1	DBE	222.75	261.29	314.71	343.35
2	MCE	487.27	547.49	730.54	783.43

Table 5. Maximum roof displacement under bidirectional loading.

S. No.	Design Level	Maximum Roof Displacement 'Δ_u' (mm)			
		Case I	Case II	Case III	Case IV
1	DBE	236.83	277.79	337.26	365.12
2	MCE	519.49	592.09	758.86	820.96

3.2. Estimation of Ductility Demand of the Buildings

According to ATC-40 [53] recommendations, the yield (Δ_y) and ultimate displacements (Δ_u) were determined by performing bilinearisation of the building capacity curve found by performing non-linear pushover analysis. This was conducted to assess the structural vulnerability curve for estimating damage and loss. The ductility demand (μ_D) was estimated from the ultimate roof displacement and yield displacement ($\mu_D = \Delta_u/\Delta_y$), which indicates the displacement ductility demand of the buildings at each design level. The ductility demand under unidirectional and bidirectional loading conditions for each building case was found (Tables 6 and 7).

Table 6. Ductility demand at each building case under unidirectional loading.

S. No.	Design Level	Ductility Demand (μ_D)			
		R = 3	R = 4	R = 5	R = 6
1	DBE	1.58	1.81	2.15	2.42
2	MCE	3.45	3.79	4.98	5.53

Table 7. Ductility demand at each building case under bidirectional loading.

S. No.	Design Level	Ductility Demand (μ_D)			
		R = 3	R = 4	R = 5	R = 6
1	DBE	1.67	1.92	2.30	2.58
2	MCE	3.67	4.10	5.18	5.80

3.3. Performance Level of the Building

The extent of the building damage caused by the earthquake determines the structure level of performance. Immediate occupancy (IO), life safety (LS) and collapse prevention (CP) are three different performance levels (Figure 9). According to the American Society of Civil Engineers [54], the member non-linearity was accounted for in the form of plastic hinges. To assess the non-linear behaviour of the structure, plastic hinges were placed on the structural parts, such as beams and columns. With the structural member non-linearity, the NLTHA was performed to assess the structural capacity and performance level. The rotational capacity and constraints of the plastic hinges form the basis of the non-linear behaviour of the structure. The study considered the plastic hinge rotation limitations of 0.01, 0.025 and 0.050 for IO, LS and CP levels, respectively.

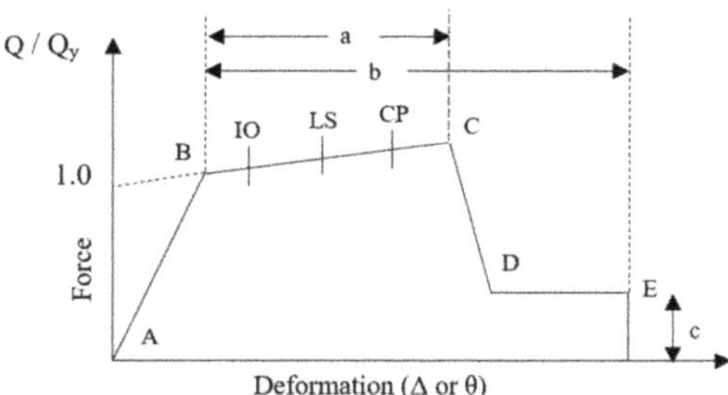

Figure 9. Generalised force deformation curve as per ASCE 41–17 [54].

From the NLTHA, the building performance level was found at each design level under both unidirectional and bidirectional loading conditions (Tables 8 and 9). The schematic representation of the performance point location of each building case with respect to MCE design level at both unidirectional and bidirectional loading conditions was shown in Figures 10 and 11, respectively.

Table 8. Performance level at each building case under unidirectional loading.

S. No.	Design Level	Performance Level			
		R = 3	**R = 4**	**R = 5**	**R = 6**
1	DBE	IO	IO	Almost LS (IO $\leq$ LS)	Almost LS (IO $\leq$ LS)
2	MCE	IO-LS	Almost CP (LS $\leq$ CP)	CP-C (CP $\leq$ C)	CP-C (CP $\leq$ C)

Table 9. Performance level at each building case under bidirectional loading.

S. No.	Design Level	Performance Level			
		R = 3	**R = 4**	**R = 5**	**R = 6**
1	DBE	IO	IO	LS (LS $\leq$ CP)	LS (LS $\leq$ CP)
2	MCE	LS	CP-C (CP $\leq$ C)	C-D	C-D

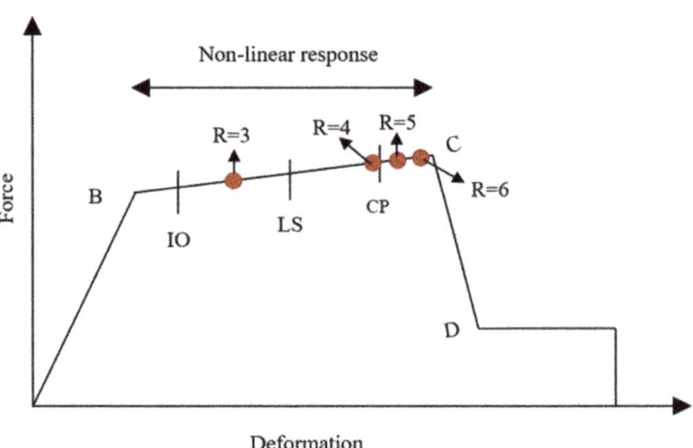

Figure 10. Schematic representation of variation in the building performance level with respect to R factors at the MCE design level under unidirectional loading.

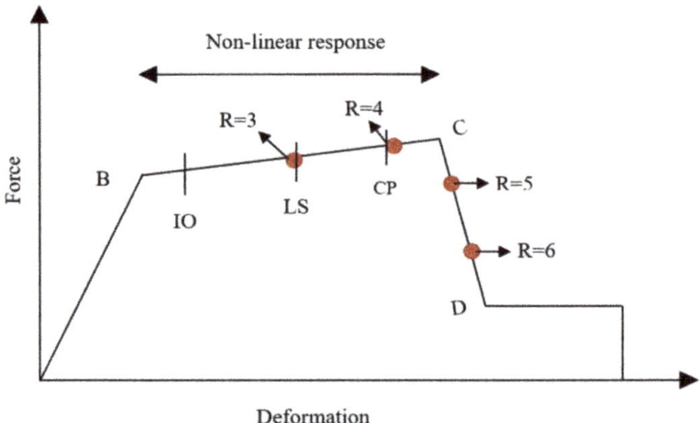

Figure 11. Schematic representation of variation in the building performance level with respect to R factors at the MCE design level under bidirectional loading.

According to Table 8, the building performance level is impacted by changes in the R factor. At the DBE design level, the building lies at the IO level at R = 3 and 4, while at R = 5 and 6 it almost reaches the LS level. At the MCE design level (Figure 10), the building reaches IO-LS, almost CP, CP-C and CP-C at R equal to 3, 4, 5 and 6, respectively. It was observed that, at each R factor, the performance level was maintained to limit the ductility demand. At higher R factors (R = 5 and 6), the building lies at the CP-C level. This demonstrates how a greater R factor influences a building performance level because a higher R value calls for more ductility.

In the case of bidirectional loading, the building performance reduces significantly when compared with unidirectional loading (Table 9). At the DBE level, the building with R equal to 3 lies at the IO level, while at R = 6 it experiences the LS (LS ≤ CP) level under bidirectional loading conditions (Figure 11). Under unidirectional loading, at the MCE design level the building reaches IO-LS at R = 3, almost CP (LS ≤ CP) at R = 4, CP-C at R = 5 and CP-C at R = 6. With bidirectional loading, the building performance level reaches the LS level at R equal to 3, the CP-C level at R = 4, the C-D level at R = 5 and the C-D level at R = 6. This behaviour of the structure was not observed in the case of unidirectional loading where the building reaches the CP-C level at R equals 5 and 6 (Table 8). It clearly indicates that the consideration of bidirectional effects decreased the building performance level significantly, especially at higher R factors at both the DBE and MCE design levels.

3.4. Vulnerability Assessment

As per the HAZUS methodology [55], all of the buildings experience four damage states, i.e., slight, moderate, extreme and collapse. By creating fragility curves, it was possible to determine the vulnerability in terms of the likelihood of exceeding each damage condition in each situation. In this study, the fragility curves were plotted against cumulative probability exceedance of each damage state and spectral displacement. Using HAZUS methodology [55], the damage probability of exceedance is given by the following expression:

$$P\left(ds \,\big/\, S_d\right) = \phi\left[\frac{1}{\beta_{ds}} \ln\left(S_d \,\big/\, S_{d,ds}\right)\right] \tag{1}$$

The median ($S_{d,ds}$) at each damage state with respect to spectral displacement was found from the proposed equations [56]. Lognormal standard deviation (β_{ds}) values that describe the variability (dispersions) of fragility curves are developed for each damage state (i.e., slight, moderate, extensive and complete). The total variability of each structural damage state (β_{ds}) was taken from the HAZUS MR4 technical manual. The spectral displacement (S_d) on the x-axis and the probability of exceedance on the y-axis were used to illustrate the vulnerability in terms of the fragility curve. The probability of exceeding the stated damage level or state is shown by the y-axis, which ranges from 0% to 100%. For each example of the construction, fragility curves were created (Figures 12–15).

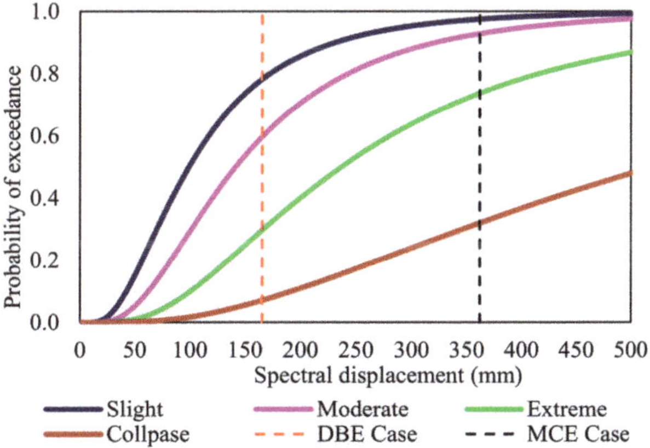

Figure 12. Fragility curve for unsymmetrical building Case I (R = 3).

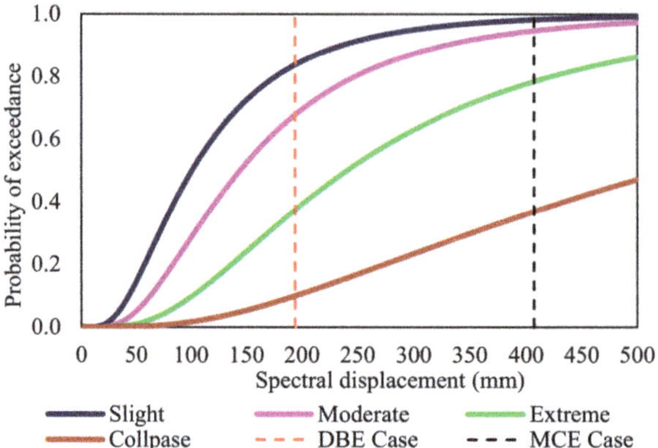

Figure 13. Fragility curve for unsymmetrical building Case II (R = 4).

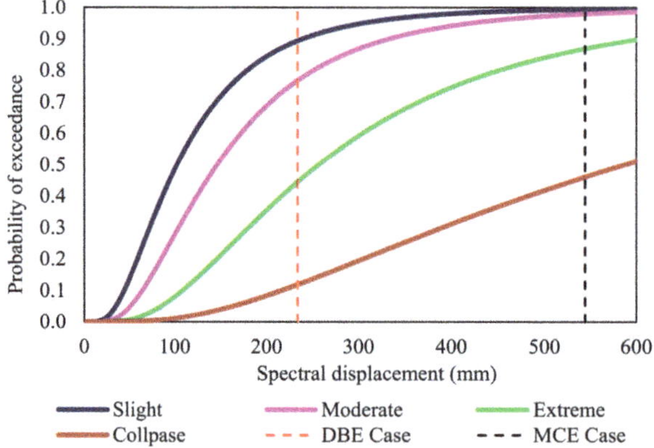

Figure 14. Fragility curve for unsymmetrical building Case III (R = 5).

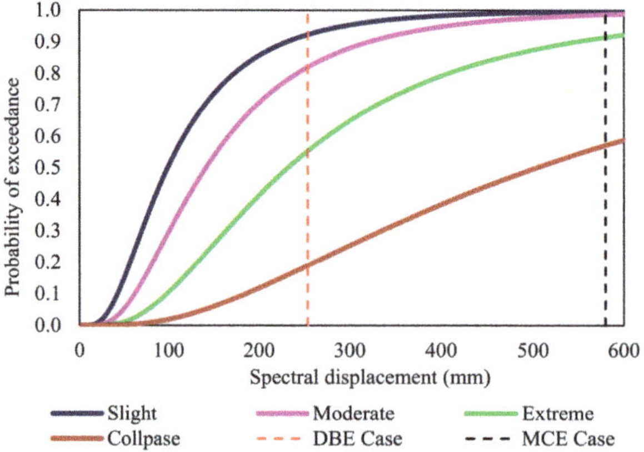

Figure 15. Fragility curve for unsymmetrical building Case IV (R = 6).

From the fragility curves, the probability of the building exceeding each damage state was evaluated. The vertical line that intersects the fragility curves of various damage states was shown with regard to the spectral displacement at each design level. The likelihood percentage of exceeding for that particular damage condition is provided by the intersection point.

From Figures 12–15, it was observed that at the DBE design level, the building with R = 5 and 6 (Cases III and IV) shows 45 to 55% certainty of experiencing extreme damage. The building in all cases shows less than a 20% probability of experiencing a collapse damage level. This shows that the building has much less collapse probability at the DBE design level. At a higher design level (MCE), it was noted that all of the building cases (Case I to IV) have more than 90% certainty to experience moderate damage and 70% certainty to experience extreme damage states. Since the collapse probability of the building ranges from 45 to 57% at higher R factors such as R = 5 and 6, it shows that the building is more prone to the collapse damage level, which affects the recovery planning of a post-disaster event. Under bidirectional loading, at the MCE design level, the building Cases I and II have more than 75–82% certainty of experiencing an extreme damage state, whereas Cases III and IV have 88–92% certainty of experiencing an extreme damage state. The slight increase in the damage probability was observed at a higher R factor with bidirectional loading.

3.5. Estimation of Damage–Loss Ratio

The estimation of damage losses due to post-disaster events is uncertain as it depends on location and environmental conditions. Based on the loss estimation, the strategic recovery plan was fixed to recover back the functionality in the control period of time. In estimating the functionality of the building, the losses are incorporated in terms of loss function, denoted as $L(I,T_{RE})$. There are two types of losses involved in finding the loss function, namely direct (L_D) and indirect (L_{ID}) economic losses. The direct damage–loss ratio was found using Equation (2).

$$L_D = \sum P_E(DS = K) \times r_K \tag{2}$$

where K is the harm state of the building, $P_E (DS = K)$ is the discrete damage probability that the building was in that state when the incident occurred and r_K is the damage ratio associated with discrete damage probability of each damage state which was estimated using the HAZUS MR4 technical manual [55]. The damage–loss ratios under both loading conditions for each building cases was found (Tables 10 and 11). The direct damage losses were estimated and then used to evaluate the building resilience.

Table 10. Direct damage–loss ratio under unidirectional loading.

S. No.	Design Level	Maximum Roof Displacement 'Δ_u' (mm)			
		R = 3	R = 4	R = 5	R = 6
1	DBE	0.390	0.461	0.522	0.589
2	MCE	0.735	0.774	0.833	0.875

Table 11. Direct damage–loss ratio under bidirectional loading.

S. No.	Design Level	Maximum Roof Displacement 'Δ_u' (mm)			
		R = 3	R = 4	R = 5	R = 6
1	DBE	0.419	0.486	0.554	0.625
2	MCE	0.758	0.795	0.843	0.885

According to these tables, the damage–loss ratio increased for each building case at each design level. Larger R factors (R = 5 and 6) result in a higher damage–loss ratio be-cause of a higher ductility demand. This increase in loss ratio has an adverse effect on the building resilience. A slight increase in the damage–loss ratio under bidirectional loading was observed, which affects the resilience of the building.

3.6. Estimation of Building Resilience from Functionality Curves

To estimate the building resilience, functionality curves for each building case were developed with respect to each design level, using various recovery functions. The analytical recovery functions, such as linear (RP-1), exponential (RP-2) and trigonometric (RP-3) recovery paths, were proposed by Bruneau et al. [57] and Kumar et al. [58]. The functionality curve in terms of resilience was found as follows:

Functionality:

$$Q(t) = 1 - \{L(I, T_{RE}) \times [H(t - t_{OE}) - H(t - (t_{OE} + T_{RE}))] \times f_{rec}(t, t_{OE}, T_{RE})\} \quad (3)$$

The time of occurrence of a seismic event is designated as t_{OE}, its recovery time is designated as T_{RE}, and its Heaviside step function is designated as $H()$. For the purposes of this study, t_{OE} is set at 50 days, and total recovery time is calculated as 65 days with a total control time period (T_{LC}) of 140 days. The analytical equations for each recovery function are the following:

$$\text{Linear function}: \ f_{rec}(t, t_{OE}, T_{RE}) = \left[1 - \frac{t - t_{OE}}{T_{RE}}\right] \quad (4)$$

$$\text{Exponential function} f_{rec}(t, t_{OE}, T_{RE}) = \exp\left[-\frac{(t - t_{OE})(\ln 200)}{T_{RE}}\right] \quad (5)$$

$$\text{Trigonometric function}: \ f_{rec}(t, t_{OE}, T_{RE}) = 0.5\left\{1 + \cos\left[\Pi\frac{(t - t_{OE})}{T_{RE}}\right]\right\} \quad (6)$$

Along with conventional, two different recovery paths named modified trigonometric path (RP-4) and combined recovery path (RP-5) were also considered.

The idea behind the parameter introduction was that, under realistic conditions, recovery could not begin after a seismic event had happened (Figure 16). Real-world situations necessitate a careful assessment of the structural damage before the recovery procedure can begin. The functionality and recovery equations were adjusted as follows in light of this.

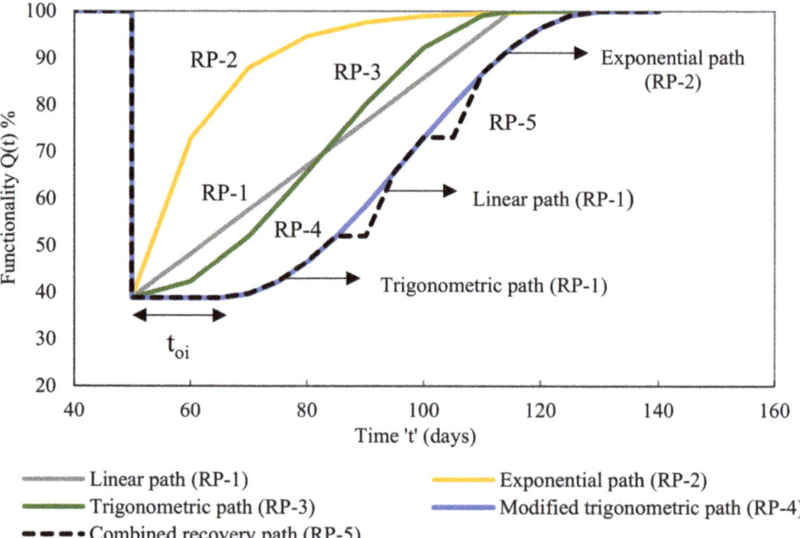

Figure 16. Various recovery paths considered (RP-1 to RP-5).

In addition to the 't_{oi}' parameter, the modified trigonometric path is the fourth recovery path (RP-4) which has an initial delay in the recovery process, whereas the combined recovery path is the fifth recovery path (RP-5) which has a certain breakdown in the recovery process (Figure 16). The functionality Equation (3) now includes the new argument 'toi'. Based on this, the functionality and recovery equations were modified as follows:

$$Q(t) = 1 - \{L(I, T_{RE}) \times [H(t - t_{OE} - t_{oi}) - H(t - (t_{OE} + T_{RE}) - t_{oi})] \times f_{rec}(t, t_{OE}, T_{RE}, t_{oi})\} \tag{7}$$

$$f_{rec}(t, t_{OE}, T_{RE}, t_{oi}) = 0.5\left\{1 + \cos\left[\Pi \frac{(t - t_{OE} - t_{oi})}{T_{RE}}\right]\right\} \tag{8}$$

3.7. Resilience of Each Building Case under Unidirectional Loading

Under conditions of unidirectional loading, functionality curves corresponding to each design level were developed (Figures 17–20) based on several recovery paths. The area below and above the functionality curves demonstrates the building resilience and loss of resilience (LOR). Utilising the Origin software, the curve area was discovered.

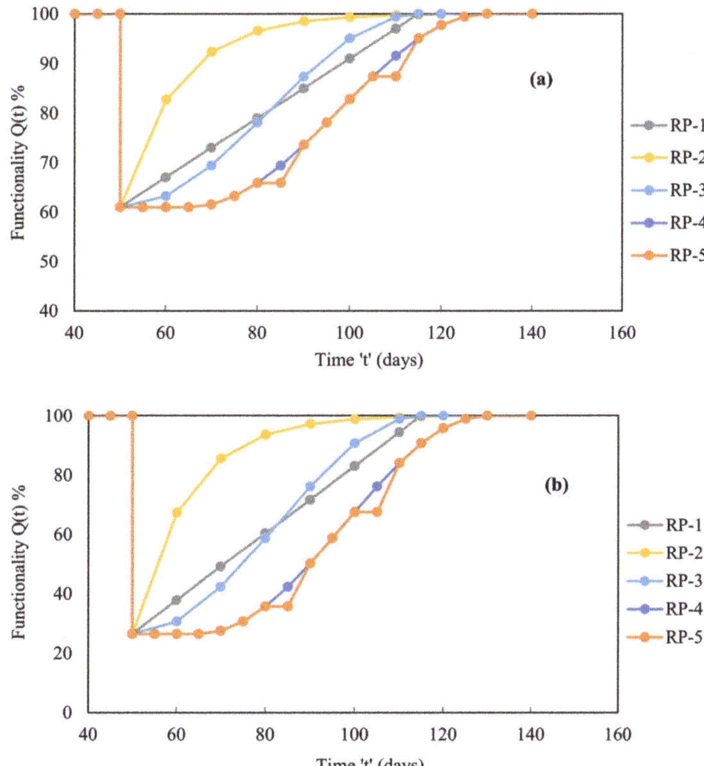

Figure 17. Functionality curves for unsymmetrical building Case I (R = 3) using various recovery paths (**a**) with respect to DBE level and (**b**) with respect to MCE level.

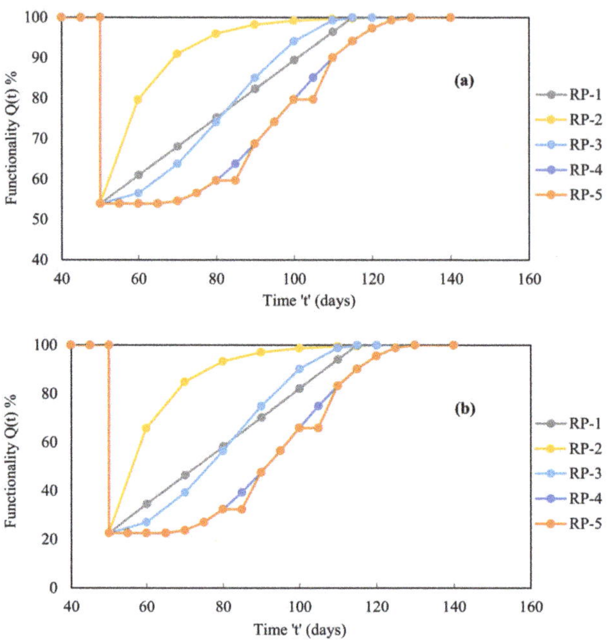

Figure 18. Functionality curves for unsymmetrical building Case II (R = 4) using various recovery paths (**a**) with respect to DBE level and (**b**) with respect to MCE level.

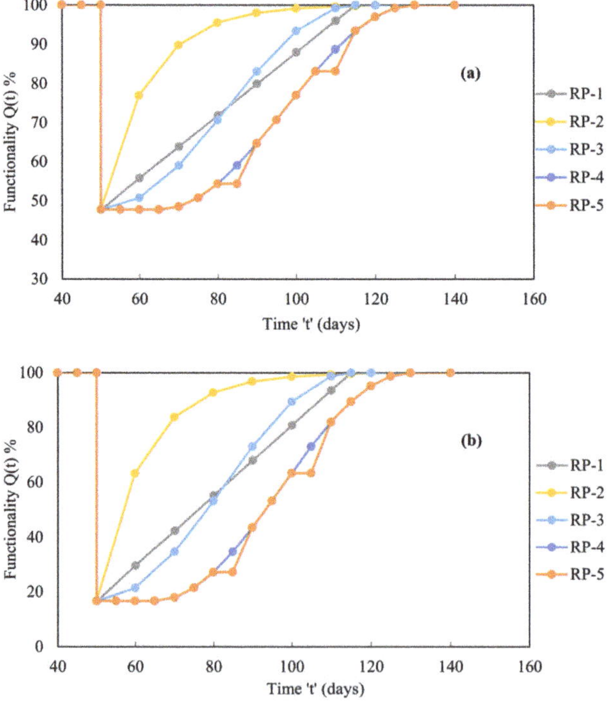

Figure 19. Functionality curves for unsymmetrical building Case III (R = 4) using various recovery paths (**a**) with respect to DBE level and (**b**) with respect to MCE level.

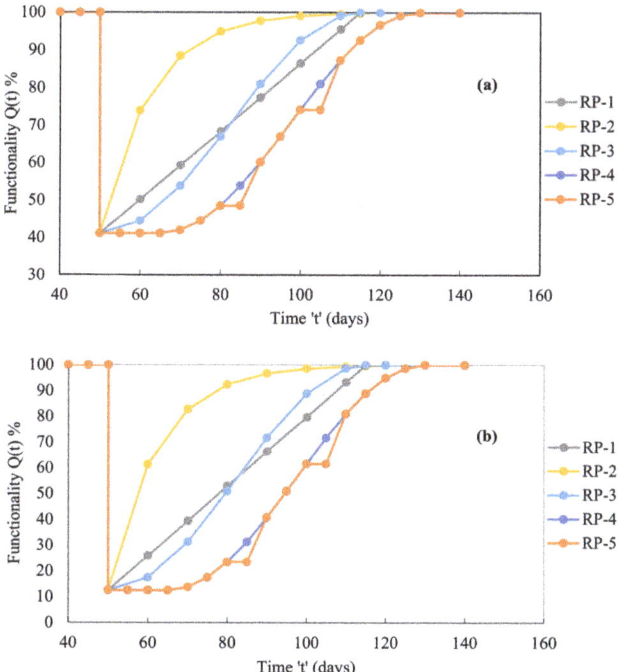

Figure 20. Functionality curves for unsymmetrical building Case IV (R = 6) using various recovery paths (**a**) with respect to DBE level and (**b**) with respect to MCE level.

The functionality curves were plotted using five different recovery paths at each design level for Case I building (Figure 17). At the time of the occurrence of a seismic event, the functionality of the building dropped down from 100% to 61.00% at the DBE design level and from 100% to 26.50% at the MCE design level. The loss in functionality was estimated to be 39.00% and 73.50% at the DBE and MCE design levels, respectively. The loss of functionality was higher at the MCE level which was due to a higher damage–loss ratio when compared with the DBE design level.

Figure 18 shows the different functionality curves at various design levels for the Case II building. The functionality drops from 100% to 54% at the DBE level and from 100% to 23% at the MCE level. The functionality loss was estimated to be 46% and 77% at the DBE and MCE design levels, respectively.

For the Case III building, the different functionality curves were developed at design levels (Figure 19). At the time of occurrence of the seismic event (on the 50th day), the functionality is reduced from 100% to 48% and 17% at the DBE and MCE design levels, respectively. The functionality loss was estimated to be 52% and 83% at the DBE and MCE design levels, respectively.

Figure 20 depicts the functioning curves for the Case IV building utilising various recovery pathways. The functionality decreases from 100% to 41% at the DBE design level and from 100% to 13% at the MCE design level at the time of the seismic event (on the 50th day). At the DBE and MCE design levels, the functionality loss was calculated to be 59% and 88%, respectively.

The region under and above the functioning curves (Figures 17–20) represents the building resilience and loss of resilience (LOR), respectively. Origin software was used to calculate the area of the curve. The functionality curves were used to calculate the resilience for each building case (Tables 12–15).

Table 12. Building resilience with respect to each design level corresponding to Case I.

S. No.	Design Level	Resilience (%)				
		RP-1	RP-2	RP-3	RP-4	RP-5
1	DBE	81	92	88	77	76
2	MCE	63	86	66	56	55

Table 13. Building resilience with respect to each design level corresponding to Case II.

S. No.	Design Level	Resilience (%)				
		RP-1	RP-2	RP-3	RP-4	RP-5
1	DBE	77	92	85	73	72
2	MCE	61	86	64	54	53

Table 14. Building resilience with respect to each design level corresponding to Case III.

S. No.	Design Level	Resilience (%)				
		RP-1	RP-2	RP-3	RP-4	RP-5
1	DBE	74	90	76	69	68
2	MCE	58	85	61	51	49

Table 15. Building resilience with respect to each design level corresponding to Case IV.

S. No.	Design Level	Resilience (%)				
		RP-1	RP-2	RP-3	RP-4	RP-5
1	DBE	71	89	73	65	64
2	MCE	56	84	59	48	47

At the DBE design level, building resilience corresponds to Case I ranging from 81% to 76%, and at the MCE design level resilience ranges from 63% to 55% (Table 12). When compared to the DBE level for RP-5, the highest reduction in resilience at the MCE level was around 28%. The recovery path RP-2 was not always realistically feasible, as was already mentioned. The remaining recovery pathways were contrasted with each other (RP-1, RP-3, RP-4 and RP-5). When compared to the standard recovery path RP-3, roughly 13% lesser resilience was seen at the DBE level and 15% less was seen at the MCE design level for the proposed recovery path RP-4. This resulted from the recuperation process breaks and the early delay in the beginning. At the DBE and MCE design levels, the maximum loss of resilience (LOR) was discovered to be approximately 24% and 45%, respectively.

Table 13 shows that, with regard to recovery paths, the resilience of the Case II building varies between 77% and 72% at the DBE design level and between 61% and 53% at the MCE design level. At the DBE and MCE design levels, respectively, it was discovered that the maximum loss of resilience (LOR) was estimated to be approximately 28% and 47%.

Building resilience in Case III (Table 14) at the DBE design level ranges from 74% to 68%, and 58% to 49% at the MCE design level. At the DBE and MCE design levels, respectively, the maximum loss of resilience (LOR) was estimated to be around 32% and 51%.

Building resilience in Case IV at the DBE design level varies from 71 to 64% (Table 15) and at the MCE design level it varies from 56 to 47%. At the DBE and MCE design levels, the maximum loss of resilience (LOR) was observed to be about 36% and 53%, respectively. It was found that the building corresponding to Cases III and IV (R = 5 and 6) exhibits less resilience than other building cases at both design levels. This resulted from the strong need for ductility at R equals 5 and 6 (Cases III and IV). It was discovered that the building resilience at both design levels significantly decreases for Cases III and IV. This was brought on by the building's increased ductility requirements at the DBE and MCE design levels in comparison with Cases III and IV.

3.8. Resilience of Each Building Case under Bidirectional Loading

The resilience of each building case under bidirectional loading was found (Tables 16–19). A comparison of resilience was made between unidirectional and bidirectional loading conditions (Figures 21 and 22).

Table 16. Seismic resilience of unsymmetrical building under bidirectional loading (Case I).

S. No.	Design Level	Resilience (%)				
		RP-1	RP-2	RP-3	RP-4	RP-5
1	DBE	79	93	81	75	74
2	MCE	62	86	65	55	54

Table 17. Seismic resilience of unsymmetrical building under bidirectional loading (Case II).

S. No.	Design Level	Resilience (%)				
		RP-1	RP-2	RP-3	RP-4	RP-5
1	DBE	76	91	77	71	70
2	MCE	60	85	63	53	52

Table 18. Seismic resilience of unsymmetrical building under bidirectional loading (Case III).

S. No.	Design Level	Resilience (%)				
		RP-1	RP-2	RP-3	RP-4	RP-5
1	DBE	72	90	74	67	66
2	MCE	58	85	61	49	48

Table 19. Seismic resilience of unsymmetrical building under bidirectional loading (Case IV).

S. No.	Design Level	Resilience (%)				
		RP-1	RP-2	RP-3	RP-4	RP-5
1	DBE	69	88	71	63	62
2	MCE	56	84	59	47	46

From Figure 21, it was observed that the variation in resilience was not significant with bidirectional loading at the DBE design level. In all of the building cases at the DBE level, the resilience under bidirectional loading was marginally less when compared with the unidirectional loading. At the MCE design level, the same variation in resilience was observed when compared with unidirectional and bidirectional loading conditions (Figure 22). The variation in resilience was not significant, which was due to the fact that the damage ratio with bidirectional loading did not vary much when compared with unidirectional loading.

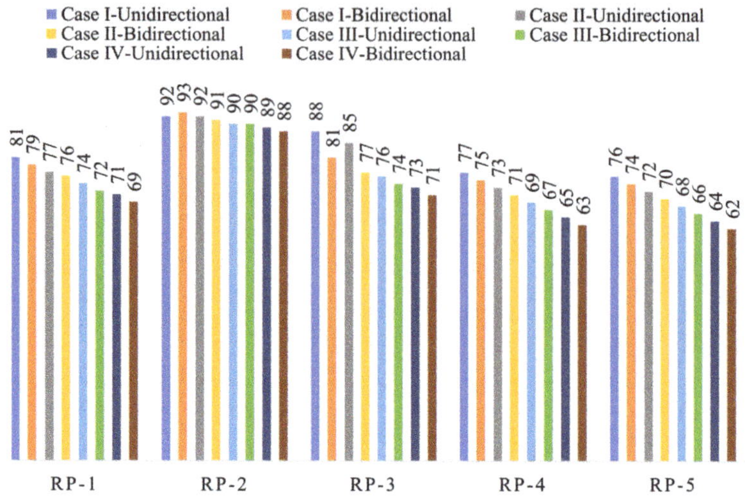

Figure 21. Comparison of unsymmetrical building resilience under unidirectional and bidirectional loading at DBE design level (in percentage).

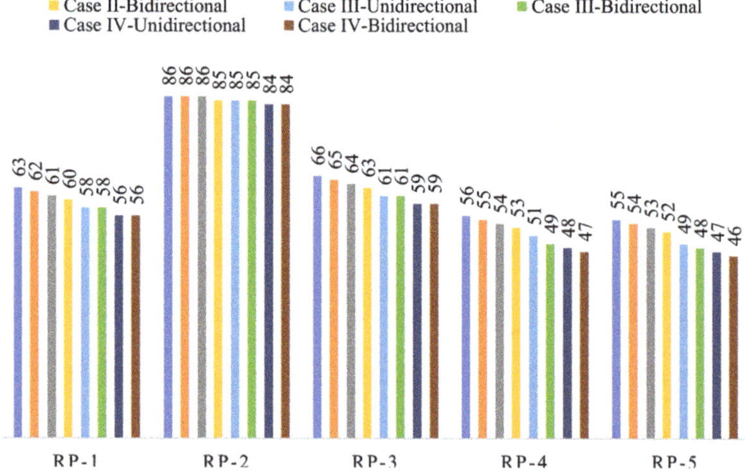

Figure 22. Comparison of unsymmetrical building resilience under unidirectional and bidirectional loading at MCE design level (in percentage).

4. Conclusions

In this study, the selection of appropriate R factors was performed for an unsymmetrical L-shaped RC building in accordance with resilience, performance level and ductility demand aspects. The study proposes a framework in the R factor selection, and it can be adapted to other irregular buildings since only the modelling and loading conditions will be varied for each type of building. The framework remains the same which proposes various recovery paths in the assessment of R factors with resilience considerations. The seismic response of the building has to be taken care of via proper selection of R factors. It is possible to consider the irregularity in the design level as did in the present study. To plan for post-disaster recovery, the building irregularity found using the R factor should be incorporated with resilience consideration. Though several codes are silent on the above aspects, the present study gives a detailed description about the basis of R factor selection in the design. In this study, the seismic performance of an unsymmetrical building subjected to unidirectional and bidirectional loadings at two design levels such as the DBE and MCE levels was assessed. Some of the major conclusions of the study are as follows.

- All of the cases of the unsymmetrical building reported a moderate ductility requirement at the DBE design level for both unidirectional and bidirectional loadings. The building almost meets the high ductility demand and corresponds to R equalling 5 and 6 at the MCE design level. This demonstrates that the ductility demand rises as the R factor increases. Buildings with lower ductility demands are generally more affordable to construct. This aids in making the right choice of R factors for the building design.
- A higher R factor has an impact on building performance levels due to higher ductility demand. The performance level at the DBE design level advances from the IO level (at R equals 3 and 4) to the IO-LS level (at R equals 5 and 6). The building performance level varied significantly (IO-LS level to CP-C level) from R = 3 to 6 at the MCE design level. Under bidirectional loading, the performance level of the building has higher R (R = 5 and 6) cases which lie at the C-D level. This shows that the building reaches a full collapse damage state without having any residual strength. This was due to the reduction in transverse member participation in structural stiffness due to a bidirectional loading effect at a higher R factor. Thus, in the case of unsymmetrical building, the bidirectional effect significantly alters the building performance level with a marginal increase in ductility demand.
- Under both unidirectional and bidirectional loading conditions, with respect to the DBE and MCE design level, the loss of resilience (LOR) was less than 30% and 50% at R = 3 and 4 (Cases I and II), respectively. At higher R factors of 5 and 6, the LOR marginally increases with more than 50% LOR at the MCE design level. At unidirectional and bidirectional loadings, though the building at R = 5 and R = 6 suffers significant functionality loss at the MCE level, the building holds almost 50% resilience. However, the recovery of the building from that level may lead to higher retrofitting costs to recover back to its target performance level.
- It was concluded that as the building reaches a higher ductility demand with a value of 4.98 at R = 5 along with a performance level at the CP-C level, the maximum R factor for the considered unsymmetrical building can be recommended up to 4 with respect to unidirectional loading. Due to asymmetry, the bidirectional loading is predominant in the case of the MCE design level, which alters the building's performance level to CP-D and increases the ductility demand to a higher level. This resulted in a reduction in maximum values of the R factor from 4 to 3, as at R equals 4, 5 and 6 the building experiences a higher ductility demand and a higher performance level with a marginal increase in loss of resilience.

The study concludes that the building resilience, performance level and ductility demand influence the selection of the R factors in accordance with the asymmetry of the building along with a consideration of directional effects.

Author Contributions: Conceptualisation, S.P. and G.G.; data curation, P.K.G.; formal analysis, S.P. and P.K.G.; investigation, S.P.; methodology, S.P. and G.G.; project administration, S.P. and P.K.G.; supervision, G.G., C.C. and L.G.; validation, S.P.; visualisation, G.G.; writing—original draft, S.P., P.K.G. and G.G.; writing—review and editing, S.P., P.K.G., G.G., C.C. and L.G. All authors have read and agreed to the published version of the manuscript.

Funding: This research received no external funding.

Institutional Review Board Statement: Not applicable.

Informed Consent Statement: Not applicable.

Data Availability Statement: All data, models and codes generated or used during the study appear in the submitted article.

Conflicts of Interest: The authors have no relevant financial or non-financial interests to disclose.

References

1. *ASCE/SEI 7-22*; Minimum Design Loads and Associated Criteria for Buildings and Other Structures. American Society of Civil Engineers: Reston, VA, USA, 2005.
2. *EN 1998-1*; Eurocode 8: Design of Structures for Earthquake Resistance. 1st ed. BSi—Brussels: Brussels, Belgium, 2004.
3. *IS: 1893*; Part-1. Indian Standard Criteria for Earthquake Resistance Design of Structures. Bureau of Indian Standards: New Delhi, India, 2016.
4. Mondal, A.; Ghosh, S.; Reddy, G.R. Performance-based evaluation of the response reduction factor for ductile RC frames. *Eng. Struct.* **2018**, *56*, 1808–1819. [CrossRef]
5. Abdollahzadeh, G.; Sadeghi, A. Earthquake recurrence effect on the response reduction factor of steel moment frame. *Asian J. Civ. Eng.* **2018**, *19*, 993–1008. [CrossRef]
6. Yahmi, D.; Branci, T.; Bouchaïr, A.; Fournely, E. Evaluating the Behaviour Factor of Medium Ductile SMRF Structures. *Period. Polytech. Civ. Eng.* **2018**, *62*, 373–385. [CrossRef]
7. Tamboli, K.; Amin, J.A. Evaluation of Response Reduction Factor and Ductility Factor of RC Braced Frame. *J. Mater. Eng. Struct.* **2015**, *2*, 120–129.
8. Nishanth, M.; Visuvasam, J.; Simon, J.; Packiaraj, J.S. Assessment of seismic response reduction factor for moment resisting RC frames. *Int. J. IOP Conf. Ser. Mater. Sci. Eng.* **2017**, *263*, 032034. [CrossRef]
9. Chaulagain, H.; Rodrigues, H.; Spacone, E.; Guragain, R.; Mallik, R.; Varum, H. Response reduction factor of irregular RC buildings in Kathmandu valley. *Earthq. Eng. Eng. Vib.* **2014**, *13*, 455–470. [CrossRef]
10. Mohsenian, V.; Mortezaei, A. Evaluation of seismic reliability and multilevel response reduction factor (R factor) for eccentric braced frames with vertical links. *Earthq. Struct.* **2018**, *14*, 537–549.
11. Patel, K.N.; Amin, J.A. Performance-based assessment of response reduction factor of RC-elevated water tank considering soil flexibility: A case study. *Int. J. Adv. Struct. Eng.* **2018**, *10*, 233–247. [CrossRef]
12. Pérez Jiménez, F.J.; Morillas, L. Effect of the Importance Factor on the Seismic Performance of Health Facilities in Medium Seismicity Regions. *J. Earthq. Eng.* **2022**, *26*, 546–562. [CrossRef]
13. Hussein, M.M.; Gamal, M.; Attia, W.A. Seismic response modification factor for RC-frames with non-uniform dimensions. *Cogent Eng.* **2021**, *8*, 1923363. [CrossRef]
14. Attia, W.A.; Irheem, M.M. Boundary condition effect on response modification factor of X-braced steel frames. *HBRC J.* **2018**, *14*, 104–121. [CrossRef]
15. Keykhosravi, A.; Aghayari, R. Evaluating response modification factor (R) of reinforced concrete frames with chevron brace equipped with steel slit damper. *KSCE J. Civ. Eng.* **2017**, *21*, 1417–1423. [CrossRef]
16. Kappos, A.J. Evaluation of behaviour factors on the basis of ductility and overstrength studies. *Eng. Struct.* **1999**, *21*, 823–835. [CrossRef]
17. Patel, B.; Shah, D. Formulation of Response Reduction Factor for RCC Framed Staging of Elevated Water tank using static pushover analysis. In Proceedings of the World Congress on Engineering, London, UK, 30 June–2 July 2010. ISSN 2078-0966.
18. Galasso, C.; Maddaloni, G.; Cosenza, E. Uncertainly Analysis of Flexural Over strength for Capacity Design of RC Beams. *J. Struct. Eng.* **2014**, *140*, 04014037. [CrossRef]
19. Abdi, H.; Hejazi, F.; Saifulnaz, R.; Karim, I.A.; Jaafar, M.S. Response modification factor for steel structure equipped with viscous damper device. *Int. J. Steel Struct.* **2015**, *15*, 605–622. [CrossRef]
20. Prasanth, S.; Ghosh, G. Effect of variation in design acceleration spectrum on the seismic resilience of a building. *Asian J. Civ. Eng.* **2021**, *22*, 331–339. [CrossRef]
21. Prasanth, S.; Ghosh, G. Effect of cracked section properties on the resilience based seismic performance evaluation of a building. *Structures* **2021**, *34*, 1021–1033. [CrossRef]
22. SeMarasco, S.; Cardoni, A.; Noori, A.Z.; Kammouh, O.; Domaneschi, M.; Cimellaro, G.P. Integrated platform to assess seismic resilience at the community level. *Sustain. Cities Soc.* **2021**, *64*, 102506. [CrossRef]
23. Hashemi, M.J.; Al-Attraqchi, A.Y.; Kalfat, R.; Al-Mahaidi, R. Linking seismic resilience into sustainability assessment of limited-ductility R.C. buildings. *Eng. Struct.* **2019**, *188*, 121–136. [CrossRef]
24. Cimellaro, G.P.; Reinhorn, A.M.; Bruneau, M. Framework for analytical quantification of disaster resilience. *Eng. Struct.* **2010**, *32*, 3639–3649. [CrossRef]
25. Cimellaro, G.P.; Reinhorn, A.M.; Bruneau, M. Seismic resilience of a hospital system. *Struct. Infrastruct. Eng.* **2010**, *6*, 127–144. [CrossRef]
26. Hudson, S.; Cormie, D.; Tufton, E.; Inglis, S. Engineering resilient infrastructure. *Proc. Inst. Civ. Eng. Civ. Eng.* **2012**, *165*, 5–12. [CrossRef]
27. Gallagher, D.; Cruickshank, H. Planning under new extremes: Resilience and the most vulnerable. *Proc. Inst. Civ. Eng. Munic. Eng.* **2015**, *169*, 127–137. [CrossRef]
28. Grigorian, M.; Kamizi, M. High-performance resilient earthquake-resisting moment frames. *Proc. Inst. Civ. Eng. Struct. Build.* **2022**, *175*, 401–417. [CrossRef]
29. Dukes, J.; Mangalathu, S.; Padgett, J.E.; DesRoches, R. Padgett and Reginald DesRoches. Development of a bridge-specific fragility methodology to improve the seismic resilience of bridges. *Earthq. Struct.* **2018**, *15*, 253–261.
30. Haselton, C.B.; Deierlein, G.G. Assessing seismic collapse safety of modern reinforced concrete moment-frame buildings. In *PEER Report 2007/08*; Pacific Engineering Research Center, University of California: Berkeley, CA, USA, 2008.

31. Haselton, C.B.; Liel, A.B.; Deierlein, G.G.; Dean, B.S.; Chou, J.H. Seismic collapse safety of reinforced concrete buildings. I: Assessment of ductile moment frames. *J. Struct. Eng.* **2011**, *137*, 481–491. [CrossRef]
32. Liel, A.B.; Haselton, C.B.; Deierlein, G.G. Seismic collapse safety of reinforced concrete buildings. II: Comparative assessment of nonductile and ductile moment frames. *J. Struct. Eng.* **2011**, *137*, 492–502. [CrossRef]
33. Dutta, S.C.; Kunnath, S.K. Effect of bidirectional interaction on seismic demand of structures. *Soil Dyn. Earthq. Eng.* **2013**, *52*, 27–39. [CrossRef]
34. Hachem, M.M.; Moehle, J.P.; Mahin, S.A. Performance of circular reinforced concrete bridge columns under bidirectional earthquake loading. In *PEER Report 2003/06*; Pacific Engineering Research Center, University of California: Berkeley, CA, USA, 2003.
35. Opabola, E.; Elwood, K.J. Accounting for the influence of bidirectional loading on deformation capacity of reinforced concrete columns. In Proceedings of the 11th Pacific Conference on Earthquake Engineering, Auckland, New Zealand, 4–6 April 2019.
36. Opabola, E.; Elwood, K.J.; Pujol, S. Influence of Biaxial Lateral Loading on Seismic Response of Reinforced Concrete Columns. *ACI Struct. J.* **2020**, *117*, 211–224.
37. Sengupta, A.; Quadery, L.; Sarkar, S.; Roy, R. Influence of bidirectional near-fault excitations on RC bridge piers. *J. Bridge Eng.* **2016**, *21*, 04016034. [CrossRef]
38. Shirmohammadi, F.; Esmaeily, A. Performance of reinforced concrete columns under biaxial lateral force/displacement and axial load. *Eng. Struct.* **2015**, *99*, 63–77. [CrossRef]
39. Gwalani, P.; Singh, Y.; Varum, H. Effect of bidirectional excitation on seismic performance of regular RC frame buildings designed for modern codes. *Earthq. Spectra.* **2022**, *38*, 950–980. [CrossRef]
40. Hussain, M.A.; Dutta, S.C. Inelastic seismic behavior of asymmetric structures under bidirectional ground motion: An effort to incorporate the effect of bidirectional interaction in load resisting elements. *Structures* **2020**, *25*, 241–255. [CrossRef]
41. Puppio, M.L.; Pellegrino, M.; Giresini, L.; Sassu, M. Effect of Material Variability and Mechanical Eccentricity on the Seismic Vulnerability Assessment of Reinforced Concrete Buildings. *Buildings* **2017**, *7*, 66. [CrossRef]
42. Puppio, M.L.; Giresini, L.; Doveri, F.; Sassu, M. Structural irregularity: The analysis of two reinforced concrete (r.c.) buildings. *Eng. Solid Mech.* **2019**, *7*, 13–34. [CrossRef]
43. Anagnostopoulos, S.A.; Kyrkos, M.T.; Stathopoulos, K.G. Earthquake induced torsion in buildings: Critical review and state of the art. *Earthq. Struct.* **2015**, *8*, 305–377. [CrossRef]
44. Cimellaro, G.P.; Giovine, T.; Lopez-Garcia, D. Bidirectional Pushover Analysis of Irregular Structures. *J. Struct. Eng.* **2014**, *140*, 04014059. [CrossRef]
45. Ruggieri, S.; Chatzidaki, A.; Vamvatsikos, D.; Uva, G. Reduced-order models for the seismic assessment of plan-irregular low-rise frame buildings. *Earthq. Eng. Struct. Dyn.* **2022**, *51*, 3327–3346. [CrossRef]
46. Pacific Earthquake Engineering Research Center. PEER Strong Motion Database. *Berkley.* 2005. Available online: https://peer.berkeley.edu/peer-strong-ground-motion-databases (accessed on 17 May 2021).
47. *SAP V22*; Integrated Software for Structural Analysis and Design. Computers and Structures. Inc.: Berkley, CA, USA, 2000.
48. *FEMA-440*; Improvement of Nonlinear Static Seismic Analysis Procedures. Federal Emergency Management Agency: Washington, DC, USA, 2005.
49. Gupta, P.K.; Ghosh, G. Effect of various aspects on the seismic performance of a curved bridge with HDR bearings. *Earthq. Struct.* **2020**, *19*, 427–444.
50. Gupta, P.K.; Ghosh, G. Effect of Bidirectional excitation on a curved bridge with lead rubber bearing. *Mater. Today Proc.* **2021**, *44*, 2239–2244. [CrossRef]
51. Gupta, P.K.; Ghosh, G.; Pandey, D.K. Parametric study of effects of vertical ground motions on base isolated structures. *J. Earthq. Eng.* **2021**, *25*, 434–454. [CrossRef]
52. Gupta, P.K.; Ghosh, G.; Kumar, V.; Paramasivam, P.; Dhanasekaran, S. Effectiveness of LRB in Curved Bridge Isolation: A Numerical Study. *Appl. Sci.* **2022**, *12*, 11289. [CrossRef]
53. *ATC-40*; Seismic Evaluation and Retrofit of Reinforced Concrete Buildings. Applied Technology Council: Redwood City, CA, USA, 1996.
54. *ASCE/SEI 41-17*; Seismic Evaluation and Retrofit of Existing Buildings. American Society of Civil Engineers: Reston, VA, USA, 2017.
55. *HAZUS-MR4 Technical Manual*; Multihazard Loss Estimation Methodology. Department of Homeland Society: Washington DC, USA, 2003.
56. Barbat, A.H.; Pujades, L.G.; Lantada, N. Seismic damage evaluation in urban areas using the capacity spectrum method: Application to Barcelona. *Soil Dyn. Earthq. Eng.* **2007**, *28*, 851–865. [CrossRef]
57. Bruneau, M.; Chang, S.E.; Eguchi, R.T.; Lee, G.C.; O'Rourke, T.D.; Reinhorn, A.M.; Shinozuka, M.; Tierney, K.; Wallace, W.A.; Von Winterfeldt, D. A framework to quantitatively assess and enhance the seismic resilience of communities. *Earthq. Spectra* **2003**, *19*, 733–752. [CrossRef]
58. Kumar, A.; Sharma, K.; Dixit, A.R. A review on the mechanical properties of polymer composites reinforced by carbon nanotubes and graphene. *Carbon Lett.* **2021**, *31*, 149–165. [CrossRef]

applied sciences

Article

A Calculation Model for Determining the Bearing Capacity of Strengthened Reinforced Concrete Beams on the Shear

Zeljko Kos [1,*], Zinovii Blikharskyi [2], Pavlo Vegera [2] and Iryna Grynyova [3]

[1] Department of Civil Engineering, University North, 42000 Varazdin, Croatia
[2] Department of Building Construction and Bridges, Lviv Polytechnic National University, Bandery Street 12, 79013 Lviv, Ukraine; zinovii.y.blikharskyi@lpnu.ua (Z.B.); pavlo.i.vehera@lpnu.ua (P.V.)
[3] Department of Architecture Structures, Odessa State Academy of Civil Engineering and Architecture, 65029 Odessa, Ukraine; grynyova@ogasa.org.ua
* Correspondence: zeljko.kos@unin.hr; Tel.: +385-9875-7989

Featured Application: This work is aimed at carrying out calculations of strengthening of reinforced concrete structures on the shear, taking into account a new factor—the loading level at which strengthening occurs. Its application field is strengthening of reinforced concrete bent structures with composite materials on the shear.

Abstract: This article presents research on the bearing capacity and methods of calculating reinforced concrete beams on the shear without internal shear reinforcement, which are strengthened with a composite FRCM system. The test samples were divided into two series: the first series—control, in which the variable parameter was the shear span (a/d = 2, a/d = 1.5, and a/d = 1); and the second series—reinforced by the FRCM system, without load, and strengthened at different load levels. The method of calculating experimental beams was tested according to the current code and data from the fib report. In this article, recommendations for determining the angle of inclined struts θ, the coefficient of the concrete shear strength $C_{Rd,c}$, and the coefficient of the load level at which strengthening is performed are proposed. The calculation with the these recommendations showed a good convergence of experimental and theoretical data in the 16–29% range, which is a much higher convergence than the calculation without these recommendations.

Keywords: reinforced concrete beam; shear; FRCM; bearing capacity; strengthening; composite materials; calculation method

Citation: Kos, Z.; Blikharskyi, Z.; Vegera, P.; Grynyova, I. A Calculation Model for Determining the Bearing Capacity of Strengthened Reinforced Concrete Beams on the Shear. *Appl. Sci.* **2023**, *13*, 4658. https://doi.org/10.3390/app13084658

Academic Editors: Mariella Diaferio and Francisco B. Varona Moya

Received: 14 March 2023
Revised: 2 April 2023
Accepted: 4 April 2023
Published: 7 April 2023

1. Introduction

The study of the bearing capacity of reinforced concrete (RC) elements on the shear is essential due to the complex stress–strain state. In codes [1–3], shear designs are calculated with a significant margin of strength, leading to excessive overspending on materials. The development and implementation of dependencies that can realistically assess the bearing capacity of RC beams on the shear that need strengthening are essential tasks for scientists. The presented experimental and theoretical research is based on the existing methodology for calculating current codes [3].

2. Analysis of Previous Research

The main reasons for the need to strengthen structures are a decrease in their bearing capacity or the need to increase their bearing capacity to manage a greater load.

Defects occur in all structures during operation, and reduce the bearing capacity of RC structures. Most common are concrete defects in reinforced concrete beams [4] or columns [5]. The defects described in these works lead to a significant decrease in the bearing capacity of such structures.

One of the reasons for the rapid spread of defects is the opening of cracks in the RC structure. In the study [6], it was established that the load level of the opening shear crack was 53% higher than the load level of the appearing bending cracks. This dependence was derived for prestressed reinforced concrete T-beams based on experimental data. An empirical expression for determining the projection of a dangerous shear crack and the necessary shear reinforcement has been obtained. The use of modern methods and devices for viewing and detecting cracks allows faster identification of their effects on the safe operation of structures with higher accuracy [7].

In the case of emergency dynamic loads, it is also important to research the shear crack resistance of concrete and the reinforced concrete elements of protective structures. The work [8] describes the influence of dynamic loads on protective structures, data about changing the shear crack resistance, and the calculation principles in this case.

The appearance of corrosion or mounting holes in the structure decreases the strength of RC beams. In this regard, experimental research was conducted focused on different RC beams with a web opening, such as beams made from ultra-high-performance concrete (UHPC) with a web opening in the support area [9].

Developing methods for strengthening RC structures requires the improvement of calculation methodologies. To determine the necessary amount of reinforcement for structures, their residual bearing capacity needs to be known [10,11]. Only by having the exact value of the residual strength is it possible to design economic strengthening of structures.

Today, structures often need to be strengthened due to the demand to restore or increase the bearing capacity of elements. Classic methods of strengthening are improving and changing. For example, geopolymer concrete (GPC) is used with improved properties for jacketing [12]. In this case, there is a need to perform experimental investigations and improve the calculation methodology.

The most effective way to increase the bearing capacity of bent elements is to strengthen the compressed zone of concrete and tensile reinforcement [13]. Strengthening was performed by using fine-grained or steel fiber concrete in the compressed area and glued carbon fibers in the tensile area. The overall strengthening effect was 32–69%, depending on the type of strengthening and the type of load (one-time or low-cycle).

A study of various types of composites (Fiber-Reinforced Polymer (FRP) and Fiber-Reinforced Cementitious Matrix (FRCM)) as systems for strengthening shear beams was conducted in the article [14]. It was established that the beams strengthened by the FRCM system showed a worse interaction than the samples strengthened by the FRP system.

Research on the implementation of U wraps of the FRCM system is relevant [15]. In this case, the possibility of the practical application of such strengthening increases. Strengthening took place along the entire length of the beam and was performed in one or more layers. The reinforcement effect ranged from 21 to 56%, depending on the characteristics of the concrete and the number of layers of reinforcement material.

The influence of the reinforcement numbers at different reinforcement ratios for longitudinal reinforcement for strengthened concrete beams by the FRP system showed the necessity to consider the specified reinforcement ratio and the importance of the quality performance of design works [16].

The combination of different types of strengthening to increase the bearing capacity has been used more and more widely [17]. Experimental studies of two-span RC beams with strengthening frames from rebars and concrete with steel fibers have been conducted. Research [18] proposed a method of strengthening the concrete properties by laying a fibrous mesh in the top (mortar) layer of the concrete surface. Basalt fiber, aramid fiber, and carbon fiber mesh were selected as reinforcing materials to study the concrete's early cracking resistance and the mortar's early cracking characteristics. Another method is installing a polymer–cement coating on wall structures [19]. The influence on the durability and destruction of the tested wall elements was described.

The need for reinforcement often arises in connection with an increase in the seismicity of the building or a change in code [20]. Research into reinforced concrete beams with different schemes and reinforcement principles for seismic impact is directed toward the most effective construction of RC beams.

A particular type of strengthening increases the seismic resistance of the joints of reinforced concrete structures. A promising direction is their reinforcement with composite materials with significant strength and deformability, which can prevent the destruction and collapse of structures [21].

Long-term effects on structures also lead to the need to strengthen them [22]. As the service life increases, the mechanical characteristics deteriorate due to the properties of filled concrete that wear out over time. Another factor that is rarely taken into account is the level of load on the structure for which strengthening is performed [23].

An important aspect of the development of methods of strengthening reinforced concrete structures is their practical application, such as the strengthening of an RC tank [24]. According to results of a technical inspection, numerous defects were discovered, and a method of using steel external bandages was developed. However, to restore tightness and improve operational characteristics by installing reinforced concrete brackets [25]. The given examples of strengthening were carried out taking into account the existing defects and the real state of the structure.

Expanding the application areas of FRCM systems requires appropriate design models. The paper [26] proposes a simplified modified compression field theory (SMCFT) for shear model calculation. According to this calculation model, the calculation accuracy was 0.96 (the ratio of theoretical data to experimental data) with a deviation of 0.14.

The calculation of strengthening concrete T-beams reinforced in shear with FRCM composites based on the ACI549.4R-13 code is relevant [27]. The results of the calculations were significantly underestimated and require further refinement.

For a high-quality calculation of RC structures, it is necessary to consider the nonlinear properties of materials and the deformation criteria for the exhaustion of the bearing capacity [28]. The same requirements apply to the calculation of constructions on the shear. The proposed method for determining the reliability of RC structures on shear allows optimal design decisions to be made, ensuring the specified level of construction reliability without overspending on materials [29]. Optimizing the design method is an urgent issue. This technique can be based on geometric parameters and structural features of systems [30]. This technique can be based on design features and systems' geometric parameters. Required reinforcements for beams with span-to-depth ratios varying from 10 to 20 were first determined in accordance with Eurocode 2. A probabilistic analysis using Monte Carlo simulations then revealed that doubling the span-to-depth ratio would not worsen the performance of the beams in terms of ultimate strength [31].

One of the effective methods of strengthening structures is their unification of prefabricated individual elements into a continuous spatial structure [32]. The model for determining the shear strength of such combined reinforced concrete beams was proposed on the basis of experimental and calculation data.

An important aspect of the FRCM application of strengthening systems is studying their carrying capacity under the influence of high temperatures [33]. It was established that heating/cooling down cycles affect the bearing capacity of FRCM systems for reinforcing reinforced concrete structures. However, it should be noted that according to an investigation of reinforced concrete beams on the FRCM (TRP) section, the systems showed better resistance to temperature effects than the FRP systems [34].

Based on the above studies, the findings do not consider the load level at which the stress-deformed state of structures changes (strengthening and/or damaging). Moreover, applying structural design models in practice is difficult since regulatory documents do not support them. This study proposes the calculation of reinforced concrete beams on the shear, without internal steel shear reinforcement, strengthened with a composite material based on the current code, considering the load level at which the reinforcement will be performed.

3. Experimental Data

In this research work, six test samples were tested. The initial dimensions of the sample's cross-section was 200 × 100, and the length was 2100 mm, but the actual dimensions had minor deviations (Table 1). Beams were reinforced with tensile reinforcement 2 Ø18 A500C, and compressed—2Ø10A500C. They were without transverse reinforcement in the supporting areas. The concrete for all beams was the same—C32/40.

Table 1. Bearing capacity on the shear without transverse reinforcement.

Beam Number	Number of Tested Support Area	The Actual Cross-Section $h \times b$ mm	Span l_0 mm	Shear Span, a/d	Shear Bearing Capacity, V_{Ed}^{exp}, kN	Average Value, V_{Ed}^{exp}, kN	$\frac{V_{ed}^{i-th}}{V_{ed}^{BO1.1}}$
BO 1.1	BO 1.1.1	201 × 106	1900	2	97	95	1.00
	BO 1.1.2		1550		93		
BO 1.2	BO 1.2.1	199 × 98	1900	1.5	139	140.5	1.48
	BO 1.2.2		1750		142		
BO 1.3	BO 1.3.1	202 × 98	1900	1	192	198	2.08
	BO 1.3.2		1650		204		

Three beams were tested as control samples in the first research stage. The variable parameter was the shear span, which acquired the following values: a/d = 1, a/d = 1.5, and a/d = 2. The criterion for exhausting the bearing capacity was obtaining limit values of the strains of compressed concrete area in the zone above the diagonal (inclined or shear) crack [1]. Each support area was tested separately, according to the developed testing methodology [35].

Exhaustion of the bearing capacity on the shear was equated to the physical destruction of the concrete compressed area. The occurrence of the limit state of control samples occurred in the following sequence:

- opening of a shear crack of maximum width (a_{crc} = 0.4 mm) on the concrete surface;
- cracking of the compressed area concrete and plastic deformation of the rebars of the reinforcing frame took place (Figure 1).
- shear reinforcement.

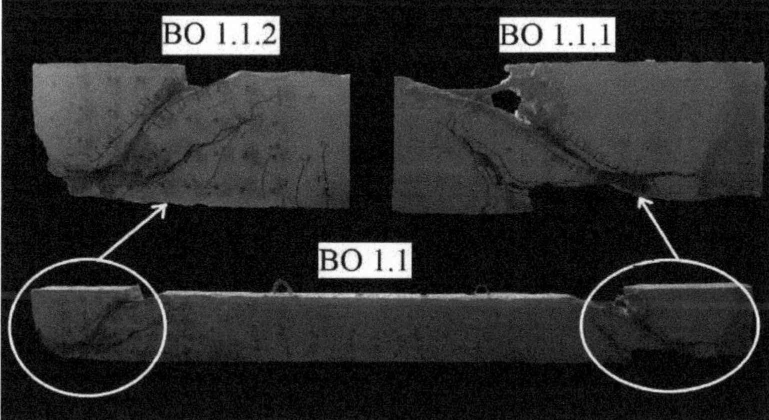

Figure 1. Tested control samples without internal shear reinforcement.

Test results of the bearing capacity on the shear without transverse reinforcement are described in Table 1.

According to them, the increase in bearing capacity on the shear was recorded at 1.48 times when the shear span was reduced from a/d = 2 to a/d = 1.5. When the shear span was a/d = 1, it increased 2.08 times. This effect is shown when the shear span is reduced, the compressive forces increase in the support zone, which are effectively absorbed by the concrete.

In the second stage, reinforced concrete beams were reinforced by sticking P.B.O. fabric in the form of vertical strips with a width of 70 mm for the possibility of fixing concrete strains in the support areas (Figure 2). Beams were tested according to the following program: BS 1.1–0, strengthened without initial load; BS 1.2–0.3 and BS 1.3–0.5—at the initial load level equal to 0.3 and 0.5 from the destructive one determined by experimental tests of unreinforced samples. The criterion for exhausting bearing capacity was adopted from unreinforced samples: the exhaustion of the shear strength was equated to the physical destruction of the compressed concrete area of the samples.

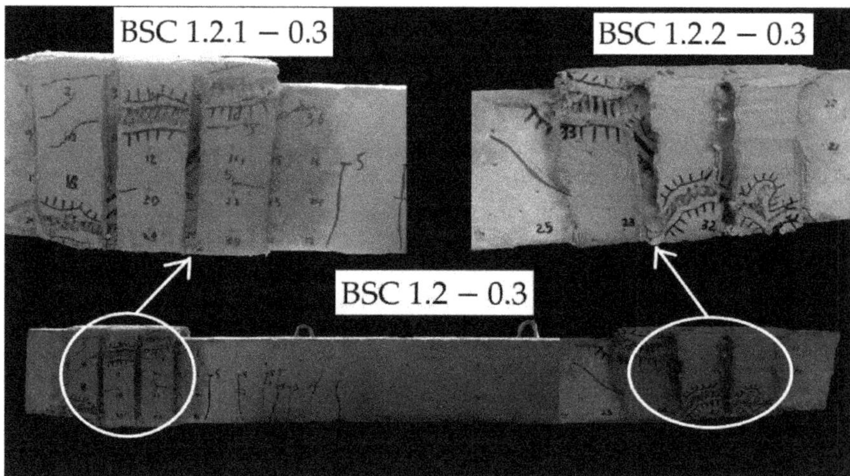

Figure 2. General view of exhausted shear strength of reinforced samples.

The collapsing of an RC beam reinforced with a composite fabric on the shear occurred in the following sequence:

- opening of a diagonal crack of maximum width (a_{crc} = 0.4 mm) on the concrete surface;
- propagation of the diagonal crack to the concrete compressed area. The appearance of a network of cracks with an opening width of a_{crc} = 0.2 mm on the surface of the strengthening system;
- collapse of concrete in the zone of action of the main tensile stresses; detachment of the reinforcement system in this area;
- plastic deformation of reinforcing bars. Destruction of the concrete compressed area and significant deformations of the reinforcing fabric can be seen due to the violation of the protective layer.

When the load was further increased, the ends of the fabric tape completely peeled off, and their anchoring failed. The exhaustion of the bearing capacity occurred at the moment of exfoliation of the concrete compressed area, together with the sharp elongation of the tape and the damage of the cover of the FRCM system in the area of propagation of the inclined crack (Figure 2).

The increase in the bearing capacity of the strengthened RC beams on the shear is given in Table 2.

Table 2. Bearing capacity of strengthened research samples on the shear without internal transverse reinforcement.

Beam Number	Number of Tested Support Area	The Actual Cross-Section $h \times b$ mm	Span l_0 mm	Shear Span, a/d	Shear Bearing Capacity, V_{Ed}^{exp}, kN	Average Value, V_{Ed}^{exp}, kN	$\frac{V_{ed}^{i-th}}{V_{ed}^{BO1.1}}$
BO 1.1	BO 1.1.1	201 × 101	1900		97	95	1.00
	BO 1.1.2		1650		93		
BS 1.1-0	BS 1.1.1-0	199 × 100	1900		130	137.5	1.45
	BS 1.1.2-0		1650		145		
BS 1.2-0.3	BS 1.2.1-0.3	200 × 100	1900	2	126	120	1.26
	BS 1.2.2-0.3		1650		117		
BS 1.3-0.5	BS 1.3.1-0.5	201 × 98	1900		116	110	1.16
	BS 1.3.2-0.5		1650		114		

The maximum effect of increasing the bearing capacity was 45% for samples reinforced without an initial load. Accordingly, as the initial load increased, the strengthening effect decreased. For a beam strengthened at the level of 0.3·, the strengthening effect was 21%, and for BS 1.3–0.5—16%. Increasing the load level to a higher value did not occur because the strengthening effect would have decreased even more, and it can be taken as a research deviation. The maximum strengthening effect is commensurate with the increase in the shear bearing capacity when the shear span is reduced from a/d = 2 to a/d = 1.5 (see Table 1).

For more details, experimental investigations of strengthened and unstrengthened RC beams on the shear are given in the articles [36,37].

4. Calculation Model for Determining the Shear Strength of the Tested Samples

4.1. Methodology of Calculating Shear Strength of the Control RC Beams

For a long time in Eastern Europe, codes [38] allowed the calculation of the bearing capacity on the shear according to the engineering method. Figure 3 shows the distribution of forces in the calculation scheme from the distributed load q (or equivalent action F) and the occurrence of forces in concrete Q_b and transverse reinforcement $R_{sw}A_{sw}$, which is located with step s. The bearing capacity of such an element is considered at the section on the length of the projection of a diagonal crack on the longitudinal axis (c—length for reinforcement, c_0—for concrete).

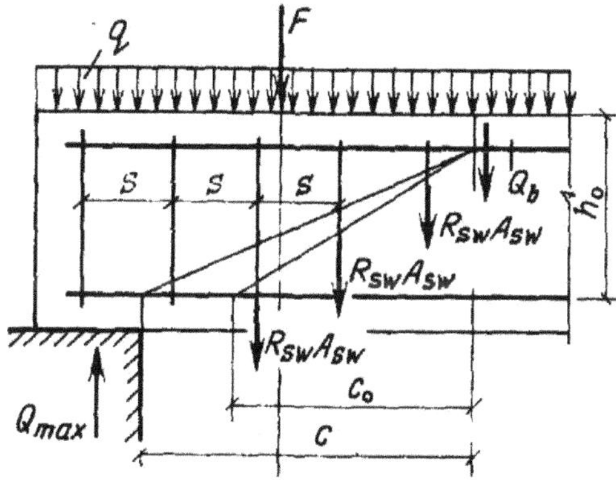

Figure 3. Scheme of the shear forces when calculating according to the engineering method [38].

The engineering method is based on the equilibrium of limit forces. The basis is the determination of the actual internal forces in the support section and their comparison with the external acting loads.

The calculation of RC elements for the action of the transverse force was carried out under the condition:

$$Q \leq Q_b + Q_{sw} \tag{1}$$

This relationship determines the total shear force as the sum of the strength of shear reinforcement and the concrete shear strength that can be borne by the support section. Since there is no transverse reinforcement in the samples, condition (1) takes the form:

$$Q \leq Q_b \tag{2}$$

The shear force perceived by concrete was calculated based on the following:

$$Q_{b,1} = \frac{\varphi_{b2} \cdot \left(1 + \varphi_f + \varphi_n\right) \cdot R_{bt} \cdot b \cdot h_0^2}{c} \tag{3}$$

However, if the value is not less than:

$$Q_{b,2} = \phi_{b3} \cdot \left(1 + \phi_f + \phi_n\right) \cdot R_{bt} \cdot b \cdot h_0. \tag{4}$$

$\varphi_{b2} = 2$; $\varphi_{b3} = 0.6$—coefficient that takes into account the influence of the type of concrete (light or heavy); $\varphi_f = 0$—a coefficient that takes into account the influence of compressed flanges in T-shaped and I-shaped elements (accepted no more than 0.5); $\varphi_n = 0$—a coefficient that takes into account the influence of longitudinal compressive forces; c—the length of the projection of the most dangerous inclined section on the longitudinal axis of the element, cm; b, h_0—respectively, the width and effective depth of the section, cm; R_{bt}—design value of concrete axial tensile strength, MPa.

Current codes [1–3] regulate shear strength calculation according to the "truss model". The bearing capacity in the zone of action of the transverse force, where is no shear reinforcement, is considered the strength of the concrete in a section at an angle of 45°.

The shear strength of reinforced concrete beams, according to current standards [1,3], is considered similar to condition (2):

$$V_{Ed} \leq V_{Rd,c} \tag{5}$$

$V_{Ed} = Q$—the calculated value of the transverse force from the external load. This is the value of the transverse force that a support area can absorb without transverse rebar and prestressed longitudinal reinforcement or axial force, according to the codes [2,3], is determined by the dependence:

$$V_{Rd,c1} = \frac{\left[C_{Rd,c} \cdot k \cdot (100 \cdot \rho_1 \cdot f_{ck})^{\frac{1}{3}}\right] \cdot b_w \cdot d}{\beta} \tag{6}$$

However, if the value is not less than:

$$V_{Rd,c2} = v_{\min} \cdot b_w \cdot d \tag{7}$$

$C_{Rd,c}$—concrete shear strength [39], codes [2,3] recommend taking 0.18; $k = 1 + \sqrt{200/d}$—coefficient, which takes into account the influence of the beam cross-section depth; $\rho_1 = \frac{A_{sl}}{b_w \cdot d}$—reinforcement ratio for longitudinal reinforcement; A_{sl}—cross-sectional area of tensile reinforcement, mm^2; f_{ck}—characteristic value of concrete compressive strength at the age of 28 days, MPa; b_w, d—respectively, the smallest width of the cross-section in the tensile area and the effective depth of the cross-section, mm; $v_{\min} = 0.035 \cdot k^{3/2} \cdot f_{ck}^{1/2}$—the minimum value of the shear strength that can lead to failure before yielding occurs in the longitudinal reinforcement, MPa.

Longitudinal reinforcement was taken into account in the calculation only if it was installed at a distance l_{bd} beyond the projection of the calculated section at an angle of $45°$. In this case, l_{bd}—the minimum required anchoring length of longitudinal reinforcement (Figure 4).

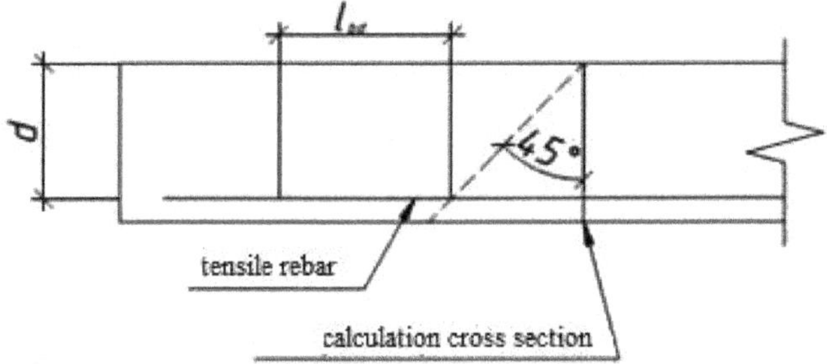

Figure 4. A model of the influencing factors on the shear strength of a support area [2,3].

The coefficient β is equal to $a_v/2d$ when loading the design elements from above within limits $0.5 \cdot d \leq a_v \leq 2 \cdot d$ (Figure 5) as a reducing coefficient of the shear force. In this model, it is proposed to consider it in the calculation by dividing the determined bearing capacity by this coefficient.

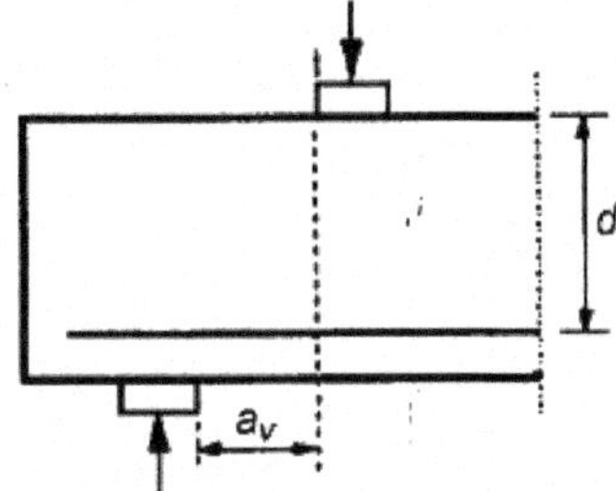

Figure 5. Case of applying load for the coefficient β [2,3].

When using the coefficient β, the following condition must be met:

$$V_{Ed} \leq 0.5 \cdot b_w \cdot d \cdot v \cdot f_{cd} \tag{8}$$

f_{cd}—design value of concrete compressive strength, MPa; v—coefficient of reduction shear strength of cracked concrete. It is recommended to determine the coefficient v according to the dependence [3]:

$$v = 0.6 \cdot \left[1 - \frac{f_{ck}}{250}\right] \tag{9}$$

Make calculations of the control samples according to the above dependencies. For convenience, we summarize the main calculation parameters in Table 3.

The results of the calculation are given in the Table 4.

According to the engineering method, the calculation showed convergence within 32–46% in the direction of exaggeration of experimental results. Such an indicator is a satisfactory result, considering the complex stress–strain state, lacking influence of longitudinal reinforcement in the methodology, and the suddenness and speed of exhaustion of the bearing capacity on the shear.

Table 3. Basic calculation parameters of beams without transverse reinforcement.

Beam Number	Tensile Reinforcement	f_{ck}, MPa	f_{ctk}, MPa	The Actual Cross-Section $h \times b$ mm	Effective Height d, mm	Shear Span, a/d	Experimental Shear Bearing Capacity. V_{Ed}, kN
BO 1.1				201×106	171	2	95
BO 1.2	2∅18	30.49	5.06	199×98	171	1.5	148.5
BO 1.3				202×98	171	1	198

Table 4. The results of determining bearing capacity on the shear without transverse reinforcement.

Beam Number	Shear Span, a/d	Experimental Shear Bearing Capacity. V_{Ed} (Q), kN	Theoretical Shear Strength according to Engineering Method [6]		$\dfrac{Q}{Q_{b,1}}$	Theoretical Shear Strength according to Current Codes [6]		$\dfrac{V_{Ed}}{V_{Rd,c1}}$
			$Q_{b,1}$, kN	$Q_{b,2}$, kN		$V_{Rd,c1}$, kN	$V_{Rd,c}$, kN	
BO 1.1	2	95	71.9	43.1	1.32	32.8	9.9	2.89
BO 1.2	1.5	140.5	95.8	43.1	1.46	43.7	9.9	3.22
BO 1.3	1	198	143.7	43.1	1.37	65.5	9.9	3.02

The overestimation of experimental data compared to calculation ones determined according to the current code is much greater. Considering the progress of the calculation, it can be established that all values, except for one, change during the calculation process. Only the value of $C_{Rd,c}$ (shear strength of concrete [39]) was constant in all calculations and was 0.18 MPa. This value is accepted as a minimum and does not depend on the change in concrete class. To determine the influence of the shear strength of concrete on the bearing capacity on the shear, it is suggested to use relationship from European codes of the 1997 edition, which considers changing the concrete shear strength from tension strength [40]:

$$C_{Rd,c} = \tau_{Rd} = 0.25 \cdot f_{ctk0.05} \tag{10}$$

When designing RC structures, the compressive strength of concrete is used more often. Therefore, we switch from the concrete tensile strength to the compressive in the dependence (10), using the relationship given in the same standards:

$$f_{ctk0.05} = 0.7 \cdot f_{ctm} \tag{11}$$

$$f_{ctm} = 0.3 \cdot f_{ck}^{2/3} \tag{12}$$

By substituting dependencies (11) and (12) into the relationship (10), we obtain the equation for determining the shear strength of concrete from the concrete compressive strength:

$$C_{Rd,c} = 0.0525 \cdot \sqrt[3]{f_{ck}^2} \tag{13}$$

After substituting the obtained values into dependence (6), we recalculate the shear strength of the control beams (Table 5).

Comparisons of the results of theoretical and experimental studies are presented graphically in Figure 6.

The results of determining the shear strength of the RC beams using the updated values of $C_{Rd,c}$, defined by the formula (13), showed a much higher convergence of results. Experimental data exceed theoretical calculation results by 16–29%., which is acceptable given the sharp nature of the destruction of such elements without transverse reinforcement.

Table 5. The bearing capacity of the control beams on the shear without transverse reinforcement.

Beam Number	Shear Span, a/d	Experimental Shear Bearing Capacity. $V_{Ed}(Q)$, kN	Theoretical Shear Strength according to Current Codes [6]		Theoretical Shear Strength according to Current Codes with $C_{Rd,c}$ Calculated by (13)	
			$V_{Rd,c1}$, kN	$\frac{V_{Ed}}{V_{Rd,c1}}$	$V_{Rd,c2}$, kN	$\frac{V_{Ed}}{V_{Rd,c2}}$
BO 1.1	2	95	32.8	2.89	81.6	1.16
BO 1.2	1.5	140.5	43.7	3.22	108.8	1.29
BO 1.3	1	198	65.5	3.02	163.2	1.21

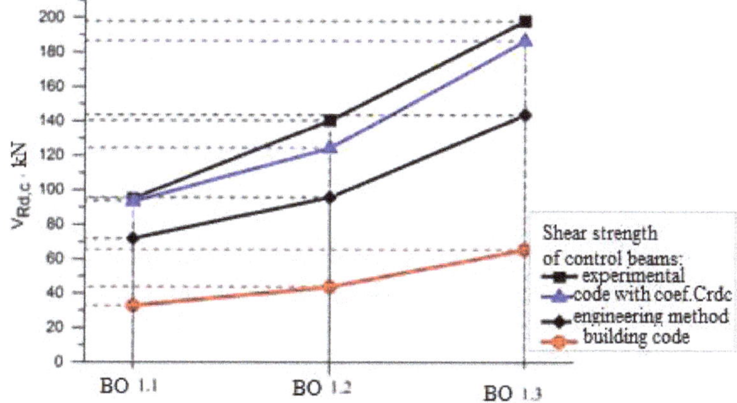

Figure 6. Experimental and theoretical data of the shear strength for control samples.

4.2. Methodology of Calculating Shear Strength of the Strengthened RC Beams by FRCM System

The bearing capacity of the RC beams on the shear, which is strengthened with composite system in the form of fabric strips, is calculated according to the method of the truss analogy [2,3]. It is calculated based on the condition that the shear force is perceived only by the shear reinforcement. In this case, the stressed–strained state of the support area is considered by the truss analogy (Figure 7). The concrete compressed zone is taken as a top, compressed chord (Figure 7A), and the working tensile reinforcement, respectively, as a bottom, tensile truss chord (Figure 7C). In this case, compressed struts of the webs are reduced to compressed strips of concrete in the support area of the beams (Figure 7B), while tensile struts of the web—shear reinforcement (Figure 7D).

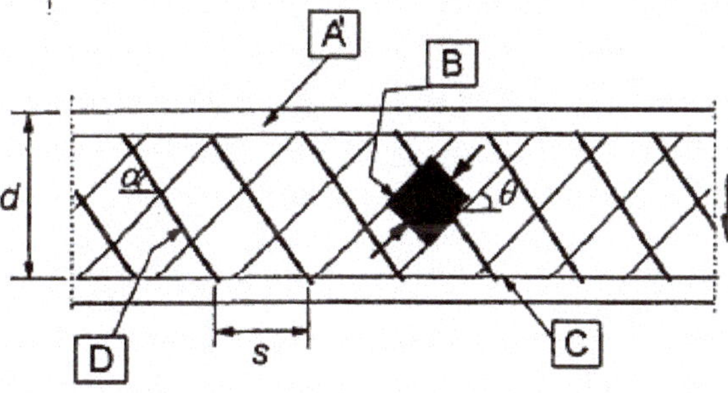

Figure 7. Truss model of the calculation state in the supporting area of RC beam [2]: A—compressed chord; B—compressed struts; C—tensile truss chord; D—tensile struts.

In Ukraine, codes do not allow the design of reinforced concrete elements strengthened with composite materials. Therefore, to calculate the bearing capacity on the shear, we accept the following prerequisites for the operation:

- there is a valid hypothesis of flat sections for the supporting areas;
- reinforcement with composite material works as additional external reinforcement;
- the reinforcement system works together with the concrete in the support area.

In this way, the external composite reinforcement is included in the design model, as with tensile elements of the truss web.

Provided that the accepted prerequisites are correct, it is proposed to calculate the shear strength of the strengthening element, such as internal shear reinforcement. Therefore, to calculate the shear strength of the strengthening system, we use an equation from codes [2,3], substituting the parameters of the composite fabric:

$$V_{Rd}^{add} = \frac{A_{sw}^{add}}{s} \cdot z \cdot f_{ywd}^{add} \cdot \cot\theta \tag{14}$$

A_{fw} = (0.00455 × 7) × 2 = 0.0637 mm^2—cross-sectional area of the composite fabric; s_f = 100 mm—step of strengthening elements.

The angle θ was determined along the diagonal crack in which the destruction of the element took place. However, it should be noted that the angle of inclined struts extends from the point of the center of force applied to the face of the element's support (Figure 8).

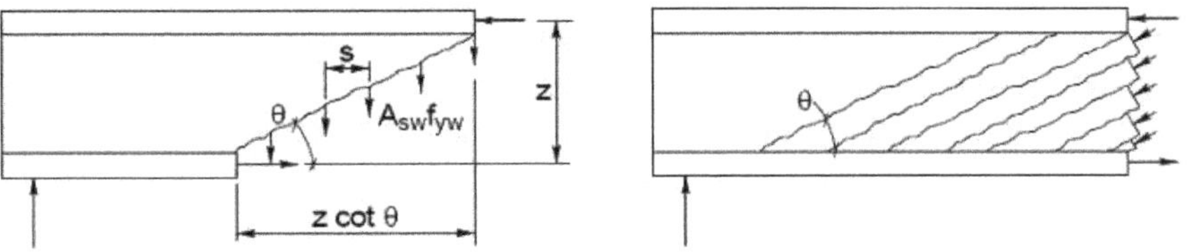

Figure 8. Location of the inclined struts [41].

With the determined bearing capacity of the additional reinforcement, the angle of inclined struts was 21.8°, which was defined from experimental data (Figure 9). The obtained value of the angle of inclination corresponds to the maximum value from the code $\cot\theta$ = 2.5 [2,3].

Figure 9. The angle of inclined struts in the sample BS 1.1.1–0.

The calculated shear strength of the composite reinforcement was determined by relationship (15), taking into account the calculation recommendations given in the report [42]:

$$f_{ywd}^{add} = k \cdot \varepsilon_{fd,e} \cdot E_{fud} = k \cdot \frac{\varepsilon_{fk,e}}{\gamma_f} \cdot 0.4 \cdot E_{fuk} \tag{15}$$

$\varepsilon_{fd,e}$—design value of limit strains for composite reinforcement; $\varepsilon_{fk,e} = \frac{\varepsilon_{f,e}}{\gamma_f}$—characteristic value of limit strains for composite reinforcement; $\gamma_f = 1.3$—partial factor by material, according to recommendations [42], which is accepted when there is a possibility of slipping through the fabric or $\gamma_f = 1.35$ if there is a possibility of tissue rupture (or the coefficient is taken equal to 1 in the case of the impossibility of the above-described conditions); $k = 0.8$ —reducing coefficient showing the dependence of the shear strength to the tensile strength of the reinforcement, described in the code [2] and Fib recommendations [42]; E_{fud}—the design value of the modulus of elasticity of the composite material, which is taken at the level of 40% of the characteristic value. According to Fib recommendations [38], the joint operation of concrete and reinforcement elements occurs only with such an underestimation of the characteristic value of the composite material modulus of elasticity.

To take into account the load level of the beam at which the strengthening took place, it is proposed to use a coefficient γ_{yw}^{add} that depends on the carrying capacity of the beam for shear strength and the load level. For the load factor of the element, it is suggested to use the dependence:

$$\gamma_{yw}^{add} = \left(1 - \frac{V_{Ed}}{V_{Rd}}\right)^{n} \tag{16}$$

V_{Ed}—external transverse force; V_{Rd}—design value of the shear strength of the RC beams; n—the index, which considered influencing from internal steel reinforcement (3/2—for a support area without transverse reinforcement; 1/2—for a support area with transverse reinforcement).

The bearing capacity is proposed to be calculated as the sum of the concrete and reinforcement composite system shear strength. This principle is proposed in Formula (1) [38], in Section 4.1 described above.

Based on the described Equations (6), (14), and (16), the following dependence of determining the shear strength of the RC bent elements without transverse (shear) reinforcement strengthened with composite fabric under action of the load was obtained:

$$V_{Rd} = V_{Rd,c} + V_{Rd}^{add} \cdot \gamma_{yw}^{add} \tag{17}$$

Analogous expressions are proposed for providing the shear strength strengthened with composite materials by many researchers and are given in Fib recommendations [42].

The above expressions were tested to calculate the bearing capacity of tested samples without internal transverse reinforcement. The value of the shear strength of the control samples is determined according to the current code, and consider the refined value $C_{Rd,c}$ determined according to Equation (13). Result of the theoretical calculation data is shown in the Table 6.

Table 6. Comparison theoretical and experimental data of the shear strength of the tested RC beams without transverse reinforcement strengthened with a composite system.

Beam Number	Theoretical Data				Experimental Results			$\frac{V_{Ed}}{V_{Rd}}$	$\frac{V_{Ed}^{add}}{V_{Rd}^{add} \cdot \gamma_{yw}^{add}}$
	$V_{Rd,c}$, kH	γ_{yw}^{add}	$V_{Rd}^{add} \cdot \gamma_{yw}^{add}$, kH	$V_{Rd'}$ kH	$V_{Ed'}$ kH	V_{Ed}^{add}, kH			
BO 1.1		–	–	81.6	95	–	1.16	–
BS 1.1-0	81.6	1.0	34.6	116.2	137.5	42.5	1.18	1.23
BS 1.2-0.3		0.7	20.3	101.9	120	25.0	1.17	1.23
BS 1.3-0.5		0.5	12.2	93.8	110	15.0	1.17	1.23

Additionally, the obtained results are shown as curves in Figure 10.

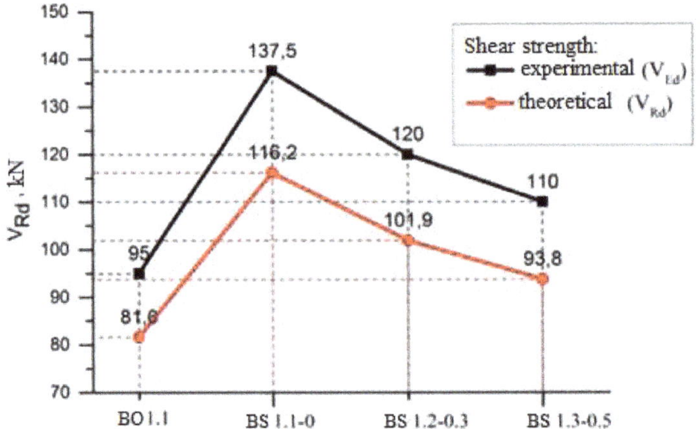

Figure 10. Comparison theoretical and experimental results of shear strength of strengthened beams.

Satisfactory convergence of results was obtained for all experimental samples, within 16–18% in the direction of overstatement of experimental data. The calculation shear strength of the composite material as a transverse reinforcement, taking into account the proposed coefficient, shows a satisfactory convergence, namely an overestimation of the experimental data by 23% for all investigated levels of the active load.

5. Conclusions

The calculation of the bearing capacity of the RC beams on the shear was tested according to the current codes with clarifications regarding the consideration of the concrete shear strength $C_{Rd,c}$. The calculation results showed a better convergence; the deviation was 16–29% in the direction of the experimental data exceeding the theoretical.

An improved methodology for calculating the bearing capacity of reinforced concrete beams without shear reinforcement strengthened with a composite FRCM system is proposed. This methodology includes the following:

- application of compatible bearing capacity of concrete and external reinforcement;
- taking into account the angle of the inclined struts θ;
- principles for calculating the external composite reinforcement are proposed, which are based on FIB recommendations and existing design code;
- the proposed coefficient takes into account the reduction in the use of composite reinforcement depending on the initial load level of the beam.

The proposed method allows the determination of the actual additional carrying capacity of the reinforcement system. A new coefficient is proposed that considers the change in the load level and, accordingly, the change in the reinforcement effect. Approbation of the proposed methodology for calculating strengthened samples without internal transverse reinforcement showed a good convergence of 23%. This methodology can be simply applied in practical design.

Author Contributions: Conceptualization Z.K.; methodology, Z.K., Z.B. and I.G.; validation, Z.B., P.V. and I.G.; formal analysis, P.V. and I.G.; resources, I.G. and P.V.; data curation, Z.K. and Z.B.; writing—original draft preparation, P.V., I.G. and Z.B.; writing—review and editing, Z.K. and I.G.; visualization, P.V. and I.G.; supervision, Z.K. and Z.B.; project administration, Z.K. and Z.B.; funding acquisition, Z.K. All authors have read and agreed to the published version of the manuscript.

Funding: This research received no external funding.

Institutional Review Board Statement: Not applicable.

Informed Consent Statement: Not applicable.

Data Availability Statement: Publicly available datasets were analyzed in this study. This data can be found here: https://www.researchgate.net/profile/Pavlo-Vegera.

Conflicts of Interest: The authors declare no conflict of interest.

References

1. *DBN V.2.6-98:2009*; Concrete and Reinforced Concrete Structures. Ministry of Regional Construction: Kyiv, Ukraine, 2011. (In Ukrainian)
2. *DSTU B.V.2.6-156:2010*; Concrete and Reinforced Concrete from Heavy Concrete. Ministry of Regional Construction: Kyiv, Ukraine, 2011. (In Ukrainian)
3. *EN 1992-1-1*; Eurocode 2: Design of Concrete Structures—Part 1-1: General Rules for Buildings. European Committee for Standartization: Brussels, Belgium, 2004.
4. Lobodanov, M.; Vegera, P.; Khmil, R.; Blikharskyy, Z. Influence of damages in the compressed zone on bearing capacity of reinforced concrete beams. *Lect. Notes Civ. Eng.* **2021**, *100*, 260–267.
5. Klymenko, Y.; Grynyova, I.; Kos, Z.; Maksiuta, O. Method for Determining the Residual Bearing Capacity of Damaged I-Beam Reinforced Concrete Columns. *Lect. Notes Civ. Eng.* **2022**, *290*, 171–184.
6. Karpiuk, V.; Somina, Y.; Karpiuk, F.; Karpiuk, I. Peculiar aspects of cracking in prestressed reinforced concrete T-beams. *Acta Polytech.* **2021**, *61*, 633–643. [CrossRef]
7. Ge, Y.; Liu, J.; Zhang, X.; Tang, H.; Xia, X. Automated Detection and Characterization of Cracks on Concrete Using Laser Scanning. *J. Infrastruct. Syst.* **2023**, *29*, 04023005. [CrossRef]
8. Karkhut, I. Improving the crack resistance of inclined cross-sections of reinforced concrete containment shells in areas of emergency loads of pushing. *East.-Eur. J. Enterp. Technol.* **2022**, *4*, 31–41.
9. Elsayed, M.; Badawy, S.; Tayeh, B.A.; Elymany, M.; Salem, M.; ElGawady, M. Shear behaviour of ultra-high performance concrete beams with openings. *Structures* **2022**, *43*, 546–558. [CrossRef]
10. Blikharskyy, Z.; Lobodanov, M.; Vegera, P. Investigation of defective reinforced concrete beams with obtained damage of compressed area of concrete. *Prod. Eng. Arch.* **2022**, *28*, 225–232. [CrossRef]
11. Kos, Z.; Klymenko, Y.; Crnoja, A.; Grynyova, I. Calculation Model for Estimation of Residual Bearing Capacity of Damaged Reinforced Concrete Slender Columns. *Appl. Sci.* **2022**, *12*, 7430. [CrossRef]
12. Sudha, C.; Sambasivan, A.K.; Kannan Rajkumar, P.R.; Jegan, M. Investigation on the performance of reinforced concrete columns jacketed by conventional concrete and geopolymer concrete. *Eng. Sci. Technol. Int. J.* **2022**, *36*, 101275.
13. Borysiuk, O.; Ziatiuk, Y. Experimental research results of the bearing capacity of the reinforced concrete beams strengthened in the compressed and tensile zones. *Lect. Notes Civ. Eng.* **2020**, *100*, 63–70.
14. Gonzalez-Libreros, J.H.; Sneed, L.H.; D'Antino, T.; Pellegrino, C. Behavior of RC beams strengthened in shear with FRP and FRCM composites. *Eng. Struct.* **2017**, *150*, 830–842. [CrossRef]
15. Loreto, G.; Babaeidarabad, S.; Leardini, L.; Nanni, A. RC beams shear-strengthened with fabric-reinforced-cementitious-matrix (FRCM) composite. *Int. J. Adv. Struct. Eng.* **2015**, *7*, 341–352. [CrossRef]
16. Sabzi, J.; Esfahani, M.R.; Ramezani, A. The effect of tensile reinforcement on the behavior of CFRP strengthened reinforced concrete beams: An experimental and analytical study. *Steel Compos. Struct.* **2023**, *46*, 115–132.
17. Babych, Y.M.; Savitskiy, V.V.; Andriichuk, O.V.; Ninichuk, M.V.; Kysliuk, D.Y. Results of experimental research of deformability and crack-resistance of two span continuous reinforced concrete beams with combined reinforcement. *Mater. Sci. Eng.* **2019**, *708*, 012043. [CrossRef]
18. Li, Y.; Xu, W.; Zhang, J.; Fan, J.; Gao, P.; Hu, C. Early cracking characteristics of concrete strengthened with fiber mesh. *J. Hohai Univ.* **2022**, *50*, 114–122.
19. Paruta, V.; Gnyp, O.; Lavrenyuk, L.; Grynyova, I. The Influence of the Structure on the Destruction Mechanism of Polymer-Cement Plaster Coating. *Key Eng. Mater.* **2020**, *864*, 93–100. [CrossRef]
20. Cansız, S. Effect of tightening zone length of reinforced concrete beams on beam capacity. *Građevinar* **2022**, *74*, 857–868.
21. Yurdakul, Ö.; Del Vecchio, C.; Di Ludovico, M.; Avşar, Ö.; Prota, A. Fragility functions for fiber-reinforced polymers strengthened reinforced concrete beam-column joints. *Eng. Struct.* **2023**, *279*, 115570. [CrossRef]
22. Han, X.; Han, B.; Xie, H.; Yan, W.; Ma, Q. Experimental and numerical study on the in-plane behaviour of concrete-filled steel tubular arches with long-term effects. *Thin-Walled Struct.* **2021**, *169*, 108507. [CrossRef]
23. Khmil, R.Y.; Tytarenko, R.Y.; Blikharskyy, Y.Z.; Vegera, P.I. Improvement of the method of probability evaluation of the failure-free operation of reinforced concrete beams strengthened under load. *Mater. Sci. Eng.* **2021**, *1021*, 012014. [CrossRef]
24. Vatulia, G.L.; Komagorova, S.D.; Opanasenko, O.V.; Lobiak, O.V. Optimal Design of a Three-Hinged Arch with Given Topology Under Constant Load. *Lect. Notes Civ. Eng.* **2020**, *47*, 501–509.
25. Kramarchuk, A.; Ilnytskyy, B.; Hladyshev, D.; Lytvyniak, O. Strengthening of the reinforced concrete tank of anaerobic purification plants with the manufacture of biogas, damaged as a result of design and construction errors. *Mater. Sci. Eng.* **2021**, *1021*, 012017. [CrossRef]
26. Wakjira, T.G.; Ebead, U. A shear design model for RC beams strengthened with fabric reinforced cementitious matrix. *Eng. Struct.* **2019**, *200*, 109698. [CrossRef]

27. Marcinczak, D.; Trapko, T.; Musiał, M. Shear strengthening of reinforced concrete beams with PBO-FRCM composites with anchorage. *Compos. Part B Eng.* **2019**, *158*, 149–161. [CrossRef]
28. Kochkarev, D.; Galinska, T. Calculation methodology of reinforced concrete elements based on calculated resistance of reinforced concrete. *MATEC Web Conf.* **2017**, *116*, 02020. [CrossRef]
29. Ahaieva, O.; Vegera, P.; Karpiuk, V.; Posternak, O. Design Reliability of the Bearing Capacity of the Reinforced Concrete Structures on the Shear. In *Proceedings of the EcoComfort 2022*; Lecture Notes in Civil Engineering; Springer: Cham, Switzerland, 2022; Volume 290, pp. 1–15.
30. Khmil, R.; Tytarenko, R.; Blikharskyy, Y.; Vegera, P. The probabilistic calculation model of RC beams, strengthened by RC jacket. *Lect. Notes Civ. Eng.* **2021**, *100*, 182–191.
31. Nguyen, H.A.T. Effect of Span-to-Depth Ratio on Strength and Deflection Reliability of Reinforced Concrete Beams. In *Lecture Notes in Mechanical Engineering: Recent Advances in Structural Health Monitoring and Engineering Structures Select Proceedings of SHM and ES 2022*; Springer Nature: Singapore, 2023; pp. 331–344.
32. Kos, Z.; Klymenko, Y.; Karpiuk, I.; Grynyova, I. Bearing Capacity near Support Areas of Continuous Reinforced Concrete Beams and High Grillages. *Appl. Sci.* **2022**, *12*, 685. [CrossRef]
33. Ombres, L.; Mazzuca, P.; Verre, S. Effects of thermal conditioning at high temperatures on the response of concrete elements confined with a pbo-frcm composite system. *J. Mater. Civ. Eng.* **2022**, *34*, 04021413. [CrossRef]
34. Tetta, Z.C.; Bournas, D.A. TRM vs. FRP jacketing in shear strengthening of concrete members subjected to high temperatures. *Compos. Part B Eng.* **2016**, *106*, 190–205. [CrossRef]
35. Blikharskyy, Z.; Vegera, P.; Vashkevych, R.; Shnal, T. Fracture toughness of RC beams on the shear, strengthening by FRCM system. *MATEC Web Conf.* **2018**, *183*, 02009. [CrossRef]
36. Blikharskyy, Z.; Khmil, R.; Vegera, P. Shear strength of reinforced concrete beams strengthened by PBO fiber mesh under loading. *MATEC Web Conf.* **2017**, *116*, 02006. [CrossRef]
37. Vegera, P.; Vashkevych, R.; Blikharskyy, Z. Fracture toughness of RC beams with different shear span. *MATEC Web Conf.* **2018**, *174*, 02021. [CrossRef]
38. *NiP 2.03.01-84**; Concrete and Reinforced Concrete Structures. NDIZB: Moscov, Russia, 1988. (In Ukrainian)
39. Amadio, C.; Macorini, L.; Sorgon, S.; Suraci, G. A novel hybrid system with RC-encased steel joints. *Eur. J. Environ. Civil Eng.* **2011**, *15*, 1433–1463. [CrossRef]
40. *DD ENV 1992-1-1*; Eurocode 2: Design of Concrete Structures—Part 1: General Rules and Rules for Buildings (Together with United Kingdom National Application Document). British Standards Institution: London, UK, 1992.
41. Bond, A.J.; Brooker, O.; Harris, A.J.; Harrison, T.; Moss, R.M.; Narayanan, R.S.; Webster, R. *How to Design Concrete Structures Using Eurocode 2*; The Concrete Centre: London, UK, 2006; p. 104.
42. Triantafillou, T.; Matthys, S.; Audenaert, K.; Balázs, G. *Externally Bonded FRP Reinforcement for RC Structures*; Technical Report; International Federation for Structural Concrete (FIB): Lausanne, Switzerland, 2001; 130p.

Article

Comparative Analyses of Selected Neural Networks for Prediction of Sustainable Cementitious Composite Subsurface Tensile Strength

Slawomir Czarnecki * and Mateusz Moj

Department of Materials Engineering and Construction Processes, Wroclaw University of Science and Technology, Wybrzeże Wyspiańskiego 27, 50-370 Wrocław, Poland
* Correspondence: slawomir.czarnecki@pwr.edu.pl

Abstract: The article assesses comparative analyses of some selected machine-learning algorithms for the estimation of the subsurface tensile strength of cementitious composites containing waste granite powder. Any addition of material to cementitious composites causes their properties to differ; therefore, there is always a need to prepare a precise model for estimating these properties' values. In this research, such a model of prediction of the subsurface tensile strength has been carried out by using a hybrid approach of using a nondestructive method and neural networks. Moreover, various topologies of neural networks have been evaluated with different learning algorithms and number of hidden layers. It has been proven by the very satisfactory results of the performance parameters that such an approach might be used in practice. The errors values (MAPE, NRMSE, and MAE) of this model range from 10 to 12%, which, in the case of civil engineering practice, proves that this model is sufficient for being used. This novel approach can be a reasonable alternative for evaluating the properties of spacious cementitious composite elements where there is a need to analyse not only the compressive strength but also its subsurface tensile strength.

Keywords: machine learning; eco cementitious composite; tensile strength; granite powder

Citation: Czarnecki, S.; Moj, M. Comparative Analyses of Selected Neural Networks for Prediction of Sustainable Cementitious Composite Subsurface Tensile Strength. *Appl. Sci.* **2023**, *13*, 4817. https://doi.org/10.3390/app13084817

Academic Editors: Francisco B. Varona Moya and Mariella Diaferio

Received: 20 March 2023
Revised: 7 April 2023
Accepted: 10 April 2023
Published: 11 April 2023

1. Introduction

A composite is defined as a material made up of at least two components with different properties such that it has superior and/or new properties over those components taken separately. However, the given definition is not generally accepted and precise, but is intended to generalize the most common descriptive definitions. On its basis, it can be assumed that a cement composite is a composite of whom one of the basic components is cement. The most popular composites in which cement acts as a binder are: grout, mortar, and concrete [1]. By the term grout, we mean the material that results from the combination of water and cement. The leaven itself is rarely used in construction. Cement mortar is used much more often, e.g., for bricklaying, plastering, or making floor underlays. Standard cement mortar is a mixture of the appropriate proportions of cement, water, and sand. The composition of the mortar can be supplemented with additives and admixtures added to give the appropriate properties to the mixture or hardened mortar. Concrete is another material. It is the most widespread cement composite in the world. In addition to water, sand, and cement, concrete contains coarse aggregate. In concrete, as in the case of mortar, we use admixtures and additives, which, in standard concretes, allow us, for example, to obtain better working parameters of the mixture with the use of less water, or to improve the physical parameters of the hardened composite [2].

An additive to concrete can be called a fine-grained component added to the concrete mix, whose task is to improve its properties or obtain the special parameters of hardened concrete. The amount of additive is generally from 5% to 15% of the weight of the cement. The content of the additive should be taken into account in subsequent iterations of the calculations of the designed mixtures. There are two main types of additives. The first type is nearly inert additives, which are in turn divided into fillers and pigments. Pigments are used to tint the concrete to the desired color. The role of fillers is to fill the pores of the concrete mixture [3]. An additive typical for this group is limestone powder. Additives with latent hydraulic or pozzolanic properties are the second type of concrete additives. Their use leads to an improvement in the properties of concrete or the acquisition of special properties. Representatives of this group are silica fume or fly ash. The most popular methods of adding fillers to a cement composite are in the form of (i) cement replacement, (ii) aggregate replacement, and (iii) grout replacement. These methods are successfully applied to waste materials in the form of quartz dust, marble powder, powdered glass, or fly ash, added to concrete and mortar as fillers or additives that modify the properties of the mixture or the hardened cement composite [4].

The mere production of cement as a basic component of concretes and mortars also has an undesirable impact on the environment. During the production of 1 ton of this material, about 900 kg of carbon dioxide are emitted; therefore, the partial replacement of cement with another component may contribute to reducing CO_2 emissions to the atmosphere [5].

Granite powder is a waste material that results from the crushing or cutting of granite rocks for industrial purposes. In recent years, an increase in the amount of granite powder waste has been observed as a result of the development of the granite industry. The powder fraction of this material alone is currently not applicable, which exacerbates the problem of the proper management of this product. The improper storage of dry material in the form of heaps, which can easily spread in the environment, causes water pollution, changes in the pH of land intended for agricultural purposes, and air pollution in the vicinity of landfills. Insisting that the effects of the environmental degradation by granite powder can be felt directly by humans, it is necessary to take the appropriate measures to reduce the amount of this production residue in the environment. One way to reduce its harmful effects is to use this by-product for the sustainable production of cement composites as the most common building material and use it as a supplementary cementitious material (SCM) [6].

However, when another material is added to cementitious composites, it is necessary to evaluate how it affects their properties. Most of the tests used for evaluating the materials' properties in civil engineering are destructive; therefore, they are costly and time-consuming, and they produce wastes. Therefore, sometimes for this purpose, the nondestructive tests are reasonable substitutes [7]. It is also important while performing such tests in buildings which are operating or which are characterized as historical heritage sites and cannot be tested destructively [8]. It was previously shown that an NDT method such as the Schmidt hammer method was successful in evaluating the compressive strength of concrete [9,10]. However, there is an issue with evaluating the other properties of cementitious composites such as subsurface tensile strength. For this purpose, very often, a method such as the pull-off test is used, but, after the test, the sample needs to be repaired, which makes this method not fully nondestructive [11]. However, it is still possible to try to correlate the compressive strength with the subsurface tensile strength. Therefore, it is possible to try to evaluate the opposite, which means that it is possible to evaluate the subsurface tensile strength using a Schmidt hammer [12]. Evaluating the subsurface tensile strength in cementitious composites such as mortars dedicated for floor substrate is very important because of the fact that, very often, these parameters are responsible for the durability of the element [13].

Recently, very often, soft computing techniques such as artificial neural networks are used for this purpose, while they are able to strengthen the results of NDT methods by improving the performance of these methods, which might be necessary in this case [14]. Such approaches have previously been successful when using the combination of a Schmidt hammer and neural networks to predict the unconfined compressive strength of rocks [15], to predict the damage analyses of concrete subjected to high temperature [16], or to assess the load-carrying capacity of concrete anchor bolts [17].

Taking the aforementioned thoughts into account, it might be reasonable to use such a combination for evaluating the subsurface tensile strength of cementitious composites using the combination of a Schmidt hammer and neural networks. Therefore, for this purpose, the authors propose the hybrid method of using the nondestructive Schmidt hammer method and neural networks for evaluating the properties of cementitious composites containing granite powder that can be used in the civil-engineering practice. This solution might be beneficial for evaluating the properties of existing floors made of mortars which are located in operating buildings.

2. Materials and Methods

2.1. Materials

In order to create the database used for ANN modeling, the research part was necessary. Four different cementitious composite mixes were prepared with varying compositions in terms of the amount of granite powder used, as shown in Table 1. The first mix, which is the reference mix designated REF, contained only Portland cement as a binder, and is typical mortar mixture used in floors. In the other mixes, Portland cement was partially replaced by granite powder, obtained from a local stone plant in Strzegom, in proportions of 10%, 20%, and 30% (designated GP10, GP20, and GP30, respectively). The amount of water was assumed to be constant for all mixes and was determined for a reference composition for a water–cement ratio of 0.5. CEM I 42.5 cement, which consisted of 95–100% Portland clinker and 0–5% secondary components, was used for the mortars. Its normal compressive strength was $\geq$42.5 MPa and $\leq$62.5 MPa. The sand used in the study had a specific density of about 1600 kg/m^3 and a water absorption of $W_{(A,24)} = 0.6\%$. The samples were made into 500 mm $\times$ 500 mm $\times$ 40 mm slabs, which were then cured for the required time (56 days or 90 days). All samples were compacted and manually trowelled. The samples were cured in air (AIR) and in a moist environment (WET), created by applying moist sponges to the surface of the samples, at a temperature of 18 $\pm$ 3 °C. The condition of the environment in which the samples were cured was maintained for 28 days, which, in the case of the wet environment, consisted of checking the water saturation of the sponges and re-wetting them if necessary. It was recommended that the sponges not be saturated, so, after wetting, they were left for 15–30 min to drain excess water.

Table 1. Compositions of individual floor mortar mixes.

Series [-]	Cement CEM I 42.5R [kg/m^3]	Granite Powder [kg/m^3]	Dry Quartz Sand [kg/m^3]	Water [kg/m^3]	W/C Ratio [-]
REF	512.0	0.0	1536.0	256.0	0.50
GP10	460.8	51.2	1536.0	256.0	0.56
GP20	409.6	102.4	1536.0	256.0	0.63
GP30	358.4	153.6	1536.0	256.0	0.71

2.2. Methods

2.2.1. Schmidt Hammer Test

In order to better evaluate the surface layer of the cement mortars made, a Schmidt hammer test was performed to correlate the results with the compressive strength of the surface layer of the specimens. The test is carried out by placing the hammer at right angles to the test surface and gradually increasing the pressure until the hammer strikes. At the moment of impact, the number of rebounds is recorded, representing a test of the hardness of the material. Adjacent test points shall not be closer than 25 mm to either each other or to the edge. If the test results in crushing or damage to the mortar at the measurement point, the result shall be rejected. The hammer was checked before the test on a calibration steel anvil—6 control measurements were taken, each indicating a reflection number of 80 $\pm$ 2. This allowed the artificial neural networks to increase the database on which they worked. The test was conducted according to the method of [18]. Twelve impact tests were performed, 3 at each pull-off strength test location.

The approximate strength f_c was taken as equal to the value of f_L. For the obtained reflection number (L), the compressive strength value was calculated from Equations (1) and (2) based on the base curve (Figure 1a):

$$f_L = 1.25L - 23 \text{ by } 20 \leq L \leq 24, \tag{1}$$

$$f_L = 1.73L - 34.5 \text{ by } 24 \leq L \leq 50, \tag{2}$$

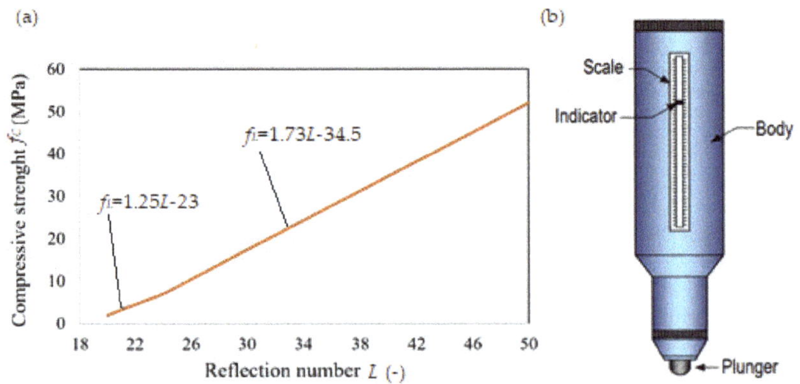

Figure 1. (**a**) Base curve according to the standard [13]; and (**b**) Schmidt hammer.

2.2.2. Pull-Off Tests

Four boreholes were drilled on the prepared panels after each curing time as shown in Figure 2a. The holes were drilled with a 50 mm inside diameter, using diamond saw to a depth of 8 mm. This will clarify the tests directly on the near-surface layer, which was defined as a range from 0 to 4 mm depth (Figure 2b). According to the standard [19], to achieve meaningful results, it was necessary to place the edge of the disc a minimum of 50 mm from the edge of the specimen. After drilling, the surface of the discs was vacuumed and cleaned with acetone without compromising the integrity of the coating. Then, in order to connect the Proceq Dy-216 test device to the specimen, the steel discs (cleaned with acetone) were glued to the wells with Poxipol two-component adhesive. The glue was applied to the surface of the puck and the surface of the sample, then positioning the puck by carefully pressing it to remove air bubbles. The adhesive had no effect on the test specimen. Once the adhesive was fully cured, the test was proceeded. The connection between the puck and the test instrument was screwed into the puck. The machine was levelled and programmed to complete the test in a maximum of 100 s at a loading rate of 0.05 MPa/s. The result of the test is recorded on the instrument along with a graph of load increase over time. After entering the relevant data into the apparatus, the pull-off strength was automatically calculated from Equation (3). Regarding the work of [20] and [21] where the pull-off strength between the substrate layer and the overlay layer is predicted using ANN, the network performed on the basis of the obtained results is supposed to allow for us to estimate the pull-off strength of the near-surface layer as the most exposed layer.

$$f_h = \frac{P}{A}, \tag{3}$$

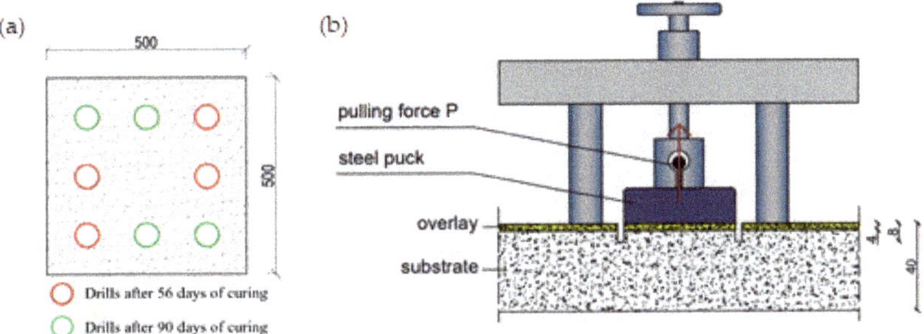

Figure 2. Pull-off test: (**a**) pull-off disc arrangement diagram; and (**b**) pull-off test.

2.2.3. Neural Networks

As we work to reduce the use of cement in the construction of subfloors, replacing it with waste granite powder, it is necessary to determine the optimal values of the components while maintaining the appropriate parameters of the final product. In order to optimize the process of designing the composition of cement mortars with cement replacements, artificial neural networks (ANN) in the form of multilayer perceptron (MLP) were used. MLP is one of the most recognized ANNs used when solving design problems in the construction industry. Karthiyaini et al. [22] predicted the mechanical strength of concrete, doped with fibers, using a multiple regression model (MRA) and an artificial neural network (ANN). The purpose of their work was to compare the two models with each other. The work of Mehdi et al. [23] describes the use of evolutionary neural networks (EANN), as a combination of artificial neural network (ANN) and evolutionary search procedures, in predicting the compressive strength of concrete. Panagiotis G. Asteris et al. [24] trained a database, in conventional machine learning models, of an artificial neural network (ANN) to predict the compressive strength of concrete. Czarnecki et al. [25] presented the application of artificial neural networks (ANNs) to nondestructively evaluate the adhesion of a repair overlay to a concrete substrate. The subject of the paper by Czarnecki et al. [26] was the prediction of the strength of a cementitious composite with the addition of ground granulated blast-furnace slag. There, they compared an artificial neural network (ANN) with a self-organizing feature map (SOFM). This tool is eagerly used by researchers to work with cementitious composites because it can handle inconsistent or "noisy" information which is a common occurrence when studying these materials. According to Figure 3, an artificial neural network in the form of an MLP consists of three layers: input, output, and one or two hidden. Each layer can have one or more neurons. Before interpreting the results, such networks should be subjected to training.

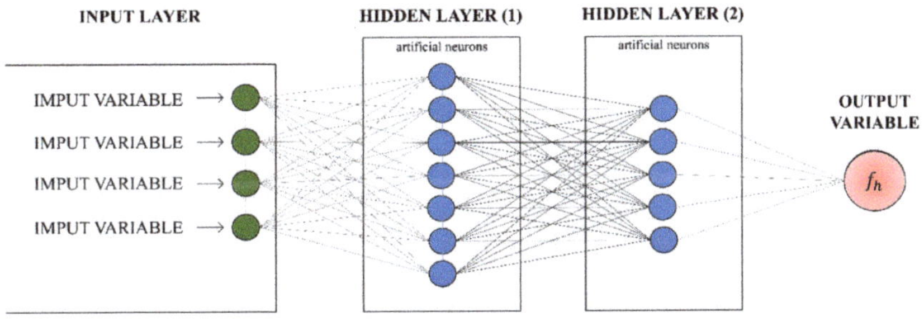

Figure 3. Block diagram of an artificial neural network.

For the purpose of this study, the MATLAB program was used for designing the neural networks' structures and learning algorithms as: gradient descent algorithm, conjugate gradient algorithm, and the Broyden–Fletcher–Goldfarb–Shanno and Levenberg–Marquardt algorithms. It also allows the use of 5 types of hidden layer activation functions and output layer activation functions. During ANN modelling, calculations were made for all possible combinations of the above-mentioned network elements. In addition, the number of hidden layer neurons was changed in the range of 1 to 15. We changed the number of layers of hidden neurons to 2 and again edited the number of neurons in layer 1 and layer 2 in the range of 1 to 15. Each of these functions processes the data using the corresponding Formulaes (4)–(8):

$$linear\ function \rightarrow f(x) = ax + b, \tag{4}$$

$$hyperbolic\ tangent\ function \rightarrow f(x) = \frac{e^x - e^{-x}}{e^x + e^{-x}}, \tag{5}$$

$$logistic\ function \rightarrow f(x) = \frac{e^x}{e^x + 1} \tag{6}$$

$$exponential\ function \rightarrow f(x) = d^x, \tag{7}$$

$$sinus\ function \rightarrow f(x) = \sin x, \tag{8}$$

Based on the database created, each input variable has 7 parameters, including: (1) the amount of cement (C), (2) the amount of sand (S), (3) the amount of water (W), (4) the amount of granite powder (GP), (5) the curing method (dry or wet), (6) the curing time, and (7) the compressive strength tested with a Schmidt hammer (fc). The output parameter consists of the tensile strength values of the floor mortar substrate measured using pull-off method.

3. Results

3.1. Schmidt Hammer Test

The samples were compared with each other based on the amount of granite powder added to partially replace the cement, as well as based on the maturation conditions of the samples. Figure 4 shows the average compressive strengths obtained when testing each of the sleepers. The basic requirement for cementitious composite substrates in terms of compressive strength is to achieve a minimum value of 20 MPa. It can be seen that all of the samples made reached the minimum value; therefore, the results obtained were used to make a database. From the graph, it can be seen that, as the amount of granite powder in the mix increases, the compressive strength of the substrate decreases, but even for the samples with the highest filler content, the substrate achieved the required strength.

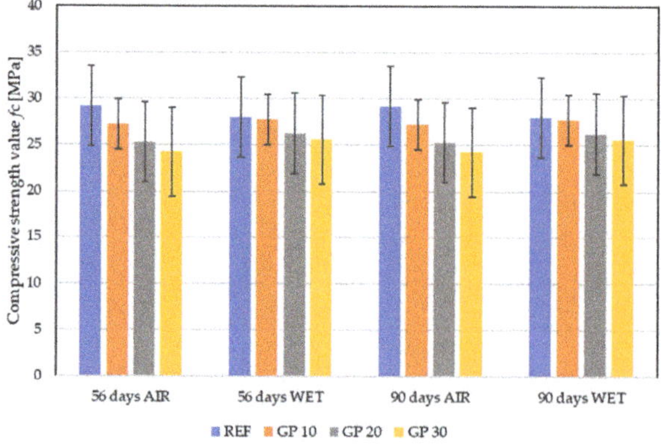

Figure 4. Average compressive strength of the near-surface layer f_c.

3.2. Pull-Off Tests

The samples were compared to each other due to the amount of granite powder added to partially replace the cement, as well as the conditions and maturation time of the samples. Figure 5 shows the average pull-off strengths of each mixture. First of all, it can be seen that there is a large difference in the results due to the different maturation conditions of the samples. Variation in testing time also affected the results. Likewise, the amount of granite powder replacing the cement made a difference. It can be noted that, with the amount of 10% filler, the obtained results for the air-matured samples increased, which suggests the validity of using this modern waste approach to mortar design.

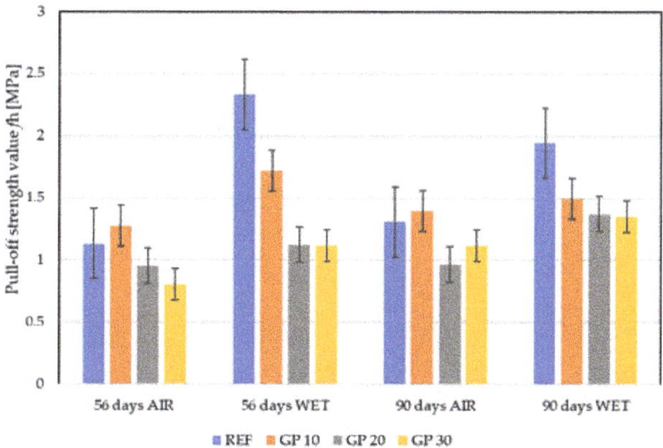

Figure 5. Average pull-off strength f_h of the near-surface layer f_c.

4. Statistical Analyses of the Collected Data

The whole dataset is presented in Appendix A. The minimum, maximum, mean, standard deviation, and range of the parameters included in the database are shown in Table 2.

Table 2. Range of input and output variables.

Descriptive Statistics	Cement [kg/m³]	Dry Quartz Sand [kg/m³]	Water [kg/m³]	Granite Powder [kg/m³]	Curing Time [days]	f_c [MPa]	f_h [MPa]
Min	358.40	1536.00	256.00	0.00	56.00	14.00	0.70
Max	512.00	1536.00	256.00	153.60	90.00	38.00	3.08
Mean	435.20	1536.00	256.00	76.80	73.00	26.69	1.33
Standard deviation	57.39	0.00	0.00	57.39	17.04	4.49	0.44
Spread	153.60	0.00	0.00	153.60	34.00	24.00	2.38

The correlation coefficients between all possible variables are shown in Table 3. High positive or negative correlation coefficient values between input variables can result in low efficiency and difficulty in determining the impact of these variables on the response. There are no significant correlations between the independent input variables, which was to be expected. As expected, it is clear that there is a strong correlation between the mortar f_h and input parameters, such as the amount of cement and granite powder, water-to-cement ratio (W/C), and curing method. In addition, Figure 6 shows the frequency histogram of the six input parameters. These figures are extremely useful because they can help identify the range of parameter values where the data are insufficient and where more data are required.

Table 3. Correlation matrix of the input and output variables.

Variables		Input Parameters						Output Parameters
		Cement	Granite Powder	W/C Ratio	Curing Method	Curing Time	f_c	Pull-Off
Input	Cement	1.00	−1.00	−0.99	-	-	0.32	0.56
	Granite Powder	−1.00	1.00	0.99	-	-	−0.32	−0.56
	W/C ratio	−0.99	−0.99	1.00	-	-	−0.31	−0.55
	Curing method	-	-	-	1.00	-	0.05	0.52
	Curing time	-	-	-	-	1.00	-	0.05
Output	f_c	0.32	−0.32	−0.31	0.05	-	1.00	0.12
	Pull-off	0.56	−0.56	−0.55	0.52	0.05	0.12	1.00

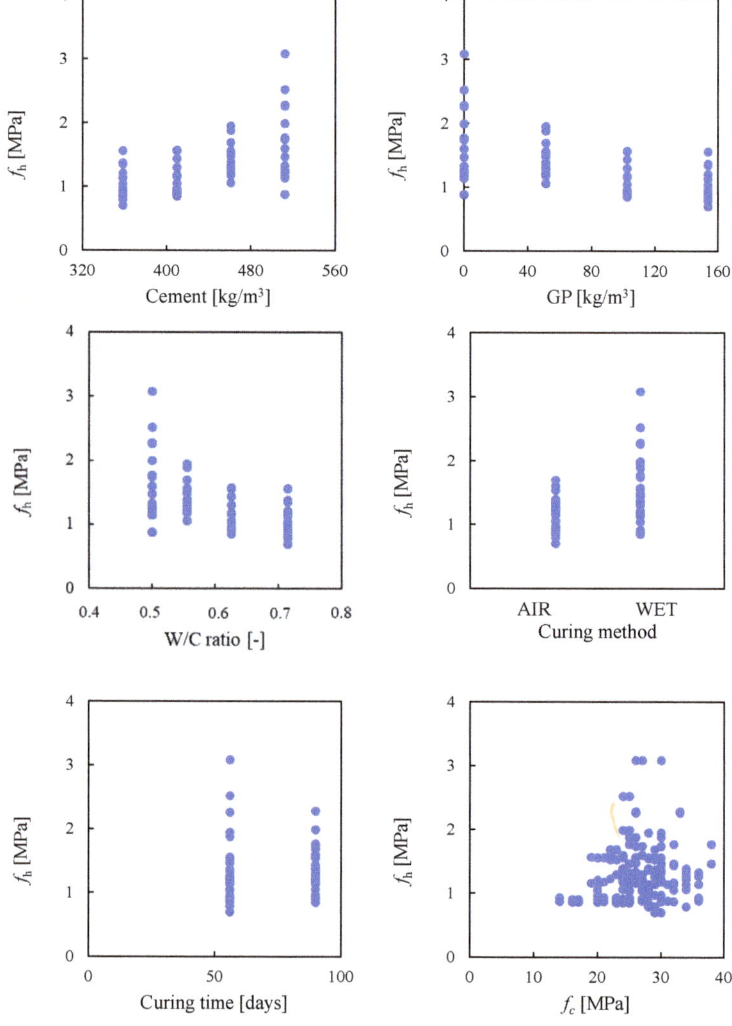

Figure 6. The relation between input variables and tensile strength.

5. Neural Network Analyses

Using numerical methods of design and prediction, it is necessary to create a database. Based on the results of the study, a database was created containing information on the material composition of the four previously mentioned mortars, how and when they were matured, and the results of the tests presented in Sections 2 and 3. To meet the requirements of the PN-EN 12504-2 standard of the Schmidt hammer tests, the authors decided to performed the four pull-off tests for each composition and maturing conditions, and, in each of these areas, the three Schmidt hammer tests were performed. Because these three tests inside one pull-off area is less than that stated in the standard, the authors have not calculated the average value out of them. With this, a database of 192 data sets was collected, which were divided into a set for the learning process (136 data sets—70%), the testing process (28 data sets—15%), and the validation process (28 data sets—15%), in a machine-learning model with artificial neural networks (ANN).

Using the ranking method, the most effective artificial neural network was determined, whose calculated values of R^2 (9), $NRMSE$ (10), MAE (11), and $MAPE$ (12) were, at the same time, closest to the ideal values of the given indicators.

$$R^2 = \frac{\sum_{i=1}^{r}(x_i - x_{mean})^2 - \sum_{i=1}^{r}(x_i - \hat{x}_i)^2}{\sum_{i=1}^{r}(x_i - x_{mean})^2} \tag{9}$$

$$NRMSE = \frac{\sqrt{\frac{1}{r}\sum_{i=1}^{r}(x_i - \hat{x}_i)^2}}{(\hat{x}_{max} - \hat{x}_{min})} \tag{10}$$

$$MAE = \frac{1}{r}\sum_{i=1}^{r}|(\hat{x}_i - x_i)| \tag{11}$$

$$MAPE = \frac{1}{r}\sum_{i=1}^{r}\left|\frac{(x_i - \hat{x}_i)}{x_i}\right| * 100 \tag{12}$$

In the given formulae, x_i is the i-th actual value obtained in the experimental part, while $\hat{x}_i$ is the i-th value predicted by the network. The total number of data is described by r, and x_{mean} is the average of the real values, i.e., the experimental value f_h of the cement composite.

Figure 7 illustrates the results of the predicted values of the pull-off strength in relation to the results obtained by the experimental method using the gradient descent learning algorithm with one layer of hidden neurons. With this algorithm, the best ability to predict the subsurface tensile strength of cementitious composite containing granite powder was obtained for a network with two hidden-layer neurons. It obtained the following network fit evaluation indices: $R^2 = 0.7807$, $NRMSE = 0.1407$, $MAE = 0.1460$, and $MAPE = 11.4751$ %. Figure 6 shows the results for the learning process (blue rhombuses), the testing process (pink triangles), and the validation process (green circles). The red solid line with the function $y = x$ represents a perfect match, while the dashed lines represent the 20% error area. About 16% of the results are outside this area.

Figure 8 illustrates the results of the predicted values of the pull-off strength in relation to the results obtained by the experimental method using the Broyden–Fletcher–Goldfarb–Shanno learning algorithm with one layer of hidden neurons. With this algorithm, the best ability to predict the subsurface tensile strength of cementitious composite containing granite powder was obtained for a network with nine hidden-layer neurons. It obtained the following network fit evaluation indices: $R^2 = 0.8399$, $NRMSE = 0.1028$, $MAE = 0.1290$, and $MAPE = 9.8658\%$. Figure 7 shows the results for the learning process (blue rhombuses), the testing process (pink triangles), and the validation process (green circles). The red solid line with the function $y = x$ represents a perfect match while the dashed lines represent the 20% error area. About 6% of the results are outside this area.

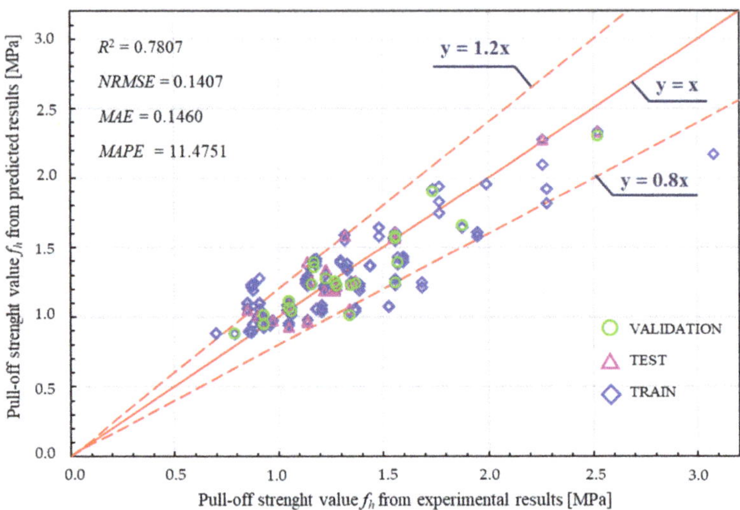

Figure 7. Relationship between the predicted value and the experimental value of the pull-off strength f_h for ANN test, train, and validation process for the GD algorithm with one layer of hidden neurons.

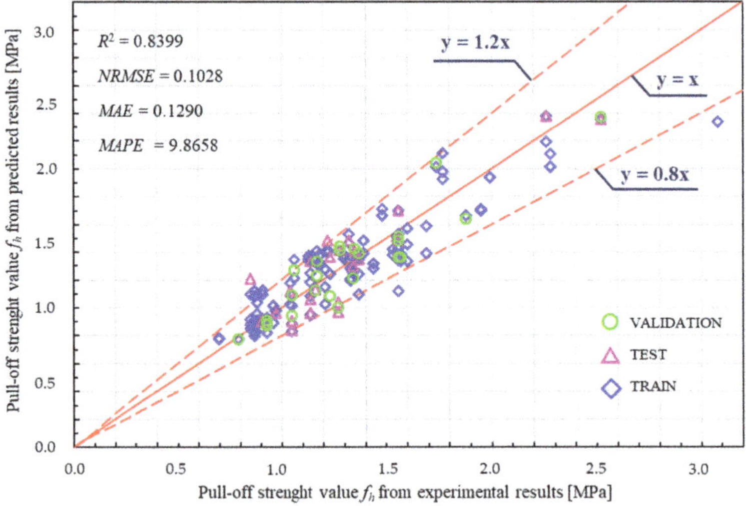

Figure 8. Relationship between the predicted value and the experimental value of the pull-off strength f_h for ANN test, train, and validation process for the BFGS algorithm with one layer of hidden neurons.

Figure 9 illustrates the results of the predicted values of the pull-off strength in relation to the results obtained by the experimental method using the Levenberg–Marquardt learning algorithm with one layer of hidden neurons. With this algorithm, the best ability to predict the subsurface tensile strength of cementitious composite containing granite powder was obtained for a network with seven hidden-layer neurons. It obtained the following network fit evaluation indices: $R^2 = 0.7934$, $NRMSE = 0.1261$, $MAE = 0.1410$, and $MAPE = 10.4077\%$. Figure 8 shows the results for the learning process (blue rhombuses), the testing process (pink triangles), and the validation process (green circles). The red solid line with the function $y = x$ represents a perfect match while the dashed lines represent the 20% error area. About 9% of the results are outside this area.

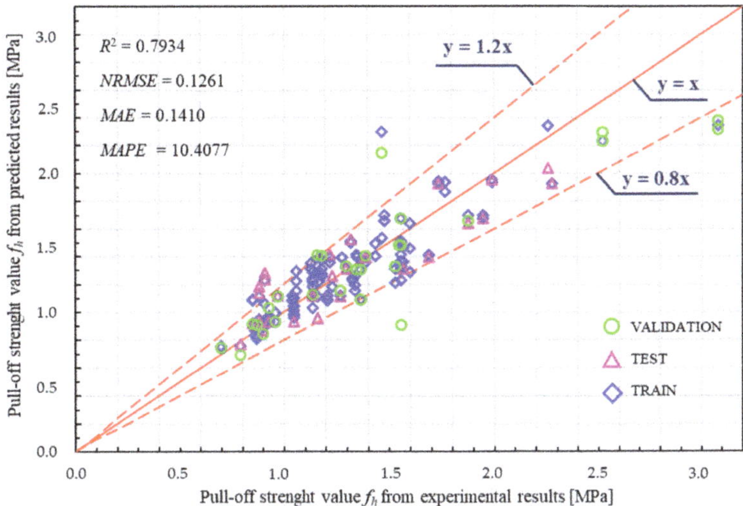

Figure 9. Relationship between the predicted value and the experimental value of the pull-off strength f_h for ANN test, train, and validation process for the LM algorithm with one layer of hidden neurons.

Figure 10 illustrates the results of the predicted values of the pull-off strength in relation to the results obtained by the experimental method using the conjugate gradient learning algorithm with one layer of hidden neurons. With this algorithm, the best ability to predict the subsurface tensile strength of cementitious composite containing granite powder was obtained for a network with three hidden-layer neurons. It obtained the following network fit evaluation indices: $R^2 = 0.8134$, $NRMSE = 0.1109$, $MAE = 0.1391$, and $MAPE = 10.3214$ %. Figure 9 shows the results for the learning process (blue rhombuses), the testing process (pink triangles), and the validation process (green circles). The red solid line with the function $y = x$ represents a perfect match while the dashed lines represent the 20% error area. About 10% of the results are outside this area.

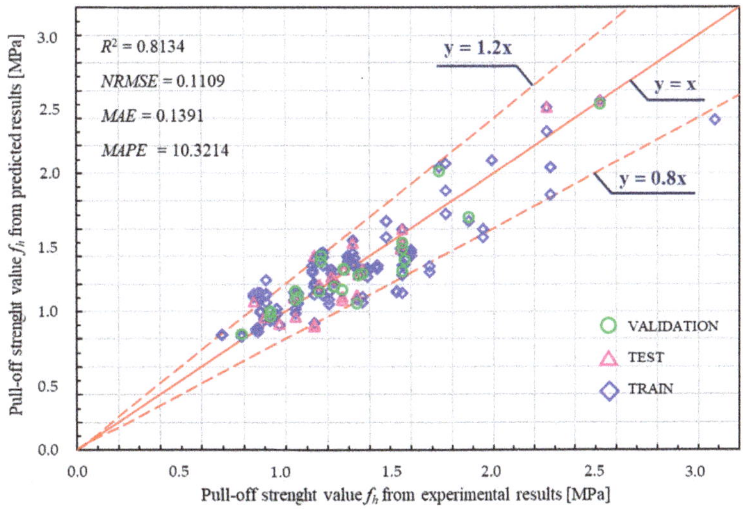

Figure 10. Relationship between the predicted value and the experimental value of the pull-off strength f_h for ANN test, train, and validation process for the CG algorithm with one layer of hidden neurons.

Figure 11 illustrates the results of the predicted values of the pull-off strength in relation to the results obtained by the experimental method using the gradient descent learning algorithm with two layers of hidden neurons. With this algorithm, the best ability to predict the subsurface tensile strength of cementitious composite containing granite powder was obtained for a network with 13 neurons in the first hidden layer and 15 neurons in the second hidden layer. It obtained the following network fit evaluation indices: $R^2 = 0.7386$, $NRMSE = 0.1262$, $MAE = 0.1582$, and $MAPE = 11.8732$ %. Figure 10 shows the results for the learning process (blue rhombuses), the testing process (pink triangles), and the validation process (green circles). The red solid line with the function $y = x$ represents a perfect match while the dashed lines represent the 20% error area. About 18% of the results are outside this area.

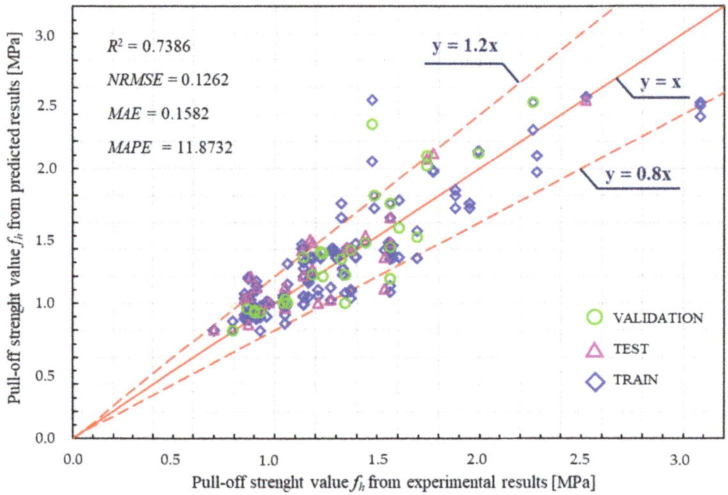

Figure 11. Relationship between the predicted value and the experimental value of the pull-off strength f_h for ANN test, train, and validation process for the GD algorithm with two layers of hidden neurons.

Figure 12 illustrates the results of the predicted values of the pull-off strength in relation to the results obtained by the experimental method using the Broyden–Fletcher–Goldfarb–Shanno learning algorithm with two layers of hidden neurons. With this algorithm, the best ability to predict the subsurface tensile strength of cementitious composite containing granite powder was obtained for a network with four neurons in the first hidden layer and three neurons in the second hidden layer. It obtained the following network fit evaluation indices: $R^2 = 0.7657$, $NRMSE = 0.1382$, $MAE = 0.1507$, and $MAPE = 11.3525\%$. Figure 11 shows the results for the learning process (blue rhombuses), the testing process (pink triangles), and the validation process (green circles). The red solid line with the function $y = x$ represents a perfect match while the dashed lines represent the 20% error area. About 11.5% of the results are outside this area.

Figure 13 illustrates the results of the predicted values of the pull-off strength in relation to the results obtained by the experimental method using the Levenberg–Marquardt learning algorithm with two layers of hidden neurons. With this algorithm, the best ability to predict the subsurface tensile strength of cementitious composite containing granite powder was obtained for a network with four neurons in the first hidden layer and 10 neurons in the second hidden layer. It obtained the following network fit evaluation indices: $R^2 = 0.8046$, $NRMSE = 0.1112$, $MAE = 0.1481$, and $MAPE = 11.0211\%$. Figure 12 shows the results for the learning process (blue rhombuses), the testing process (pink triangles), and the validation process (green circles). The red solid line with the function

$y = x$ represents a perfect match while the dashed lines represent the 20% error area. About 8% of the results are outside this area.

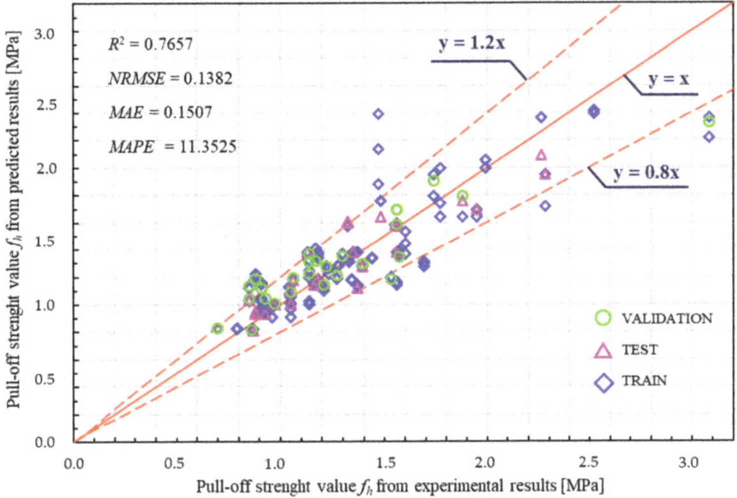

Figure 12. Relationship between the predicted value and the experimental value of the pull-off strength f_h for ANN test, train, and validation process for the BFGS algorithm with two layers of hidden neurons.

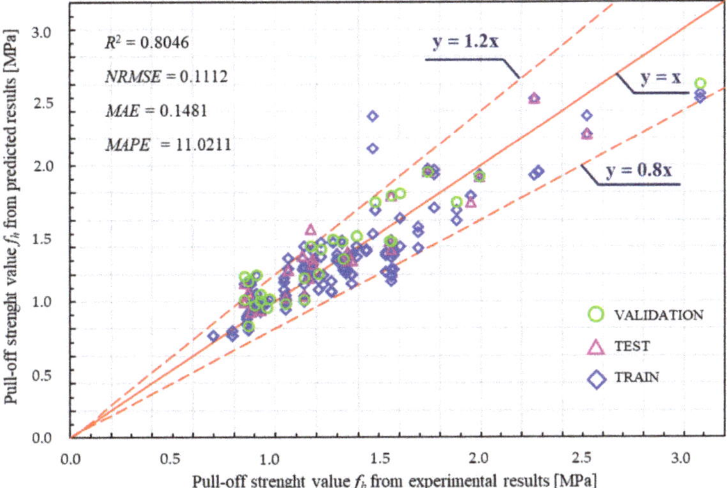

Figure 13. Relationship between the predicted value and the experimental value of the pull-off strength f_h for ANN test, train, and validation process for the LM algorithm with two layers of hidden neurons.

Figure 14 illustrates the results of the predicted values of the pull-off strength in relation to the results obtained by the experimental method using the conjugate gradient learning algorithm with two layers of hidden neurons. With this algorithm, the best ability to predict the subsurface tensile strength of cementitious composite containing granite powder was obtained for a network with 11 neurons in the first hidden layer and eight neurons in the second hidden layer. It obtained the following network fit evaluation indices:

$R^2 = 0.7734$, $NRMSE = 0.1332$, $MAE = 0.1488$, and $MAPE = 10.8728$ %. Figure 13 shows the results for the learning process (blue rhombuses), the testing process (pink triangles), and the validation process (green circles). The red solid line with the function $y = x$ represents a perfect match while the dashed lines represent the 20% error area. About 11% of the results are outside this area.

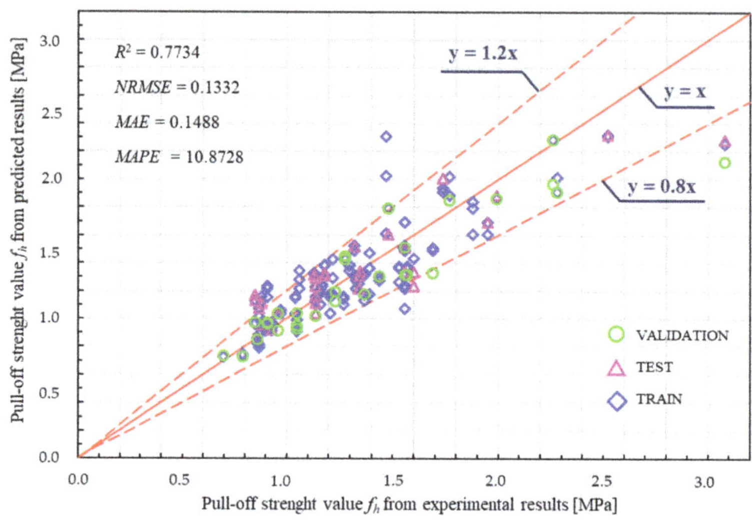

Figure 14. Relationship between the predicted value and the experimental value of the pull-off strength f_h for ANN test, train, and validation process for the CG algorithm with two layers of hidden neurons.

6. Comparison of the Models

The network allows us to estimate the strength based on the material composition of the floor mortar and the compressive strength value determined by the Schmidt hammer. This allows the use of a nondestructive method to evaluate the peel strength, thereby not exposing the tested surface to the repairs necessary after using the pull-off method. Subsequently, using the Schmidt hammer, it is possible to obtain both the compressive strength and pull-off strength. These are the two most important characteristics of floor substrates.

The machine-learning algorithms used in the article are accurate in predicting the subsurface tensile strength of green cementitious composites. This can be concluded from the values of the parameters determining their accuracy, which were R^2, $NRMSE$, MAE, and $MAPE$. Figures 15–19 and Table 4 compare the above algorithms. It can be seen that all the models tested predict the tensile strength of green cement composites well. The BFGS algorithm with one layer of hidden neurons performed best. However, since none of the models were perfectly accurate (R^2 would be equal to 1; and $NRMSE$, MAE, $MAPE$, and relative error would be equal to 0), there is still room for improvement in the algorithms by building other databases or using other algorithms.

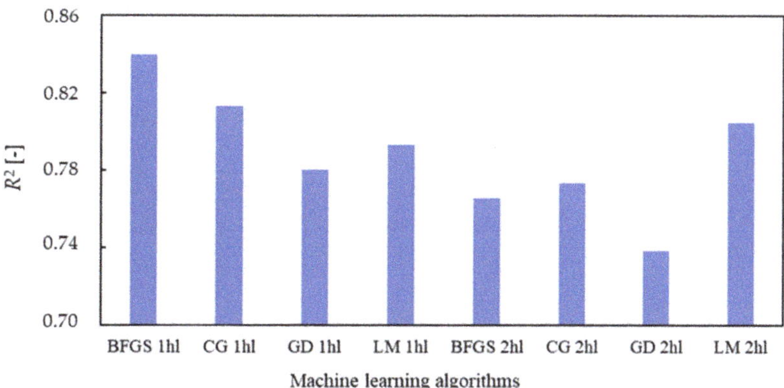

Figure 15. Comparison of models for predicting the pull-off strength of the surface layer f_h due to the value of R^2.

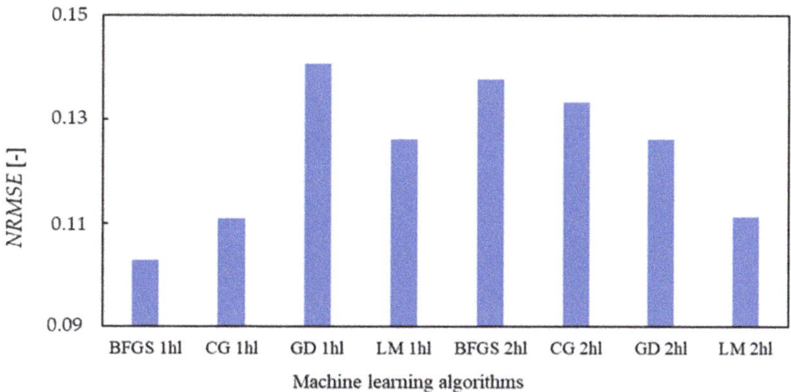

Figure 16. Comparison of models for predicting the pull-off strength of the surface layer f_h due to the value of *NRMSE*.

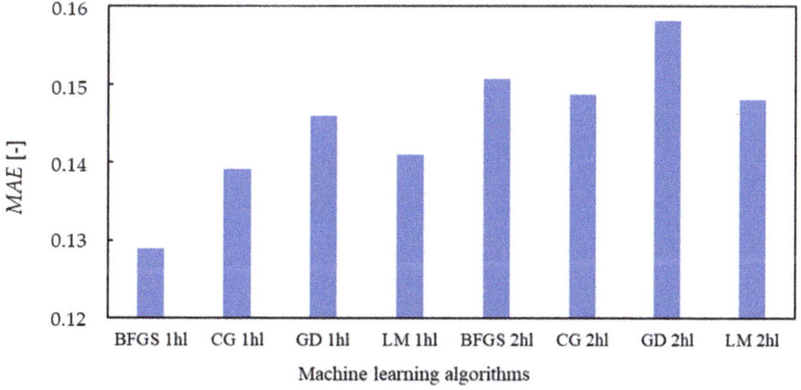

Figure 17. Comparison of models for predicting the pull-off strength of the surface layer f_h due to the value of *MAE*.

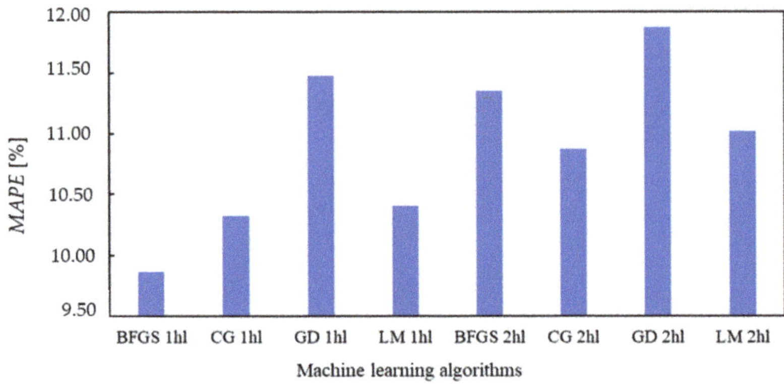

Figure 18. Comparison of models for predicting the pull-off strength of the surface layer f_h due to the value of *MAPE*.

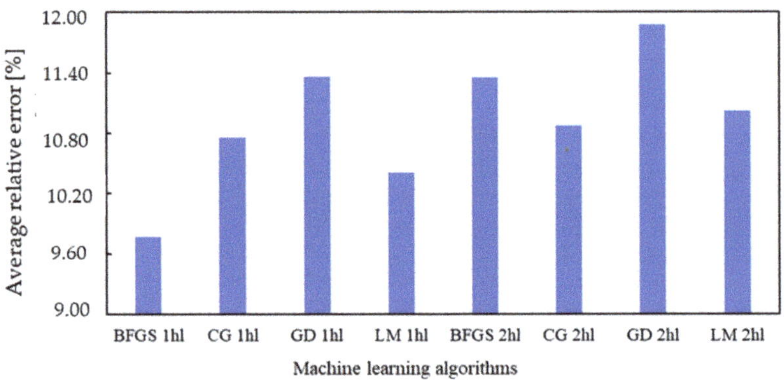

Figure 19. Comparison of models for predicting the pull-off strength of the surface layer f_h due to the value of average relative error.

Table 4. Comparison of accuracy of compared neural networks.

Learning Algorithm	Number of Hidden Layres	Number of Hidden Neurons	R^2	*NRMSE*	*MAE*	*MAPE*
gradient descent	1	2	0.7807	0.1408	0.1460	11.47
gradient descent	2	13–15	0.7386	0.1262	0.1582	11.87
conjugate gradient	1	3	0.8134	0.1109	0.1391	10.32
conjugate gradient	2	11–8	0.7734	0.1332	0.1488	10.87
Broyden–Fletcher–Goldfarb–Shanno	1	9	0.8399	0.1028	0.1290	9.87
Broyden–Fletcher–Goldfarb–Shanno	2	4–3	0.7657	0.1382	0.1507	11.35
Levenberg–Marquardt	1	7	0.7934	0.1261	0.1410	10.41
Levenberg–Marquardt	2	4–10	0.8046	0.1112	0.1481	11.02

The network based on the BFGS algorithm with nine hidden-layer neurons performed best in predicting the pull-off strength of cement floor underlayments. It obtained the following network fit evaluation indices: $R^2 = 0.8399$, $RMSE = 0.1637$, $MAE = 0.1290$, and $MAPE = 9.8658$ %. According to the article [10], the values of determination coefficients R^2 achieved by the selected network are within the range of a very good match with the results obtained experimentally because $R^2 > 0.81$. The MAPE absolute percentage error values of up to 10% are defined as a good fit. As can be seen in Figure 6, the presented network correctly predicts the pull-off strength of the surface layer. The strength ranging from 0.8 MPa to 2.6 MPa is predicted with similar accuracy. The result obtained for the experimental strength above 3 MPa significantly deviates from the rest of the results. This may be due to insufficient batch data from the 2.8 MPa to 3.2 MPa range. It should be noted that the prediction accuracy in the learning, testing, and validation processes of the given network was similar.

7. Conclusions

In this work, the authors proposed a hybrid neural–nondestructive model of subsurface tensile strength prediction in cementitious composites containing granite powder. For this purpose, the experimental tests using a Schmidt hammer were performed, combined with numerical analyses using neural networks. It has been proven that it is possible to design such accurate models using a neural network with the Broyden–Fletcher–Goldfarb–Shanno learning algorithm. The performance was very good, which was shown by the very high value of the coefficient of determination R^2, equal to 0.8399, and very low values of the errors presented as $NRMSE$, MAE, and $MAPE$, equal to 0.1028, 0.1290, and 9.8658%, respectively. According to the results of the correlation between the input variables and output parameter, it can be seen that the amount of cement content and curing conditions contributed the most to the output parameter, whether the samples were kept in a dry environment or wet.

One of the advantages of this method is the fact that it can be used in existing structures, without a need to obtain the samples and destroy them. Moreover, it can still be used after almost 100 days of concreting. Because different curing conditions were taken into account, this method is more universal in comparison to standardized equations.

This novel approach can be a reasonable alternative for evaluating the properties of spacious mortar elements in which there is a need to analyse not only the compressive strength but also its subsurface tensile strength. Moreover, it is worth emphasizing that using a Schmidt hammer during the tests allows us to evaluate the compressive strength, which is beneficial in order to shorten the time of analysing the cementitious composite element.

However, there are some limitations which should be considered before using it in practice. Firstly, it has not been tested yet in the field; therefore, there is need for experimental verification. Second, it has not been used to evaluate the subsurface tensile strength of composites containing admixtures other than granite powder, which might be interesting from a cognitive point of view. Moreover, even though the mortars have been tested, there is still an open question if this model is suitable for concrete containing core aggregates.

Taking the above into account, the authors are still investigating the possibility of using these methods for different cementitious composite mixtures containing such admixtures as fly ash and blast-furnace slag. Moreover, the authors are trying different algorithms such as SVM, linear regression, or random forest, which might be more accurate than those presented in the article. Because this is still an open problem, the authors submit the data to the article for other researchers to use.

Author Contributions: Conceptualization, S.C. and M.M.; methodology, S.C.; software, S.C. and M.M.; validation, S.C.; formal analysis, S.C. and M.M.; investigation, S.C. and M.M.; resources, S.C. and M.M.; data curation, S.C. and M.M.; writing—original draft preparation, S.C. and M.M.; writing—review and editing, S.C.; visualization, M.M.; supervision, S.C.; project administration, S.C.; funding acquisition, S.C. All authors have read and agreed to the published version of the manuscript.

Funding: The authors received funding for preparing the samples from the project, supported by the National Centre for Research and Development, Poland [grant no. LIDER/35/0130/L-11/19/NCBR/2020 "The use of granite powder waste for the production of selected construction products."].

Institutional Review Board Statement: Not applicable.

Informed Consent Statement: Not applicable.

Data Availability Statement: Data available upon request.

Conflicts of Interest: The authors declare no conflict of interest.

Appendix A

Table A1. Data collected during studies.

Sample No.	Cement [kg/m^3]	Dry Quartz Sand [kg/m^3]	Water [kg/m^3]	Granite Powder [kg/m^3]	Curing Conditions	Curing Time [days]	f_c [MPa]	f_h [MPa]
1	512	1536	256	0	AIR	56	22	1.23
2	512	1536	256	0	AIR	56	24	1.23
3	512	1536	256	0	AIR	56	27	1.23
4	512	1536	256	0	AIR	56	29	1.27
5	512	1536	256	0	AIR	56	34	1.27
6	512	1536	256	0	AIR	56	36	1.27
7	512	1536	256	0	AIR	56	32	1.14
8	512	1536	256	0	AIR	56	27	1.14
9	512	1536	256	0	AIR	56	25	1.14
10	512	1536	256	0	AIR	56	31	0.88
11	512	1536	256	0	AIR	56	36	0.88
12	512	1536	256	0	AIR	56	27	0.88
13	460.8	1536	256	51.2	AIR	56	25	1.18
14	460.8	1536	256	51.2	AIR	56	26	1.18
15	460.8	1536	256	51.2	AIR	56	26	1.18
16	460.8	1536	256	51.2	AIR	56	23	1.53
17	460.8	1536	256	51.2	AIR	56	29	1.53
18	460.8	1536	256	51.2	AIR	56	22	1.53
19	460.8	1536	256	51.2	AIR	56	24	1.06
20	460.8	1536	256	51.2	AIR	56	28	1.06
21	460.8	1536	256	51.2	AIR	56	34	1.06
22	460.8	1536	256	51.2	AIR	56	30	1.34
23	460.8	1536	256	51.2	AIR	56	34	1.34
24	460.8	1536	256	51.2	AIR	56	26	1.34
25	409.6	1536	256	102.4	AIR	56	23	0.88
26	409.6	1536	256	102.4	AIR	56	24	0.88
27	409.6	1536	256	102.4	AIR	56	25	0.88
28	409.6	1536	256	102.4	AIR	56	32	1.05
29	409.6	1536	256	102.4	AIR	56	20	1.05
30	409.6	1536	256	102.4	AIR	56	27	1.05
31	409.6	1536	256	102.4	AIR	56	28	0.96
32	409.6	1536	256	102.4	AIR	56	20	0.96
33	409.6	1536	256	102.4	AIR	56	20	0.96
34	409.6	1536	256	102.4	AIR	56	23	0.93
35	409.6	1536	256	102.4	AIR	56	25	0.93
36	409.6	1536	256	102.4	AIR	56	36	0.93
37	358.4	1536	256	153.6	AIR	56	21	0.87
38	358.4	1536	256	153.6	AIR	56	23	0.87
39	358.4	1536	256	153.6	AIR	56	14	0.87
40	358.4	1536	256	153.6	AIR	56	20	0.86
41	358.4	1536	256	153.6	AIR	56	17	0.86
42	358.4	1536	256	153.6	AIR	56	16	0.86
43	358.4	1536	256	153.6	AIR	56	30	0.7
44	358.4	1536	256	153.6	AIR	56	29	0.7
45	358.4	1536	256	153.6	AIR	56	30	0.7

Table A1. *Cont.*

Sample No.	Cement [kg/m^3]	Dry Quartz Sand [kg/m^3]	Water [kg/m^3]	Granite Powder [kg/m^3]	Curing Conditions	Curing Time [days]	f_c [MPa]	f_h [MPa]
46	358.4	1536	256	153.6	AIR	56	34	0.79
47	358.4	1536	256	153.6	AIR	56	29	0.79
48	358.4	1536	256	153.6	AIR	56	28	0.79
49	512	1536	256	0	WET	56	24	2.52
50	512	1536	256	0	WET	56	24	2.52
51	512	1536	256	0	WET	56	25	2.52
52	512	1536	256	0	WET	56	26	2.26
53	512	1536	256	0	WET	56	26	2.26
54	512	1536	256	0	WET	56	33	2.26
55	512	1536	256	0	WET	56	38	1.47
56	512	1536	256	0	WET	56	32	1.47
57	512	1536	256	0	WET	56	25	1.47
58	512	1536	256	0	WET	56	27	3.08
59	512	1536	256	0	WET	56	30	3.08
60	512	1536	256	0	WET	56	26	3.08
61	460.8	1536	256	51.2	WET	56	26	1.48
62	460.8	1536	256	51.2	WET	56	30	1.48
63	460.8	1536	256	51.2	WET	56	26	1.48
64	460.8	1536	256	51.2	WET	56	25	1.88
65	460.8	1536	256	51.2	WET	56	26	1.88
66	460.8	1536	256	51.2	WET	56	30	1.88
67	460.8	1536	256	51.2	WET	56	30	1.95
68	460.8	1536	256	51.2	WET	56	28	1.95
69	460.8	1536	256	51.2	WET	56	28	1.95
70	460.8	1536	256	51.2	WET	56	28	1.56
71	460.8	1536	256	51.2	WET	56	28	1.56
72	460.8	1536	256	51.2	WET	56	28	1.56
73	409.6	1536	256	102.4	WET	56	19	1.16
74	409.6	1536	256	102.4	WET	56	26	1.16
75	409.6	1536	256	102.4	WET	56	29	1.16
76	409.6	1536	256	102.4	WET	56	23	0.91
77	409.6	1536	256	102.4	WET	56	24	0.91
78	409.6	1536	256	102.4	WET	56	27	0.91
79	409.6	1536	256	102.4	WET	56	21	1.56
80	409.6	1536	256	102.4	WET	56	24	1.56
81	409.6	1536	256	102.4	WET	56	30	1.56
82	409.6	1536	256	102.4	WET	56	32	0.87
83	409.6	1536	256	102.4	WET	56	30	0.87
84	409.6	1536	256	102.4	WET	56	30	0.87
85	358.4	1536	256	153.6	WET	56	24	0.85
86	358.4	1536	256	153.6	WET	56	28	0.85
87	358.4	1536	256	153.6	WET	56	24	0.85
88	358.4	1536	256	153.6	WET	56	30	1.37
89	358.4	1536	256	153.6	WET	56	29	1.37
90	358.4	1536	256	153.6	WET	56	24	1.37
91	358.4	1536	256	153.6	WET	56	25	1.04
92	358.4	1536	256	153.6	WET	56	20	1.04
93	358.4	1536	256	153.6	WET	56	27	1.04
94	358.4	1536	256	153.6	WET	56	20	1.21
95	358.4	1536	256	153.6	WET	56	30	1.21
96	358.4	1536	256	153.6	WET	56	26	1.21
97	512	1536	256	0	AIR	90	22	1.6
98	512	1536	256	0	AIR	90	24	1.6
99	512	1536	256	0	AIR	90	27	1.6
100	512	1536	256	0	AIR	90	29	1.33

Table A1. *Cont.*

Sample No.	Cement [kg/m^3]	Dry Quartz Sand [kg/m^3]	Water [kg/m^3]	Granite Powder [kg/m^3]	Curing Conditions	Curing Time [days]	f_c [MPa]	f_h [MPa]
101	512	1536	256	0	AIR	90	34	1.33
102	512	1536	256	0	AIR	90	36	1.33
103	512	1536	256	0	AIR	90	32	1.17
104	512	1536	256	0	AIR	90	27	1.17
105	512	1536	256	0	AIR	90	25	1.17
106	512	1536	256	0	AIR	90	31	1.14
107	512	1536	256	0	AIR	90	36	1.14
108	512	1536	256	0	AIR	90	27	1.14
109	460.8	1536	256	51.2	AIR	90	25	1.28
110	460.8	1536	256	51.2	AIR	90	26	1.28
111	460.8	1536	256	51.2	AIR	90	26	1.28
112	460.8	1536	256	51.2	AIR	90	23	1.69
113	460.8	1536	256	51.2	AIR	90	29	1.69
114	460.8	1536	256	51.2	AIR	90	22	1.69
115	460.8	1536	256	51.2	AIR	90	24	1.39
116	460.8	1536	256	51.2	AIR	90	28	1.39
117	460.8	1536	256	51.2	AIR	90	34	1.39
118	460.8	1536	256	51.2	AIR	90	30	1.22
119	460.8	1536	256	51.2	AIR	90	34	1.22
120	460.8	1536	256	51.2	AIR	90	26	1.22
121	409.6	1536	256	102.4	AIR	90	23	0.85
122	409.6	1536	256	102.4	AIR	90	24	0.85
123	409.6	1536	256	102.4	AIR	90	25	0.85
124	409.6	1536	256	102.4	AIR	90	32	1.05
125	409.6	1536	256	102.4	AIR	90	20	1.05
126	409.6	1536	256	102.4	AIR	90	27	1.05
127	409.6	1536	256	102.4	AIR	90	28	1.05
128	409.6	1536	256	102.4	AIR	90	20	1.05
129	409.6	1536	256	102.4	AIR	90	20	1.05
130	409.6	1536	256	102.4	AIR	90	23	0.91
131	409.6	1536	256	102.4	AIR	90	25	0.91
132	409.6	1536	256	102.4	AIR	90	36	0.91
133	358.4	1536	256	153.6	AIR	90	21	0.93
134	358.4	1536	256	153.6	AIR	90	23	0.93
135	358.4	1536	256	153.6	AIR	90	14	0.93
136	358.4	1536	256	153.6	AIR	90	20	0.9
137	358.4	1536	256	153.6	AIR	90	17	0.9
138	358.4	1536	256	153.6	AIR	90	16	0.9
139	358.4	1536	256	153.6	AIR	90	30	0.97
140	358.4	1536	256	153.6	AIR	90	29	0.97
141	358.4	1536	256	153.6	AIR	90	30	0.97
142	358.4	1536	256	153.6	AIR	90	34	1.14
143	358.4	1536	256	153.6	AIR	90	29	1.14
144	358.4	1536	256	153.6	AIR	90	28	1.14
145	512	1536	256	0	WET	90	24	1.99
146	512	1536	256	0	WET	90	24	1.99
147	512	1536	256	0	WET	90	25	1.99
148	512	1536	256	0	WET	90	26	2.28
149	512	1536	256	0	WET	90	26	2.28
150	512	1536	256	0	WET	90	33	2.28
151	512	1536	256	0	WET	90	38	1.77
152	512	1536	256	0	WET	90	32	1.77
153	512	1536	256	0	WET	90	25	1.77
154	512	1536	256	0	WET	90	27	1.74
155	512	1536	256	0	WET	90	30	1.74

Table A1. *Cont.*

Sample No.	Cement [kg/m^3]	Dry Quartz Sand [kg/m^3]	Water [kg/m^3]	Granite Powder [kg/m^3]	Curing Conditions	Curing Time [days]	f_c [MPa]	f_h [MPa]
156	512	1536	256	0	WET	90	26	1.74
157	460.8	1536	256	51.2	WET	90	26	1.56
158	460.8	1536	256	51.2	WET	90	30	1.56
159	460.8	1536	256	51.2	WET	90	26	1.56
160	460.8	1536	256	51.2	WET	90	25	1.32
161	460.8	1536	256	51.2	WET	90	26	1.32
162	460.8	1536	256	51.2	WET	90	30	1.32
163	460.8	1536	256	51.2	WET	90	30	1.56
164	460.8	1536	256	51.2	WET	90	28	1.56
165	460.8	1536	256	51.2	WET	90	28	1.56
166	460.8	1536	256	51.2	WET	90	28	1.55
167	460.8	1536	256	51.2	WET	90	28	1.55
168	460.8	1536	256	51.2	WET	90	28	1.55
169	409.6	1536	256	102.4	WET	90	19	1.57
170	409.6	1536	256	102.4	WET	90	26	1.57
171	409.6	1536	256	102.4	WET	90	29	1.57
172	409.6	1536	256	102.4	WET	90	23	1.3
173	409.6	1536	256	102.4	WET	90	24	1.3
174	409.6	1536	256	102.4	WET	90	27	1.3
175	409.6	1536	256	102.4	WET	90	21	1.18
176	409.6	1536	256	102.4	WET	90	24	1.18
177	409.6	1536	256	102.4	WET	90	30	1.18
178	409.6	1536	256	102.4	WET	90	32	1.44
179	409.6	1536	256	102.4	WET	90	30	1.44
180	409.6	1536	256	102.4	WET	90	30	1.44
181	358.4	1536	256	153.6	WET	90	24	1.37
182	358.4	1536	256	153.6	WET	90	28	1.37
183	358.4	1536	256	153.6	WET	90	24	1.37
184	358.4	1536	256	153.6	WET	90	30	1.35
185	358.4	1536	256	153.6	WET	90	29	1.35
186	358.4	1536	256	153.6	WET	90	24	1.35
187	358.4	1536	256	153.6	WET	90	25	1.13
188	358.4	1536	256	153.6	WET	90	20	1.13
189	358.4	1536	256	153.6	WET	90	27	1.13
190	358.4	1536	256	153.6	WET	90	20	1.56
191	358.4	1536	256	153.6	WET	90	30	1.56
192	358.4	1536	256	153.6	WET	90	26	1.56

References

1. Danish, A.; Ozbakkaloglu, T. Greener cementitious composites incorporating sewage sludge ash as cement replacement: A review of progress, potentials, and future prospects. *J. Clean. Prod.* **2022**, *371*, 133364. [CrossRef]
2. Nadim Hasoun, M.; Al-Manaseer, A. *Structural Concrete: Theory and Design*, 7th ed.; Wiley: Hoboken, NJ, USA, 2020; ISBN 978-1-119-60512-6.
3. Herath, C.; Gunasekara, C.; Law, D.W.; Setunge, S. Performance of high volume fly ash concrete incorporating additives: A systematic literature review. *Constr. Build. Mater.* **2020**, *258*, 120606. [CrossRef]
4. Czarnecki, S. Modelling of Mechanical Properties of Eco-Friendly Cementitious Composites Used in Floors: State of the Art and Research Gaps. *Chem. Eng. Trans.* **2022**, *94*, 403–408.
5. Chajec, A.; Chowaniec, A.; Królicka, A.; Sadowski, Ł.; Żak, A.; Piechowka-Mielnik, M.; Savija, B. Engineering of green cementitious composites modified with siliceous fly ash: Understanding the importance of curing conditions. *Constr. Build. Mater.* **2021**, *313*, 125209. [CrossRef]
6. Chajec, A. Towards the Cleaner Production of Cementitious Materials with the Synergistic Addition of Granite Powder Waste and Fly Ash. *Chem. Eng. Trans.* **2022**, *94*, 289–294.
7. Malla, P.; Khedmatgozar Dolati, S.S.; Ortiz, J.D.; Mehrabi, A.B.; Nanni, A.; Dinh, K. Feasibility of Conventional Non-Destructive Testing Methods in Detecting Embedded FRP Reinforcements. *Appl. Sci.* **2023**, *13*, 4399. [CrossRef]
8. Łątka, D. Prediction of Mortar Compressive Strength Based on Modern Minor-Destructive Tests. *Materials* **2023**, *16*, 2402. [CrossRef] [PubMed]

9.   Kazemi, M.; Madandoust, R.; De Brito, J. Compressive strength assessment of recycled aggregate concrete using Schmidt rebound hammer and core testing. *Constr. Build. Mater.* **2019**, *224*, 630–638. [CrossRef]

10. *ASTM, C805*; Standard Test Method for Rebound Number of Hardened Concrete. ASTM International: West Conshohocken, PA, USA, 2019.

11. Cacciari, P.P.; Futai, M.M. Assessing the tensile strength of rocks and geological discontinuities via pull-off tests. *Int. J. Rock Mech. Min. Sci.* **2018**, *105*, 44–52. [CrossRef]

12. Pereira, E.; De Mereiros, M.H.F. Pull Off test to evaluate the compressive strength of concrete: An alternative to Brazilian standard techniques. *Ibracon Struct. Mater. J.* **2021**, *5*, 757–780. [CrossRef]

13. Kos, Ž.; Kroviakov, S.; Kryzhanovskyi, V.; Hedulian, D. Strength, Frost Resistance, and Resistance to Acid Attacks on Fiber-Reinforced Concrete for Industrial Floors and Road Pavements with Steel and Polypropylene Fibers. *Materials* **2022**, *15*, 8339. [CrossRef]

14. Martini, R.; Carvalho, J.; Arede, A.; Varum, H. Validation of nondestructive methods for assessing stone masonry using artificial neural networks. *J. Build. Eng.* **2021**, *42*, 102469. [CrossRef]

15. Le, T.-T.; Skentou, A.D.; Mamou, A.; Asteris, P.G. Correlating the Unconfined Compressive Strength of Rock with the Compressional Wave Velocity Effective Porosity and Schmidt Hammer Rebound Number Using Artificial Neural Networks. *Rock Mech. Rock Eng.* **2022**, *55*, 6805–6840. [CrossRef]

16. Almasaeid, H.H.; Suleiman, A.; Alawneh, R. Assessment of high-temperature damaged concrete using non-destructive tests and artificial neural network modelling. *Case Stud. Constr. Mater.* **2022**, *16*, e01080. [CrossRef]

17. Saleem, M. Assessing the load carrying capacity of concrete anchor bolts using non-destructive tests and artificial multilayer neural network. *J. Build. Eng.* **2020**, *30*, 101260. [CrossRef]

18. *PN-EN 12504-2*; Testing of Concrete in Structures—Part 2: Non-Destructive Testing—Determination of Reflection Number. Polish Committee for Standardization: Warsaw, Poland, 2021.

19. *ASTM, D 4541*; Standard Test Method for Pull-Off Strength of Coatings Using Portable Adhesion Testers. ASTM International: West Conshohocken, PA, USA, 2017.

20. Sadowski, Ł.; Hoła, J. New nondestructive way of identifying the values of pull-off adhesion between concrete layers in floors. *J. Civ. Eng. Manag.* **2014**, *20*, 561–569. [CrossRef]

21. Czarnecki, S.; Sadowski, Ł.; Hoła, J. Evaluation of interlayer bonding in layered composites based on non-destructive measurements and machine learning: Comparative analysis of selected learning algorithms. *Autom. Constr.* **2021**, *132*, 103977. [CrossRef]

22. Karthiyaini, S.; Senthamaraikannan, K.; Priyadarshini, J.; Gupta, K.; Shanmugasundaram, M. Prediction of mechanical strength of fiber admixed concrete using multiple regression analysis and artificial neural network. *Adv. Mater. Sci. Eng.* **2019**, *2019*, 4654070. [CrossRef]

23. Nikoo, M.; Torabian Moghadam, F.; Sadowski, Ł. Prediction of concrete compressive strength by evolutionary artificial neural networks. *Adv. Mater. Sci. Eng.* **2015**, *2015*, 849126. [CrossRef]

24. Asteris, P.G.; Skentou, A.D.; Bardhan, A.; Samui, P.; Pilakoutas, K. Predicting concrete compressive strength using hybrid ensembling of surrogate machine learning models. *Cem. Concr. Res.* **2021**, *145*, 106449. [CrossRef]

25. Czarnecki, S.; Sadowski, Ł.; Hoła, J. Artificial neural networks for non-destructive identification of the interlayer bonding between repair overlay and concrete substrate. *Adv. Eng. Softw.* **2020**, *141*, 102769. [CrossRef]

26. Czarnecki, S.; Shariq, M.; Nikoo, M.; Sadowski, Ł. An intelligent model for the prediction of the compressive strength of cementitious composites with ground granulated blast furnace slag based on ultrasonic pulse velocity measurements. *Measurement* **2021**, *172*, 108951. [CrossRef]

applied sciences

Article

Comparison of the Effectiveness of Reducing the Leaching of Formaldehyde from Immobilized Wool in Geopolymer and Cement Mortar

Beata Łaźniewska-Piekarczyk [1,*], Dominik Smyczek [1,2] and Monika Czop [3]

[1] Department of Building Processes and Building Physics, Faculty of Civil Engineering, The Silesian University of Technology, Akademicka 5, 44-100 Gliwice, Poland; dominik.smyczek@saint-gobain.com
[2] Saint Gobain Construction Products Polska Sp z o.o., ul. Okrężna 16, 44-100 Gliwice, Poland
[3] Department of Technologies and Installations for Waste Management, Faculty of Energy and Environmental Engineering, The Silesian University of Technology, Konarskiego 18, 44-100 Gliwice, Poland
* Correspondence: beata.lazniewska@polsl.pl; Tel.: +48-32-237-15-67

Abstract: Innovative building materials should also be pro-environmental. This article discusses the environmental footprint of geopolymer and cement-based mortars. It describes the methodology for preparing geopolymer and cement mortars using mineral wool waste. The phenol–formaldehyde resin used in mineral wool is a source of phenol and formaldehyde emissions to the environment. The prepared mortar samples were subjected to durability tests to assess the correlation between the amount of mineral wool and the flexural and compressive strength of the samples. The key element of the paper is to test whether immobilisation of mineral wool in the geopolymer will reduce leaching of phenol and formaldehyde into the environment. The results revealed that cements prepared with mineral wool showed higher compressive strength, whereas geopolymer samples had better flexural strength. The study also proved that immobilisation of the wool in the geopolymer reduces phenol and formaldehyde leaching significantly.

Keywords: mineral wool; stone wool; metakaolin; leaching of contaminations; geopolymer; formaldehyde; production waste

Citation: Łaźniewska-Piekarczyk, B.; Smyczek, D.; Czop, M. Comparison of the Effectiveness of Reducing the Leaching of Formaldehyde from Immobilized Wool in Geopolymer and Cement Mortar. *Appl. Sci.* **2023**, *13*, 4895. https://doi.org/10.3390/app13084895

Academic Editors: Francisco B. Varona Moya and Mariella Diaferio

Received: 9 March 2023
Revised: 6 April 2023
Accepted: 6 April 2023
Published: 13 April 2023

1. Introduction

This paper discusses the problem of leaching of hazardous substances from mineral-wool-based geopolymer and cement mortars. The partial substitution of wool waste with metakaolin and CEM II cement is investigated as a method to prevent leaching. The earliest information regarding geopolymers goes back to ancient times, when it had various applications. Some sources state that the ancient Egyptians used similar solutions in the construction of the pyramids [1]. The concept of geopolymers was introduced into the literature by Professor Joseph Davidovits in 1978 [2]. What exactly is a geopolymer? Geopolymers are synthetic, nonorganic, i.e., not containing carbon atoms in their structures, polymers of aluminosilicates. The basic reaction behind the formation of this material is the synthesis of silicon and aluminum. In other words, geopolymers are synthetic aluminosilicate polymers characterized by an amorphous internal structure. The definition geopolymer can be presented by analysing its two parts: "geo" and "polymer", the former indicating that t is a material with a structure that imitates raw materials found in nature, hence, the 'geo' in the name [3]. Furthermore, like a polymer, it is formed of multiples of repetitive minor particles, hence, 'polymer'.

The base of a geopolymer composition is a powdered material whose chemical composition is dominated by aluminosilicates. The other major component of a geopolymer is an activator—a chemical mixture whose task is to create an alkaline environment in which the binding reaction will take place [3]. Geopolymer is an innovative environmentally friendly building material, commonly referred to as 'green concrete'. According to the sources, the production of geopolymer uses up to three times less energy compared with classic Portland cement concrete. The same is true for CO_2 emissions. Geopolymers are regarded as the ecological successor to Portland cement as classical concrete production generates several times more harmful greenhouse gases. The production of one tonne of Portland cement emits approximately 1 tonne of CO_2 into the atmosphere. The synthesis of geopolymer generates 4–8 times less CO_2 [4,5]. Mineral wool is a widely used construction material, performing the function of cold, fire, and noise insulation. Waste from this building material is generated during both the demolition and production process [6]. Mineral wool waste poses a significant environmental challenge. In the age of thermomodernisation of buildings, massive amounts of this waste are being generated, reaching an estimated annual of 2.5 million tonnes in the European Union. Due to the difficulty of recycling, most of this ends up in landfills. The management of mineral wool waste is a difficult issue. Due to the specific characteristics of the waste, it is stored in industrial landfills and, thus, contributes to environmental degradation. Nevertheless, due to its low bulk density, high volume, poor compressibility, and the consequential high elasticity, mineral wool causes problems, even in landfilling. Its characteristics adversely affect the landfill stability [7]. The synthesis of geopolymer may offer a method to counteract this occurrence. Geopolymers can be obtained from various raw materials, including fly ashes [8,9], bottom ashes [10], glass powder [11], natural rocks [12], and mineral wool [13] and metakaolin [14], which are the subjects of research in this article. A problem that the production of wool-based geopolymers can generate is the leaching of environmentally hazardous substances. This issue was covered in a previous article of the team [15], where acceptable concentrations of sodium and strong alkalinity were found to be exceeded. An additional challenge is the reduced leachability of phenol and formaldehyde. This paper addresses the issue found in that article regarding the exceedance of permissible leaching standards for geopolymer mortars. The replacement of mineral wool with metakaolin aims to reduce the leaching of exceeded parameters. The article also investigates the effect of using metakaolin on the compressive strength of the geopolymer and determining the ratio of wool to MK that will result in the best durability.

2. Materials and Methods

2.1. Materials

The materials used for the research were recycled stone wool waste and metakaolin, which acted as bonding agents. The following materials were used to bond the mortar: in three series of samples, geopolymer activation was used. The other three samples were bonded with Portland cement (CEM II) multicomponent cement. The final geopolymer mass was created and examined using these materials.

2.1.1. Mineral Wool

The stone wool waste sample was sourced from production waste, so it is characterised by its homogeneity and low moisture content. The chemical composition of the wool was supplemented with alumina to equalize its chemical composition to metakaolin, which is aluminium-rich. Ten weight percent Al_2O_3 was added to the wool. The fraction of wool before milling is presented in Figure 1.

(**a**)   (**b**)

Figure 1. Mineral wool before milling: (**a**) upon collection and (**b**) before milling.

The stone wool waste sample was pulverised using a Los Angeles type ball mill according to EN 1097-2 [16]. The drum was set at 3000 rpm. After milling, the product was the residue as presented in Figure 2.

(**a**)   (**b**)

Figure 2. Mineral wool after milling: (**a**) inside the ball mill and (**b**) before being mixed into the mortar.

The granulometry of the ground wool was examined. The granulometric curve is illustrated in Figure 3.

The modal value of the milled sample was 21.20 μm, and the median was 18.31 μm. The measured specific surface area of the wool after milling was 25,700 cm^2/g. Determination of specific surface area was carried out on a Gemini 2360 apparatus from Micromeritics. The principle of measurement is to measure the adsorption of gas (nitrogen) on the surface of the adsorbate.

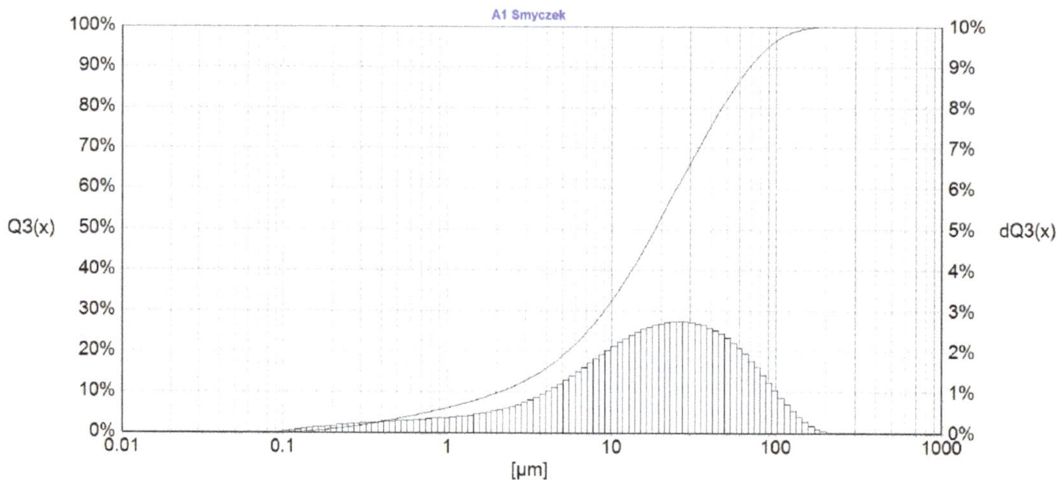

Figure 3. The granulometry of the ground wool.

2.1.2. Metakaolin

Metakaolin is a product, or more precisely a by-product, created by firing kaolin at high temperatures. Chemically, it is an aluminosilicate. It is a very efficient pozzolanic material that can be successfully used in mortars [16]. In contrast to kaolin, which has a crystalline structure, metakaolin is amorphous. Due to its amorphous structure, metakaolin is very reactive when in contact with an alkaline solution [17]. For this reason, metakaolin is ideally applicable as a raw material for the production of geopolymer [18,19] The metakaolin used for this study (Figure 4) comes from the producer ASTRA MK-40 [20]. The relative density of the metakaolin used in the study is 2.30–2.80 g/cm^3. The specific surface area of the material, as specified by the supplier, is 20,000 cm^2/g.

Figure 4. Metakaolin used for mortar preparation.

2.1.3. Comparison of Stone Wool and Metakaolin

When comparing the chemical compositions of mineral wool and metakaolin, similarities are observed. Table 1 presents a comparison of the chemical compositions of the most commonly used raw materials, i.e., stone wool and metakaolin, for the preparation of the geopolymer.

Table 1. Comparison of chemical compositions [15].

Chemical Composition	SiO_2	Al_2O_3	Fe_2O_3	TiO_2	CaO	K_2O	MgO	SO_3	Na_2O	P_2O_5
Mineral wool	43.8	16.4	5.4	0.5	21.83	0.21	9.7	0.0	1.9	0.1
Metakaolin	52.00	45.00	0.50	1.50	0.40	0.05	0.20	0	0.15	0

By analysing the chemical composition of metakaolin, it can be concluded that it is an ideal aluminosilicate [21]. There are high contents of silicon oxide and aluminium oxide with very low proportions of residual oxides. In the case of stone wool, a dominance of silicon oxide, calcium oxide, and aluminium oxide is observed. However, a lower proportion of alumina than that in metakaolin can cause difficulties in geopolymer synthesis, which is why the stone wool was enriched during the study by adding 10% alumina in powder form. The chemical composition after enrichment is shown in Table 2.

Table 2. Chemical composition of stone wool after enrichment.

Chemical Composition	SiO_2	Al_2O_3	Fe_2O_3	TiO_2	CaO	K_2O	MgO	SO_3	Na_2O	P_2O_5
Mineral wool	33.9	29.9	6	0.5	19.4	0.18	6.8	0.5	2.1	0.1

2.1.4. Geopolymer Mortar and Cement Mortar for Continued Research

Six recipes were prepared as described in Table 3. In all six, the base was norm sand (NSD), according to PN EN 196-1, consisting mainly of quartz with granulation of 0.05–2 mm. The individual recipes contained different proportions of ground stone wool (MSW) and metakaolin (MK): three where the binder was sodium hydroxide (NaOH) and three where the binder was cement (CEM II). The recipes were designed on the basis of increasing proportions of mineral wool against decreasing proportions of MK and CEM II. The method enabled the determination of the amount of leachate from the substances in the mortar depending on the proportion of mineral wool. It also served to determine which better retains leaching contaminants—MK or CEM II.

Table 3. Summary of recipes prepared for testing.

		NSD (g)	MSW (g)	MK (g)	CEM II (g)	NaOH (g)	H_2O (g)
Geopolymer mortars	WM1	1350	135	315	0	212	212
	WM2	1350	225	225	0	187	187
	WM3	1350	315	135	0	162	162
Cement mortars	WM4	1350	135	0	315	0	225
	WM5	1350	225	0	225	0	225
	WM6	1350	315	0	135	0	225

Mortar samples were then prepared on the basis of the above formulations. Geopolymers made with stone wool waste were tested (Figure 5).

Figure 5. Mortar samples prepared for testing.

The purpose of preparing the mortars was to further test them for compressive and flexural strength and leaching tests [22]. The geopolymer was forged in $40 \times 40 \times 160$ mm moulds and heat-treated for 48 h at 70 °C for geopolymer mortars and for 72 h at 45 °C for cement mortars. The test samples were then cured under laboratory conditions for an additional 26 days at an air temperature of approximately 20 °C and a humidity of approximately 50%. After this time, flexural and compressive strength tests were performed on the crushed samples (Figure 6).

Figure 6. Crushed mortar samples prepared for leaching test.

2.2. Methods

2.2.1. Compressive Strength Test of Mortars

Compressive strength testing is an overall assessment of mortar performance [23]. This parameter is determined by many factors, such as the amount of water, cement, various additives, and their proportions [24]. Durability tests of the flexural and compressive strength of the wool-based geopolymers and mortars were tested according to EN 196-1:2016 [25]. A Controls Model 65-L27C12 Cement Compression and Flexural Machine was used for the tests. The tests were carried out after 28 days of hardening under air conditions of approx. 20 °C and humidity of approx. 50%.

2.2.2. The Leachability of the Wool-Based Geopolymer and Wool-Based Concrete

The test procedure involved preparing aqueous extracts from the prepared geopolymer mortars and estimating the leaching rates from them. This was followed by a comparison with the limits specified in the regulation [26,27]. The procedure for the preparation of water extracts from solid waste was carried out in accordance with PN-EN 12457-2:2006 [28]. Water extracts for waste were prepared at a liquid to solid ratio of L/S = 10 L/kg (basic test). The washing liquid was distilled water with pH 7.1 and specific electrical conductivity of 61.18 µS/cm. The prepared water extracts were shaken on a laboratory shaker for 24 h, after which time, the obtained extracts were filtered. Analysis of the water extracts from the waste included a number of indications. The pH of the aqueous extracts was determined using the potentiometric method in accordance with PN-EN ISO 10523:2012 [29]. Concentrations and contents were determined by a variety of methods, as detailed in the following: Concentration and elemental content of Zn, Cd, Cu, Pb, Cr, Ni, Ba, As, Mo, Se, Sb, and Na by inductively coupled plasma atomic emission spectrometry (ICP-OES) according to EN ISO 11885:2009 [30]; concentration and content of mercury by atomic absorption spectrometry with cold vapour generation (CVAAS) according to EN ISO 12846:2012 [31]; concentration and content of anions: fluorides, chlorides, and sulphates by ion chromatography (IC) according to PN-EN ISO 10304-1:2009+AC:2012 [32]; dissolved organic carbon (DOC) concentration and content by high-temperature combustion with infrared (IR) detection according to PN-EN 1484:1999 [33]; dissolved substances (TDS) by weight method according to PN-EN 15216:2010 [34]; phenol index (volatile phenols) by spectrophotometric method according to PN-ISO 6439:1994 [35]; and formaldehyde concentration spectrophotometric method according to EPA Method 316:2020 [36].

3. Results

The results of the research work are described below. Table 4 presents the results of the compressive strength, and Tables 5–9 presents the comparison of the results of the leaching of environmental pollutants from the samples tested with emphasis on the emission of formaldehyde.

Table 4. Compressive and flexural test strength results.

Sample No.	Flexural Strength (Mpa)		Compressive Strength (Mpa)	
	Average Value	Standard Deviation	Average Value	Standard Deviation
WM1	2.96	0.97	11.34	1.73
WM2	3.12	0.27	9.24	1.28
WM3	3.68	0.40	8.53	1.79
WM4	1.48	0.51	17.78	1.77
WM5	2.58	0.14	17.07	1.07
WM6	1.90	0.18	5.99	1.16

Table 5. Leachability of selected contaminants from geopolymer mortar, expressed as mg/kg dry weight compared with maximum acceptable waste disposal rates.

| | | | | | Criteria for Waste Landfilling Acceptance [26] | | |
					Neutral	Other Than Neutral and Hazardous	Hazardous
PL	Unit	WM1	WM2	WM3			
Arsenic, As	mg/kg	0.42	0.39	0.42	0.5	2	25
Barium, Ba	mg/kg	0.040	0.020	<0.01	20	100	300
Cadmium, Cd	mg/kg	<0.01	<0.01	<0.01	0.04	1	5
Chromium, Cr	mg/kg	0.24	0.19	0.13	0.5	10	70
Copper, Cu	mg/kg	2.54	1.59	0.99	2	50	100
Mercury, Hg	mg/kg	0.003	0.003	0.009	0.01	0.2	2
Molybdenum, Mo	mg/kg	<0.20	0.22	0.3	0.5	10	30
Nickel, Ni	mg/kg	0.49	0.58	0.46	0.4	10	40
Lead, Pb	mg/kg	4.08	2.03	0.92	0.5	10	50
Antimony, Sb	mg/kg	<0.20	<0.20	0.21	0.06	0.7	5
Selenium, Se	mg/kg	0.24	<0.20	<0.20	0.1	0.5	7
Zinc, Zn	mg/kg	0.13	0.050	0.040	4	50	200
Chlorides, Cl^-	mg/kg	17.0	15	13.8	800	15,000	25,000
Fluorides, F^-	mg/kg	28.4	20.6	13.1	10	150	500
Sulphates, SO_4^{2-}	mg/kg	117	45.3	175	1000	20,000	50,000
Dissolved organic carbon, DOC	mg/kg	633	965	1000	500	800	1000
Total dissolved solids (TDS)	mg/kg	72,760	70,000	67,540	4000	60,000	100,000
Phenolic index	mg/kg	0.870	1.35	0.620	1	-	-

Table 6. Leachability of selected contaminants from cement mortar, expressed as mg/kg dry weight compared with maximum acceptable waste disposal rates.

| | | | | | Criteria for Waste Landfilling Acceptance [26] | | |
					Neutral	Other Than Neutral and Hazardous	Hazardous
	Unit	WM4	WM5	WM6			
Arsenic, As	mg/kg s·m	<0.01	<0.01	<0.01	0.5	2	25
Barium, Ba	mg/kg s·m	1.80	2.36	2.52	20	100	300
Cadmium, Cd	mg/kg s·m	<0.01	<0.01	<0.01	0.04	1	5
Chromium, Cr	mg/kg s·m	3.23	2.04	1.33	0.5	10	70
Copper, Cu	mg/kg s·m	0.14	0.15	0.17	2	50	100
Mercury, Hg	mg/kg s·m	0.0004	0.0004	0.0002	0.01	0.2	2
Molybdenum, Mo	mg/kg s·m	0.73	0.41	0.37	0.5	10	30
Nickel, Ni	mg/kg s·m	<0.01	<0.01	<0.01	0.4	10	40
Lead, Pb	mg/kg s·m	<0.01	<0.01	<0.01	0.5	10	50
Antimony, Sb	mg/kg s·m	1.03	0.91	0.94	0.06	0.7	5
Selenium, Se	mg/kg s·m	<0.20	<0.20	<0.20	0.1	0.5	7
Zinc, Zn	mg/kg s·m	<0.01	<0.01	0.01	4	50	200
Chlorides, Cl^-	mg/kg s·m	31	31	46.6	800	15,000	25,000
Fluorides, F^-	mg/kg s·m	2.50	2.50	3.40	10	150	500
Sulphates, SO_4^{2-}	mg/kg s·m	1870	1600	1550	1000	20,000	50,000
Dissolved organic carbon, DOC	mg/kg s·m	37.4	38.5	44.8	500	800	1000
Total dissolved solids (TDS)	mg/kg s·m	5400	4210	4040	4000	60,000	100,000
Phenolic index	mg/kg s·m	0.870	1.35	0.620	1	-	-

Table 7. Leachability of selected contaminants from geopolymer and cement mortar, expressed as mg/kg dry weight compared with maximum acceptable waste disposal rates.

PL	Unit	WM1	WM2	WM3	WM4	WM5	WM6
Arsenic, As	mg/kg	0.42	0.39	0.42	<0.01	<0.01	<0.01
Barium, Ba	mg/kg	0.040	0.020	<0.01	1.80	2.36	2.52
Cadmium, Cd	mg/kg	<0.01	<0.01	<0.01	<0.01	<0.01	<0.01
Chromium, Cr	mg/kg	0.24	0.19	0.13	3.23	2.04	1.33
Copper, Cu	mg/kg	2.54	1.59	0.99	0.14	0.15	0.17
Mercury, Hg	mg/kg	0.003	0.003	0.009	0.0004	0.0004	0.0002
Molybdenum, Mo	mg/kg	<0.20	0.22	0.3	0.73	0.41	0.37
Nickel, Ni	mg/kg	0.49	0.58	0.46	<0.01	<0.01	<0.01
Lead, Pb	mg/kg	4.08	2.03	0.92	<0.01	<0.01	<0.01
Antimony, Sb	mg/kg	<0.20	<0.20	0.21	1.03	0.91	0.94
Selenium, Se	mg/kg	0.24	<0.20	<0.20	<0.20	<0.20	<0.20
Zinc, Zn	mg/kg	0.13	0.050	0.040	<0.01	<0.01	0.01
Chlorides, Cl^-	mg/kg	17.0	15	13.8	31	31	46.6
Fluorides, F^-	mg/kg	28.4	20.6	13.1	2.50	2.50	3.40
Sulphates, SO_4^{2-}	mg/kg	117	45.3	175	1870	1600	1550
Dissolved organic carbon, DOC	mg/kg	633	965	1000	37.4	38.5	44.8
Total dissolved solids (TDS)	mg/kg	72,760	70,000	67,540	5400	4210	4040
Sodium, Na	mg/kg	29,710	27,330	28,940	826	346	249
pH	-	12.7	12.7	12.7	11.2	11.1	11.1
(T)	(°C)	(22.3)	(22.4)	(22.5)	(23)	(22.9)	(22.5)
Formaldehyde	mg/kg	5.56	8.30	12.6	5.18	7.22	8.41
Phenolic index	mg/kg	0.870	1.35	0.620	0.210	0.350	0.470

Table 8. Leachability of phenol and formaldehyde from geopolymer, cement mortar, and mineral wool, expressed as mg/kg dry weight compared with maximum acceptable waste disposal rates.

	Unit	WM1	WM2	WM3	WM4	WM5	WM6	Mineral Wool
pH	°C	12.7	12.7	12.7	11.2	11.1	11.1	8.9
(Temperature)		(22.3)	(22.4)	(22.5)	(23)	(22.9)	(22.5)	(21.9)
Formaldehyde	mg/kg	5.56	8.30	12.60	5.18	7.22	8.41	578
Phenolic index	mg/kg	8.70	1.35	0.62	0.21	0.35	0.47	7.80

Table 9. Correlation between wool quantity and formaldehyde leachate expressed as mg/kg dry weight.

Sample No.	Wool Quantity (%)	Formaldehyde (mg/kg)
WM1	7.5	5.56
WM2	12.5	8.30
WM3	17.5	12.60
WM4	7.5	5.18
WM5	12.5	7.22
WM6	17.5	8.41
Mineral wool	100	578

3.1. Compressive and Flexural Strength Test Results

The results of the durability test showed that the highest compressive strengths were those of the WM 4 series of samples based on the CEM II cement binder, in which the amount of cement predominated relative to the amount of wool. Their average compressive strength was 17.78 MPa. The series of samples with the lowest compressive strength was the series also based on cement binder (CEM II); in this case, however, the highest intake of wool was accompanied by the lowest intake of cement. Regarding the flexural strength, the samples based on geopolymer mortar performed best with a high proportion of mineral wool. The series of samples with the lowest compressive strengths was the WM6 series, with a high proportion of mineral wool and a low proportion of CEM II cement, the latter being insufficient to adequately consolidate the material. The research led to the conclusion that the fiber structure of the wool improved the compressive strength of the mortar only with alkaline activation. Another important aspect was the very high ratio of flexural strength to compressive strength in geopolymer formulations of up to 1:2. In the case of cementitious mortars designated WM4, WM5, and WM6, the flexural strength was low. In this respect, geopolymer mortar had an advantage over cement mortar.

3.2. Results of Geopolymer Mortar Leaching Tests Compared with the Limit Values

Waste generated from the geopolymer formulation materials is not qualified for landfills for neutral waste due to exceeded acceptable concentrations of copper, nickel, lead, fluoride and dissolved organic carbon. For the WM2 recipe, the phenolic indicator concentration was also exceeded. In addition, waste from the WM1, WM2, and WM3 geopolymer formulations may not be disposed of in landfills for other than neutral and hazardous waste due to exceeded concentrations of total dissolved solids (TDS). Waste from geopolymer formulation materials may be disposed of only in landfills designated for hazardous waste.

3.3. Results of Cement Mortar Leaching Tests Compared with the Limit Values

Waste from cementitious formulation materials, such as geopolymer materials, cannot be disposed of in landfills for neutral waste because concentrations of chromium, molybdenum, antimony, sulphates, and dissolved substances are exceeded. On the other hand, they can be successfully disposed of at other than neutral and hazardous landfills for non-hazardous or inert waste, as they meet all the requirements.

3.4. Comparison of the Results of Geopolymer and Cement Mortar Leaching Tests

The WM1 geopolymer recipe and WM4 cement recipe had in their composition the least mineral wool and the most bonding material metakaolin and CEMII cement, respectively. In contrast, the WM3 and WM6 recipes had the highest wool content ratio to the bonding material (MK, CEMII). The proportional increase in mineral wool content was visible in the leaching results. In fact, for both geopolymer and cementitious formulations, the concentration of formaldehyde increased in direct proportion to the amount of wool in the formulation. The main differences that could be observed between the geopolymer WM1-3 and the cement WM4-6 recipes were the sodium and sulphate concentrations. Due to the use of sodium hydroxide in the activating solution, the sodium concentration in the geopolymer formulations was very high relative to the cement formulations. On the other hand, due to the presence of cement in the cement formulations, the sulphate concentration was very high.

3.5. Comparison of Phenol and Formaldehyde Leaching Results from Geopolymer, Cement Mortar, and Mineral Wool

The summary shows the differences between the leaching of formaldehyde and phenol from the mortars tested and the mineral wool waste sample. The study showed that the use of geopolymer or cement mortar significantly stopped phenol and formaldehyde leaching from the wool.

4. Discussion

The test results presented in Table 4 for the compressive and flexural strengths of the mortars showed the properties of the mortars and the differences between them. The cement-based recipes with the highest cement-to-wool-content ratio showed the highest compressive strength. The combination of high cement content and low wool content resulted in the samples presenting the highest compressive strength. A logical conclusion is that the lower the amount of cement and the higher the amount of mineral wool, the lower the compressive strength. A similar trend can be observed with geopolymer mortars.

Recipes in which mineral wool dominated over aluminosilicate (MK) had lower compressive strengths. It is also important to note the relation in which geopolymer mortars have a better flexural strength compared with cement mortars. The durability of the geopolymer is also influenced by the homogenisation and granulation of the wool [37]. The purity of the wool waste used for recycling is also a significant factor; it is crucial that it is uncontaminated by gypsum, adhesive mortar, or other construction contaminants [38]. Emphasis should also be placed on balancing the aluminium/silicon ratio. Therefore, aluminium oxide was added to the mineral wool sample to raise the proportion of aluminium to silicon. The results of leaching tests on cement mortar samples in Table 6 showed that waste from these materials cannot be disposed of in inert landfills. Nevertheless, they can be disposed of in landfills for non-hazardous and neutral waste, as can mineral wool [15]. In the case of geopolymer mortars, the concentrations of copper, nickel, lead, and fluoride were exceeded in relation to the limits set out in the regulation on landfill of inert waste. In addition, the dissolved organic carbon limit was exceeded in relation to the limits set out in the regulation on landfills for non-hazardous and inert waste [26,27]. The comparison made in Table 8 is the most important observation resulting from the study. By comparing the leaching of formaldehyde and the phenolic index from mineral wool and geopolymer mortars, it is concluded that the geopolymer mortar effectively inhibits the leaching of hazardous substances from the wool contained in the mortar. The table below shows the effect of the amount of wool on the formaldehyde content of the water extract.

The above results indicate a correlation between the amount of wool and the amount of formaldehyde in the water extract. On the basis of the tests described in the article, it is not possible to conclude whether the geopolymer or cement mortar is a better solution for formaldehyde immobilisation. The leaching of formaldehyde with both geopolymer and cement mortar is inhibited to a comparable, impressive significant level.

5. Conclusions

The identified research problem was the leaching of formaldehyde from mineral wool. It is common to use phenol–formaldehyde resin in the production of mineral wool [39]. Over the years, major manufacturers have set targets for VOC reduction, which is derived from the use of phenol–formaldehyde resin. A number of patents have been issued, with the goal of reducing these substances, based mainly on two methods of scavenging formaldehyde molecules during heating in the curing oven and the development of new resin formulations that are not based on formaldehyde [40]. These issues are an area of research, while formaldehyde leaching is high in the case of wool with phenolic formaldehyde binders. The method used by the team to immobilise the resin-soaked mineral wool fiber was successful. The leaching of formaldehyde for the mineral wool sample was 578 mg/kg, while the maximum value for the geopolymer mortar was 12.60 mg/kg. The difference is significant. The results clearly state that the use of wool in geopolymer mortar significantly reduces its leaching into the environment. Given the increasing environmental awareness in society and the tightening of environmental requirements, the research team unanimously believes that the building materials of the future must be ecological. A research niche that is emerging from the work described in this article is undoubtedly the re-use of geopolymer waste to produce geopolymer to close the mineral wool cycle.

Author Contributions: Conceptualisation, B.Ł.-P., D.S. and M.C.; methodology, B.Ł.-P., D.S. and M.C.; validation B.Ł.-P., D.S. and M.C.; formal analysis B.Ł.-P., D.S. and M.C.; writing—original draft preparation B.Ł.-P., D.S. and M.C.; writing—review and editing, B.Ł.-P., D.S. and M.C. All authors have read and agreed to the published version of the manuscript.

Funding: The research reported in this paper was co-financed by the European Union from the European Social Fund in the framework of the project "Silesian University of Technology as a Centre of Modern Education based on research and innovation" POWR.03.05.00-00-Z098/17 and co-financed by The Center for Incubation and Technology Transfer of the Silesian University of Technology project "PŚ-08".

Institutional Review Board Statement: Not applicable.

Informed Consent Statement: Not applicable.

Data Availability Statement: Not applicable.

Conflicts of Interest: The authors declare no conflict of interest. The funders did not play any role in the design of the study plan; in collecting the resulting data, analysing its results, or interpreting data; formatting the content of the manuscript; or in the decision to publish the results of the study.

References

1. Davidovits, J. *Why the Pharaohs Built the Pyramids with Fake Stones*; Geopolymer Institute: Saint-Quentin, France, 2009.
2. Davidovits, J. Soft mineralurgy and geopolymers. In Proceedings of the geopolymer 88 International Conference, Universite de Technologie, Compiegne, France; 1988.
3. Davidovits, J. *Geopolymer Chemistry and Applications*; Geopolymer Institute: Saint-Quentin, France, 2008.
4. Błaszczyński, T.Z.; Król, M.R. *Geopolymers in Construction/Zastosowanie Geopolimerów W Budownictwie*; Civil And Environmental Engineering Reports: Koszalin, Poland, 2015. [CrossRef]
5. Baszczyski, T.; Król, M. Durability of cement and geopolimer composites. *IOP Conf. Ser. Mater. Sci. Eng.* **2017**, *251*, 012005. [CrossRef]
6. Väntsi, O.; Kärki, T. Mineral wool waste in Europe: A review of mineral wool waste quantity, quality, and current recycling methods. *J. Mater. Cycles Waste Manag.* **2014**, *16*, 62–72. [CrossRef]
7. Sattler, T.; Pomberger, R.; Schimek, J.; Vollprecht, D. Mineral wool waste in Austria, associated health aspects and recycling options. *Detritus* **2020**, *09*, 174–180. [CrossRef]
8. Piyush, G.; Garima, N. *Flyash Based Geopolymer*; Beni-Suef University Journal of Basic and Applied Scinces: Beni-Suef, Egipt, 2021.
9. Morshed, P.D.K.; Abdulrehman, M.A. Colored Geopolymer Concrete Based on Fly Ash. *Int. J. Eng.* **2022**, *35*, 1752–1758.
10. Ibrahim, W.M.W.; Al Bakri Abdullah, M.M.; Ahmad, R.; Tahir, M.F.M.; Mortar, N.A.M.; Azli, M.A.A.N. Mechanical and physical properties of bottom ash/fly ash geopolymer for pavement brick application. *IOP Conf. Ser. Mater. Sci. Eng.* **2020**, *743*, 012029. [CrossRef]
11. Dilan, P.; Mustafa, G. Processing and characterization of geopolymer and sintered geopolymer foams of waste glass powders. *Constr. Build. Mater.* **2021**, *300*, 124259. [CrossRef]
12. Zribi, M.; Baklouti, S. Phosphate-based geopolymers: A critical review. *Polym. Bull.* **2022**, *79*, 6827–6855. [CrossRef]
13. Yliniemi, J.; Luukkonen, T.; Kaiser, A.; Illikainen, M. Mineral wool waste-based geopolymers. *IOP Conf. Ser. Earth Environ. Sci.* **2019**, *297*, 012006. [CrossRef]
14. Appiah-Kubi, E.; Yalley, P.P.; Sam, A. Strength behaviour of Corn Husk Ash polymer concrete reinforced with coconut fibre. *Cogent Eng.* **2021**, *8*, 1993511. [CrossRef]
15. Łaźniewska-Piekarczyk, B.; Czop, M.; Smyczek, D. The Comparison of the Environmental Impact of Waste Mineral Wool and Mineral in Wool-Based Geopolymer. *Materials* **2022**, *15*, 2050. [CrossRef] [PubMed]
16. Jindal, B.B.; Alomayri, T.; Assaedi, H.; Kaze, C.R. Geopolymer concrete with metakaolin for sustainability: A comprehensive review on raw material's properties, synthesis, performance, and potential application. *Environ. Sci. Pollut. Res.* **2023**, *30*, 25299–25324. [CrossRef] [PubMed]
17. Fernández-Jiménez, A.; Monzó, M.; Vicent, M.; Barba, A.; Palomo, A. Alkaline Activation of Metakaolin-Fly Ash Mixtures: Obtain of Zeoceramics and Zeocements. *Microporous Mesoporous Mater.* **2008**, *108*, 41–49. [CrossRef]
18. Al-Jaberi, L.A.; Ibrahim, Y.K. Microstructure, Strength, and Physical Properties of Metakaolin- Based Geopolymer Mortar. *Int. J. Eng. Artif. Intell.* **2021**, *2*, 13–26.
19. Faza, Y.; Harmaji, A.; Takarini, V.; Hasratiningsih, Z.; Cahyanto, A. Synthesis of Porous Metakaolin Geopolymer as Bone Substitute Materials. Key Engineering Materials. *Key Eng. Mater.* **2019**, *829*, 182–187. [CrossRef]
20. Available online: https://astra-polska.com/wp-content/uploads/2020/11/karta-teczhniczna-MK40-1-2.pdf (accessed on 20 March 2023).
21. Zhan, J.; Li, H.; Li, H.; Cheng, Z.; Fu, B. Composition design and characterization of alkali-activated Slag–Metakaolin materials. *Front. Built Environ.* **2022**, *8*, 1020217. [CrossRef]
22. Council Decision of 19 December 2002 Establishing Criteria and Procedures for the Acceptance of Waste at Landfills Pursuant to Article 16 of and Annex II to Directive 1999/31/EC. Available online: https://eur-lex.europa.eu/legal-content/PL/TXT/PDF/?uri=CELEX:32003D0033&from=EN (accessed on 10 February 2023).
23. Hasan, M.; Kabir, A. Prediction of compressive strength of concrete from early age test result. In Proceedings of the 4th Annual Paper Meet and 1st Civil Engineering Congress, Dhaka, Bangladesh, 22–24 December 2011. [CrossRef]
24. Faluyi, F.; Arum, C.; Ikumapayi, C.M.; Alabi, S.A. A Review of the Compressive Strength Predictor Variables of Geopolymer Concrete. *FUOYE J. Eng. Technol.* **2022**, *7*, 404–414. [CrossRef]
25. *EN 196-1:2016*; Methods of Testing Cement Determination of Strength. European Committee for Standardization (CEN): Brussels, Belgium, 2016.

26. Regulation of the Minister of Maritime Economy and Inland Navigation of 12 July 2019 on Substances Particularly Harmful to the Aquatic Environment and the Conditions to Be Met When Discharging Sewage intoWaters or Soil, as Well as Discharging Rainwater or Meltwater into Waters or for Water Devices (Journal of Laws 2019, Item 1311). Available online: https://isap.sejm.gov.pl/isap.nsf/DocDetails.xsp?id=WDU20190001311 (accessed on 10 February 2023).
27. The Regulation of the Minister of Economy of 16 July 2015 on the Acceptance of Waste at landfilling. (Journal of Laws 2015, Item 1227). Available online: https://isap.sejm.gov.pl/isap.nsf/DocDetails.xsp?id=WDU20150001277 (accessed on 10 February 2023).
28. *EN 12457-2:2006*; Characterisation of Waste—Leaching—Compliance Test for Leaching of Granular Waste Materials and Sludges. Part 2: One Stage Batch Test at a Liquid to Solid Ratio of 10 L/kg for Materials with Particle Size below 4 mm (without or with Size Reduction). European Committee for Standardization (CEN): Brussels, Belgium, 2006.
29. *EN ISO 10523:2012*; Determination of pH. European Committee for Standardization (CEN): Brussels, Belgium, 2012.
30. *PN-EN ISO 11885:2009*; Determination of Selected Elements by Inductively Coupled Plasma Optical Emission Spectrometry (ICP-OES). Polish Committee for Standardization: Warszawa, Poland, 2009. Available online: https://sklep.pkn.pl (accessed on 20 February 2023).
31. *ISO 12846:2012*; Water Quality—Determination of Mercury—Method Using Atomic Absorption Spectrometry (AAS) with and without Enrichment. International Organization for Standardization: London, UK, 2012. Available online: https://sklep.pkn.pl/ (accessed on 20 March 2023).
32. *EN ISO 10304-1:2009/AC:2012*; Water Quality—Determination of Dissolved Anions by Liquid Chromatography of Ions—Part 1: Determination of Bromide, Chloride, Fluoride, Nitrate, Nitrite, Phosphate and Sulfate—Technical Corrigendum. European Committee for Standardization (CEN): Brussels, Belgium, 2012. Available online: https://sklep.pkn.pl (accessed on 20 February 2023).
33. *PN-EN 1484:1999*; Water Analysis—Guidelines for the Determination of Total Organic Carbon (TOC) and Dissolved Organic Carbon (DOC). Polish Committee for Standardization: Warszawa, Poland, 1999. Available online: https://sklep.pkn.pl (accessed on 20 February 2023).
34. *PN-EN 15216:2010*; Charakteryzowanie Odpadów—Oznaczanie Całkowitej Substancji Rozpuszczonej (TDS) W Wodzie I Eluatach. Polish Committee for Standardization: Warszawa, Poland, 2010. Available online: https://sklep.pkn.pl (accessed on 20 February 2023).
35. *PN-ISO 6439:1994*; Water Quality pDetermination of Phenolic Index γSpectrometric Methods with 4-Aminoantypirin after Distillation. Polish Committee for Standardization: Warszawa, Poland, 1994. Available online: https://sklep.pkn.pl (accessed on 20 February 2023).
36. *EPA Method 316:2020*; Method 316—Sampling and Analysis for Formaldehyde Emissions From Stationary Sources in the Mineral Wool and Wool Fiberglass Industries 10-7-2020.pdf. United States Environmental Protection Agency: Washington, DC, USA, 2020. Available online: https://sklep.pkn.pl (accessed on 20 March 2023).
37. Yliniemi, J.; Laitinen, O.; Kinnunen, P.; Illikainen, M. Pulverization of fibrous mineral wool waste. *J. Mater. Cycles Waste Manag.* **2018**, *20*, 1248–1256. [CrossRef]
38. Trejbal, J.; Zobal, O.; Domonkos, M.; Prošek, Z. New possibilities for recycling of mineral wool separated from thermal insulation waste. *Acta Polytech. CTU Proc.* **2022**, *34*, 122–126. [CrossRef]
39. Bennett, T.; Allan, J.; Garden, J.; Shaver, M. Low Formaldehyde Binders for Mineral Wool Insulation: A Review. *Glob. Chall.* **2022**, *6*, 2100110. [CrossRef] [PubMed]
40. Neuhaus, T.; Oppl, R.; Clausen, A. *Formaldehyde Emissions from Mineral Wool in Building Constructions into Indoor Air*; Indoor Air: Copenhagen, Denmark, 2008.

Article

Cow Dung Ash in Mortar: An Experimental Study

Muluken Alebachew Worku [1], Woubishet Zewdu Taffese [2,3,*], Behailu Zerihun Hailemariam [3]
and Mitiku Damtie Yehualaw [3,*]

[1] Department of Construction Technology and Management, Woldia Institute of Technology, Woldia University, Woldia 7220, Ethiopia
[2] School of Research and Graduate Studies, Arcada University of Applied Sciences, Jan-Magnus Jansson Aukio 1, 00560 Helsinki, Finland
[3] Faculty of Civil and Water Resource Engineering, Bahir Dar Institute of Technology, Bahir Dar University, Bahir Dar 6000, Ethiopia
* Correspondence: woubishet.taffese@arcada.fi (W.Z.T.); mtkdmt2007@gmail.com (M.D.Y.)

Abstract: This study investigated the impact of using cow dung ash (CDA) as a partial replacement for ordinary Portland cement (OPC) in mortar. Mortar mixes are prepared by replacing OPC with CDA at varying levels: 5%, 10%, 15%, 20%, 25%, and 30%. The chemical composition of CDA shows that it is composed primarily of SiO_2, Al_2O_3, and Fe_2O_3, with a significant amount of loss of ignition. The workability, hardened properties, and microstructure of CDA-containing mortars are also analyzed. The increasing CDA content in mortar reduces workability and, beyond 5%, it causes high water absorption due to CDA's porous nature and unremoved organic compounds. This impacts the density and compressive strength of the hardened mortar as well as compromising its homogeneous characteristics. When using 5% CDA, the bulk density and compressive strength of the mortar are comparable to those of the control mixes. Nonetheless, as the proportion of CDA increases, both the bulk density and compressive strength of the mortar diminish. The thermal stability of mortar mixes with 10%, 20%, and 30% CDA is unaffected at temperatures between 500 °C and 600 °C. The Fourier-transform infrared spectroscopy (FTIR) analysis reveals the presence of unreacted particles and wide stretched C–S–H gels in the mortar samples. In general, the results suggest that CDA can be utilized as a substitute for OPC at a ratio of up to 10% in the manufacturing of mortar and can serve as a feasible alternative cementitious material.

Keywords: cow dung ash; mortar; microstructure; workability; durability; sustainability; supplementary cementitious materials; carbon footprint

Citation: Worku, M.A.; Taffese, W.Z.; Hailemariam, B.Z.; Yehualaw, M.D. Cow Dung Ash in Mortar: An Experimental Study. *Appl. Sci.* **2023**, *13*, 6218. https://doi.org/10.3390/app13106218

Academic Editors: Francisco B. Varona Moya and Mariella Diaferio

Received: 17 April 2023
Revised: 15 May 2023
Accepted: 15 May 2023
Published: 19 May 2023

1. Introduction

The construction sector is a vital and rapidly expanding industry worldwide, playing a crucial part in the economic growth and development of nations. As construction projects continue to grow across numerous developing nations, there will be a consistent rise in the need for concrete, since no other material compares to it in terms of strength and accessibility. However, the concrete industry is one of the largest consumers of natural resources and contributes significantly to global anthropogenic carbon dioxide (CO_2) emissions. According to the Global Cement and Concrete Association (GCCA), 14 billion cubic meters of concrete are produced every year for use in the construction of roads, bridges, tunnels, houses, dams, and flood defenses [1]. The primary source of high CO_2 levels in the concrete industry is the production of Portland cement, which is the main binder and accounts for approximately 8% of global CO_2 emissions with an annual production of more than four billion tons [2,3]. Taking this into account, cement and concrete manufacturers are working to accelerate the transition to greener concrete by committing to reducing CO_2 emissions by 25% by 2030 and meeting a target aligned with the Paris Agreement to limit global warming to 1.5 °C [1].

Reducing carbon emissions from the cement industry is a significant challenge, but there are several approaches that can be taken. One of the most practical, economical, and sustainable approaches to reducing the CO_2 emissions caused by the concrete industry is the use of a sustainable system loop that can transform waste resources into valuable products. This approach provides a range of advantages that go beyond reducing clinker content and associated carbon emissions. It enables a reduction in the use of conventional cement ingredients such as limestone, thereby conserving valuable natural resources. Moreover, repurposing waste materials that would otherwise end up in landfills promotes waste reduction and diversion, thereby fostering more sustainable waste management practices.

In the past two decades, several studies have already explored supplementary cementitious materials (SCMs) derived from industrial and agricultural byproducts [4–11]. These include silica fume, fly ash, blast furnace slag, rice husk ash, coffee husk ash, ground nut ash, bamboo leaves ash, banana leaves ash, sugarcane bagasse ash, corncob ash, animal bone ash, tobacco waste ash, glass powders, coal bottom ash, and eggshell ash. Several studies have also investigated the potential of waste materials to replace various ingredients used in concrete production, including the use of recycled aggregates, coconut fibers, glass, marble, and rubber tree seeds [12–18]. The objective is to conserve natural resources and address the environmental impact caused by these waste materials. By substituting conventional ingredients for waste materials, the concrete industry can contribute to resource conservation and mitigate the burden on the environment.

In recent years, researchers have studied the potential of cow dung ash (CDA) as a supplementary cementitious material. Cow dung, a byproduct of cattle farming, is known for being a rich source of silicon dioxide, and thus has pozzolanic properties [19]. It also contains organic matter, including fibrous materials that have passed through the cow's digestive system. The chemical composition of cow dung typically consists of carbon, nitrogen, hydrogen, oxygen, and phosphorus, as well as potassium and calcium [20]. Ekasila et al. [21] investigated the effectiveness of using 10% CDA and 20% rice husk ash (RHA) in concrete under varying curing conditions. The researchers concluded that incorporating both SCMs in the concrete resulted in greater strength. Another study was conducted to evaluate the performance of concrete with 15% CDA that was exposed to water [22]. The specimens were cured for 28 days and then exposed to fresh water for different durations of time: 56, 90, 180, and 365 days. The results showed that the concrete with CDA had superior pH, compressive and split tensile strength, and durability compared to normal concrete. The researchers also observed a lower bacterial density in the CDA-containing concrete.

According to Statista, Ethiopia has a significant population of cattle, ranking fifth in the world and holding the top spot in Africa for total count. The primary use of cow dung in Ethiopia is as a source of fertilizer, and it is also dried to serve as fuel for cooking purposes. With the Ethiopian government's ambitious plan to provide electricity to all citizens by 2025 [23], the rural population of the country will eventually shift towards using electricity for energy, potentially resulting in wasted cow dung. Therefore, it is necessary to explore alternative plans for utilizing this free and valuable resource. Considering that cement is extensively used in building construction in Ethiopia and is responsible for the highest embodied energy and CO_2 emissions [24], incorporating cow dung ash as a supplementary cementitious material presents a promising solution for addressing the environmental challenges stemming from both the cement industry and cow dung waste.

This paper aims to investigate the viability of using CDA as a partial replacement for ordinary Portland cement (OPC) in mortar production. This study stands out from other research on the topic by conducting a thorough investigation of the impact of CDA on various properties of mortar when used as a partial replacement for OPC. Unlike previous studies that focused on a limited set of factors, this work evaluated a range of parameters, including workability, water absorption, bulk density, compressive strength, homogeneity, surface attack resistance, thermal decomposition, and mineralogical composition. Additionally, as the properties of cow dung can be affected by multiple factors, the findings of this study are significant and novel, as no prior research has been carried out on this topic in the particular region where the study was conducted.

2. Materials and Methods

2.1. Materials

2.1.1. Cement

The OPC used in this study had a 42.5R grade and was sourced from Dangote Cement Plc. The quality of the cement was assessed as per ASTM C1084 [25]. The grading and physical properties were in conformity with the requirements necessitated by standard specifications of ASTM C150/C150M [26].

2.1.2. Fine Aggregate

Fine aggregate was obtained from impurity-free river sand in Lalibela, Ethiopia. The sand was subjected to various tests to ensure its quality met the ASTM standard. Figure 1 shows the fine aggregate gradation curve result of the sieve analysis test as per the standard ASTM C117 [27]. In addition to this, other physical property evaluations were conducted on the sand samples, and the undertaken test results are outlined in Table 1, with reference to their respective test standards.

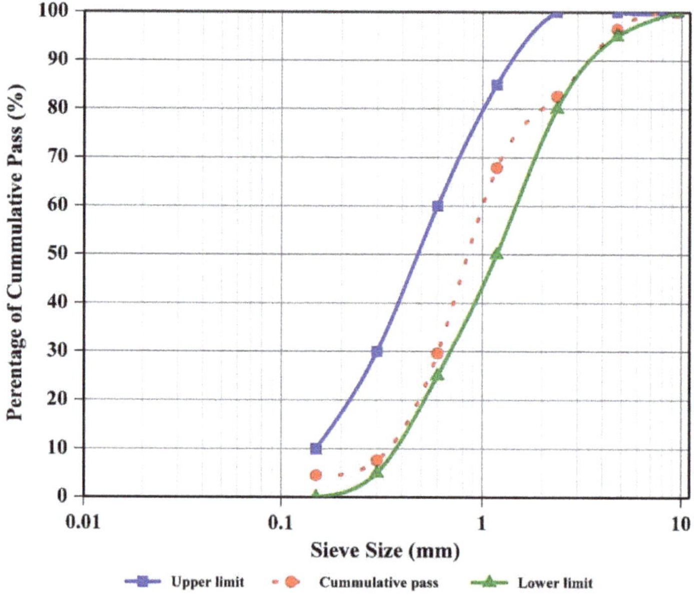

Figure 1. Fine aggregate gradation curve.

Table 1. Test results of the fine aggregate.

No.	Test Type	Test Standards	Test Result	Unit
1	Fineness modulus	ASTM C117	3.11	[-]
2	Loose bulk density	ASTM C29/C29M	1779.63	[Kg/m^3]
3	Compacted bulk density	ASTM C29/C29M	1885.46	[Kg/m^3]
4	Voids	ASTM C29/C29M	35.60	[%]
5	Specific gravity	ASTM C128	2.70	[-]
6	Water absorption	ASTM C128	3.36	[%]
7	Free moisture content	ASTM C566	2.50	[%]
8	Silt content	ASTM C136	3.50	[%]

2.1.3. Cow Dung Ash

The cow dung samples were obtained from the cattle production area surrounding the city of Bahir Dar, Ethiopia. The samples were sun-dried for a week and subsequently calcined in a muffle furnace at 800 °C for 2 h. This particular temperature was chosen as it was found to yield CDA of superior quality [28]. Once burned, the samples were allowed to cool and then ground using a milling machine. Samples that passed through a sieve size of 150 μm were used for cement replacement. The resulting cow dung ash had a dark gray appearance, as shown in Figure 2.

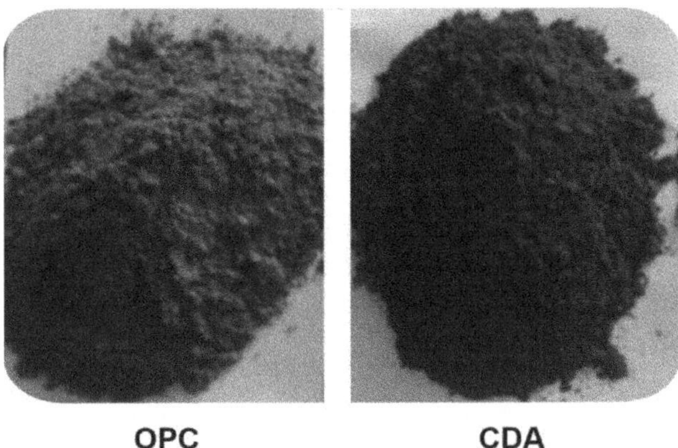

Figure 2. A visual representation depicting the OPC and CDA utilized in the study.

Table 2 compares the results of the X-ray fluorescence (XRF) test conducted on the CDA with the chemical requirements specified by ASTM C150/C150M for Portland cement. The test results indicated that the CDA contained 16.66% SiO_2, 6.15% Al_2O_3, and 7.4% Fe_2O_3, with a negligible amount of SO_3. The sum of the silicious oxides was 30.21%, which did not meet the minimum requirements specified by the standard. The content of MgO, a chemical component found in CDA, falls within the range specified by ASTM C150/C150M for OPC. The loss of ignition (LOI) in the CDA was found to be over 34%, whereas the maximum limit specified by the ASTM is 5% for OPC. This suggests that there are high amounts of unburned inorganic materials present in the CDA, which could result in reduced density and strength.

Table 2. CDA composition analysis and its correlation with the ASTM standard for OPC.

Chemical Compositions	Oxide Content [%]	
	CDA	OPC (ASTM C150/C150M)
CaO	8.18	61–67
SiO_2	16.66	19–23
SO_3	0	0–5.35
Al_2O_3	6.15	2.5–6
Fe_2O_3	7.4	0.1–5
MgO	4.72	1–5
LOI	34.13	-
Na_2O	4	-
K_2O	5.92	-
MnO	0.26	-
P_2O	7.59	-

2.1.4. Mortar Mix Preparation

The mortar mixes were prepared using a water-to-binder ratio of 0.51 and a volumetric ratio of 1:2.75 for cement and sand. As the main objective of our study was to explore the effects of partially replacing OPC with CDA, we examined different levels of CDA content, ranging from 5% to 30%, in increments of 5% (i.e., 5%, 10%, 15%, 20%, 25%, and 30%). A total of 336 cubes, three samples for each mix, were produced specifically for the purpose of performing tests on the hardened mortar properties. Table 3 illustrates the mix codes and the percentages of OPC and CDA used, along with their respective quantities. The table also presents the quantity of potable tap water used to mix the mortar for each mix series.

Table 3. Proportions and quantities of materials in mortar mixtures.

Mix Code	OPC Content		CDA Content		Water	Sand
	[%]	[kg]	[%]	[kg]	[kg]	[kg]
CDA0	100	3.219	-	0	1.631	8.85
CDA5	95	3.058	5	0.161	1.631	8.85
CDA10	90	2.892	10	0.321	1.628	8.84
CDA15	85	2.725	15	0.481	1.625	8.82
CDA20	80	2.56	20	0.64	1.621	8.8
CDA25	75	2.395	25	0.798	1.618	8.78
CDA30	70	2.231	30	0.956	1.615	8.76

2.2. Methods

After conducting multiple tests on the material characteristics of cement, fine aggregates, and CDA, the workability, hardened, and microstructural properties of the control and CDA-containing mortar specimens were assessed using the test methods and types presented in Table 4. These methods adhere to ASTM standards. The properties of hardened mortar consist of water absorption, bulk density, compressive strength, homogeneity, and resistance to sulfate attack. To carry out these tests, the mortar mixes were molded into 50 mm × 50 mm × 50 mm cubes. After molding, the cubes were covered with plastic sheets and stored at room temperature for 24 h. The cubes were then removed from the molds and submerged in water for curing until the time of the test, which occurred at 3, 7, 28, 56, and 91 days. To analyze the microstructural properties of mortar specimens containing CDA, selected samples underwent thermogravimetric analysis (TGA) and Fourier-transform infrared (FTIR) spectroscopy.

Table 4. The fresh, hardened, and microstructural properties test results of the mortars.

Test Category	Properties	Test Standards	Examined Samples	Curing Ages
Fresh	Workability	ASTM C1437	All	-
Hardened	Water absorption Bulk density Compression strength Homogeneity Sulfate attack resistance	ASTM C1403 ASTM C642 ASTM C109/C109M ASTM C597 ASTM C1012	All	3, 7, 28 and 56 days
Microstructure	Thermal decomposition Mineralogical composition	- -	CDA0, CDA10, CDA30	28 days 7 and 28 days

3. Results and Discussion

This section covers the test results and discusses the impact of cow dung ash as a partial replacement for cement in the mortar specimens. Our focus is on how CDA influences the workability, hardened properties, and microstructure of the mortar specimens compared to the reference specimens that used only ordinary Portland cement.

3.1. Effects of CDA on Workability

The workability or fresh property of the plastic mortar was determined by measuring its consistency using the flow table method in accordance with ASTM C1437. The result of this test is shown in Figure 3. As can be observed from this figure, the slump of the mortar decreased significantly with increasing levels of CDA replacement. The greatest reductions in slump were observed for CDA replacement levels of 25% and 30%, with slumps decreasing by 68% and 90%, respectively, compared to the reference mix (CDA0). Similar findings were reported by [29,30]. This phenomenon is primarily attributed to the higher porosity of CDA particles in comparison to cement. The porous nature of CDA particles leads to an increased absorption of mixing water within the mixture, especially as the proportion of CDA in the mixture increases. Consequently, compensating for the reduced workability resulting from CDA utilization necessitates a higher quantity of superplasticizer to maintain the desired workability of the mortar. According to ASTM C1437, the initial flow of the mortar should be $110 \pm 5\%$. Hence, only the fresh mortar containing 15% CDA met this requirement.

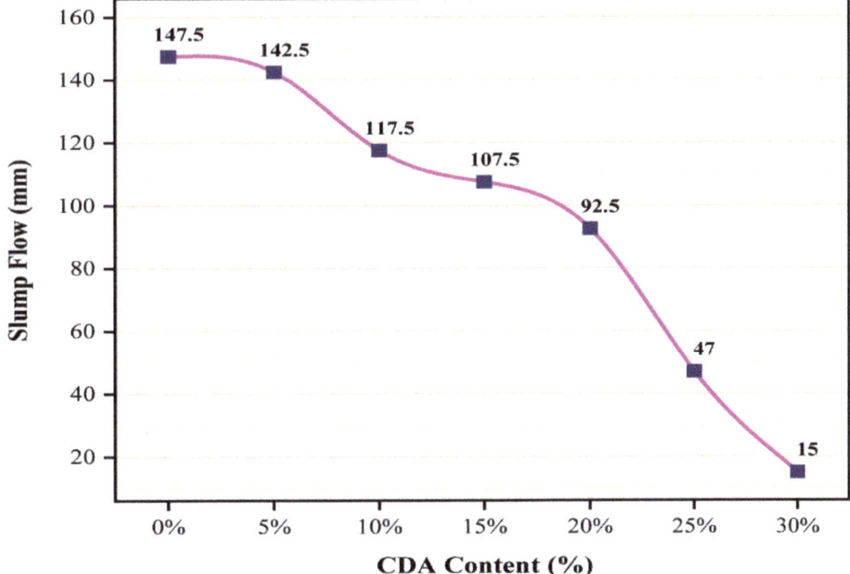

Figure 3. Slump flow of fresh mortar mixes.

3.2. Effects of CDA on Hardened Mortar Properties

3.2.1. Water Absorption

The water absorption test results are presented in Table 5, which reveal that the water absorption of the mortar specimens increased as the percentage of CDA content increased at all curing ages. The increased water absorption can be ascribed to the porous characteristics of CDA particles, as these particles possess a significant capacity to absorb water. This effect becomes particularly pronounced as the content of CDA in the mixture increases. The findings regarding water absorption align well with the observed workability of the mixture. However, it should be noted that the absorption capacity of the CDA-blended mortar decreased with longer curing periods, as reported by [31]. At a curing age of 56 days, the highest water absorption of 11.34% was observed when the CDA content was 30%, resulting in an increase of approximately 36% compared to the reference mix (8.35%). On the other hand, the lowest water absorption was 9.71% at a CDA content of 5%. The water absorption of the mortar samples with CDA content beyond 15% exceeded the maximum allowable limit of 10% [32]. It is important to note that lower water absorption is associated with lower porosity percentage and better compressive strength of the mortar paste [33].

Table 5. Water absorption of the reference and CDA-containing mortar specimens.

CDA Content	Water Absorption [%]			
	3rd Day	7th Day	28th Day	56th Day
0%	10.03	9.90	9.74	8.35
5%	10.64	10.54	10.14	9.71
10%	11.05	10.96	10.20	9.75
15%	11.38	11.25	11.12	9.78
20%	11.90	11.49	11.46	11.04
25%	12.15	11.34	11.30	11.22
30%	12.30	12.10	11.80	11.34

3.2.2. Bulk Density

The bulk density test results are presented in Table 6. From the presented table, it is evident that the bulk density of all mortar specimens incorporating CDA decreases as the percentage of CDA used as a partial replacement for cement increases. This decline in bulk density can be attributed to the utilization of CDA, which hampers the formation of dense and compacted microstructures within the mortar. This observation is supported by the results of microstructural tests discussed in Section 3.3. The bulk density of all cube specimens, including the control mixes, was found to be highest at 56 days of curing and lowest at 3 days of curing. It can be observed that the bulk density of the mortar improved as the curing ages increased from day 3 to day 56. The bulk density of the mortar blended with 5% CDA at 28 days was approximately similar to that of the control mix. Moreover, all sample cubes met the minimum bulk density requirement of 1500 kg/m^3. A similar investigation was also reported by [20].

Table 6. The bulk density of the reference and CDA-containing mortar specimens.

CDA Content	Bulk Density [Kg/m^3]			
	3rd Day	7th Day	28th Day	56th Day
0%	2568.80	2585.60	2589.86	2594.40
5%	2560.00	2582.00	2588.00	2590.40
10%	2493.34	2558.67	2561.34	2576.00
15%	2489.34	2498.00	2555.08	2560.00
20%	2466.66	2480.00	2513.20	2544.00
25%	2452.00	2464.00	2492.50	2504.00
30%	2428.00	2460.00	2480.96	2488.00

3.2.3. Compressive Strength

Figure 4 illustrates the cube compressive strength of mortar specimens at different curing periods with varying percentages of cow dung ash. It can be observed from Figure 4 that there was a linear decrease in the compressive strength of the mortar with an increase in the percentage of CDA used, and the greatest reduction was observed for the specimens containing 30% CDA across all curing periods. The observed reduction in compressive strength is expected, given that the water absorption increased and bulk density decreased with increasing CDA content. The slow rate of strength development can be attributed to the low CaO content present in CDA. As a result, the formation of the C-S-H gel is hindered due to the reduction in cement content. Additionally, the presence of pozzolanas, known for reducing early strength and slowing down hydration, contributes to this delayed strength development. Regardless, in all cases, the strength of the mortar improved as the curing period progressed. The highest compressive strength, measuring 40.79 MPa, was achieved after 56 days, while the lowest value of 15.07 MPa was recorded after 3 days of curing. Based on the findings, there was no significant difference in the compressive strength of specimens containing 5% CDA and those in the control mix, and this result is consistent with the findings reported in [30]. Moreover, even at 15% replacement of cement with CDA, the reduction in compressive strength at the age of 56 days was not found to be significant. In comparison to the control mix sample, it was observed that at a curing time of 28 days, the compressive strength of the mortar specimens containing 20%, 25%, and 30% CDA decreased by 5.66 MPa, 6.51 MPa, and 8.87 MPa, respectively. As per the Indian Standard, IS 2250 [34], the cement and sand mortar should have a minimum compressive strength of 11.5 MPa and 17.5 MPa at the end of 3 and 7 days, respectively. It can be observed that all the mortars containing CDA samples met the specified standards.

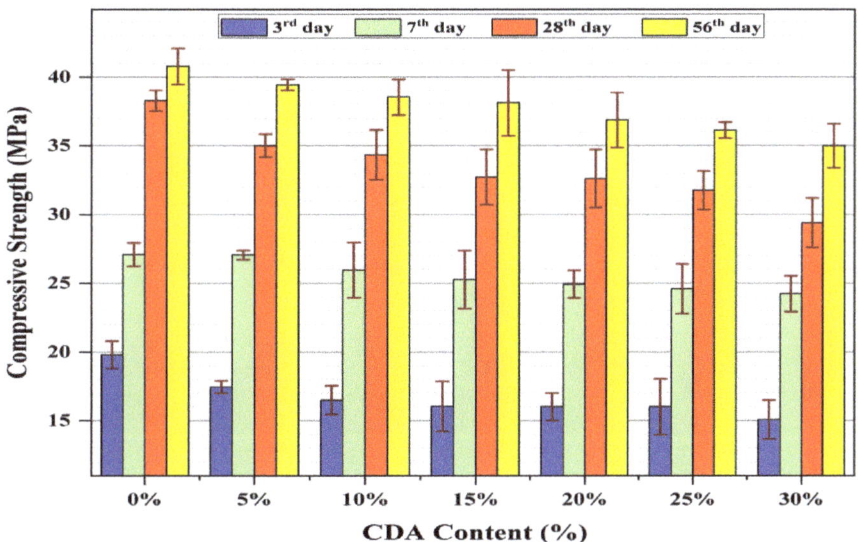

Figure 4. The compressive strength of mortar cubes.

3.2.4. Homogeneity

Figure 5 displays the results of the ultrasonic pulse velocity (UPV) test conducted to evaluate the homogeneity of the reference and CDA-containing mortar specimens. It is evident from Figure 5 that the ultrasonic pulse velocity of the mortar decreased with an increase in the percentage of CDA. The results suggest that the homogeneous characteristics of the mortar are compromised with an increase in the percentage of CDA. The trend observed in the results of the UPV test is consistent with the findings of the compressive strength tests. The experimental results indicate that the UPV values for CDA percentages of 0% to 30% were 2900–1814 m/s, 2900–954 m/s, 3100–1954 m/s, and 3280–2742 m/s after 3, 7, 28 and 56 days of curing age, respectively. As observed from the figure, there were no significant differences in the UPV values between the 3rd and 7th day results; however, the UPV values changed remarkably after later ages. The 28th day UPV results were 4.5–11.5% higher than the respective 7th day results, and the 56th day UPV results were 5–28.7% higher than the respective 28th day results. According to IS 13311 (Part 1): 1992 [35], the quality of the mortar in terms of uniformity was medium strength up to 25% CDA cement substitute.

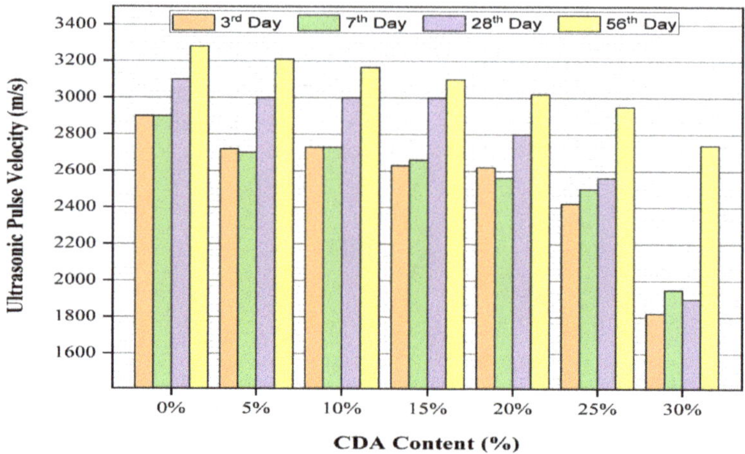

Figure 5. Ultrasonic pulse velocity (UPV) of mortar.

3.2.5. Sulfate Attack Resistance

All of the mortar specimens' ability to withstand sulfate attack was determined by analyzing the variations in their compressive strength following exposure to a sodium sulfate (Na_2SO_4) solution. Figure 6 illustrates the results of the test depicting the variation in compressive strength of mortar containing CDA, subjected to sulfate attack, in comparison to the control mortar sample, at various curing periods. It is evident that the mortar specimens containing up to 10% CDA are able to withstand sulfate attack without much impact on their compressive strength, as the reduction in strength is negligible. Furthermore, it was observed that the compressive strength starts to decrease beyond 15% CDA content, which means that at CDA20 to CDA30, the sodium sulfate significantly affected the strength of the mortar specimens. This phenomenon can be attributed to the high quantity of CaO and the formation of a gel, particularly when the CDA replacement ratio exceeds 10%, as demonstrated by the microstructural tests discussed in Section 3. Consequently, the density of mortar samples decreases, allowing for the diffusion of sulphate ions into the mortar pores. This accelerates the chemical reaction between the hydration products of cement, such as Na_2SO_4 and sulphate ions, with $Ca(OH)_2$ and monosulfate, resulting in the formation of gypsum and ettringite (crystal needles) within the mortar pores [36]. These factors collectively have a negative impact on the mortar's ability to resist sulfate.

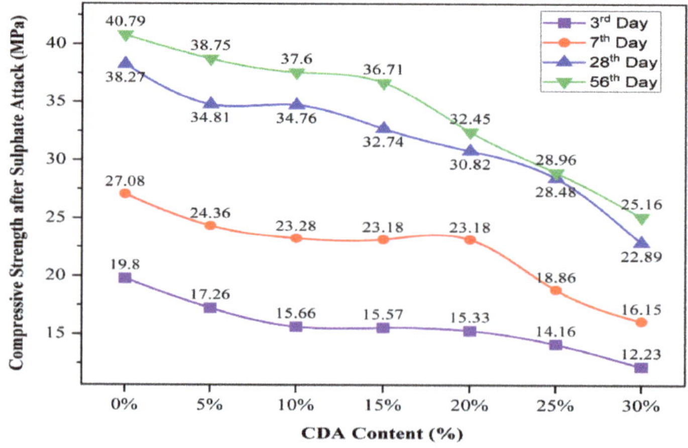

Figure 6. Compressive strength of the mortar specimens after exposure to sulfate attack.

3.3. Effects of CDA on Microstructure of Mortar

3.3.1. Thermal Decomposition

Figures 7–9 illustrate and compare the thermogravimetric (TGA) analyses of mortar samples containing 10%, 20%, and 30% CDA, respectively, after being cured for 28 days. In all figures, the weight loss was determined by subtracting the level of the lower line y-axis from that of the upper line y-axis. According to Figure 7, after 28 days of curing, the mortar sample containing 10% CDA underwent a weight loss of 23% at a heating temperature of 450 °C due to the decomposition of CaO and water from C–H. The weight loss decreased by 8% after an increase in temperature to 500 °C, and then increased again up to a heating temperature of 600 °C. From 600 to 750 °C, the weight loss of the mortar sample significantly decreased by 15% due to the decomposition of water and the C–S–H gel, as well as the carbonation of CO_2 from CO_3. In the case of the 28-day cured mortar sample containing 20% CDA (Figure 8), the trend of thermal stability was almost similar to that of the 10% CDA sample. The mortar samples containing 10% CDA revealed that the $Ca(OH)_2$ dehydrated and lost approximately 25% its weight at a temperature of 450 °C, with only a 7% weight reduction up to 500 °C. Additionally, the figure elucidates that there was no further weight loss when the heating temperature was increased from 500 °C to 550 °C. However, both $CaCO_3$ and the C–S–H gel released CO_2 and water, respectively, and lost 13% of their weight as the temperature increased from 500 °C to 700 °C. Further increases in temperature did not appear to affect the thermal stability of the mortar. The percentage weight loss for the mortar sample containing 30% CDA, which was cured for 28 days, is presented in Figure 9. It can be noticed that the weight loss of the mortar continuously decreased with an increase in temperature up to 450 °C due to the evaporation of water from the C–H gel. It can also be observed that it exhibited a weight loss of 11% due to the evaporation of water from the C–H gel, which was relatively smaller than that of the 10% and 20% CDA mortar samples. Furthermore, the weight loss due to carbonation and dehydroxylation was insignificant compared to the other samples, with a combined weight loss of about 8% at a heating temperature between 450 °C and 700 °C. This indicates that either there were no CaO, CO_2, and H_2O molecules in the C–S–H gel and $CaCO_3$, as all these oxides were released at a temperature of 450 °C, or the heating temperature might not have been sufficient in order to remove the oxides. Generally, for all the mortar samples containing 10%, 20%, and 30% CDA, the thermal stability of 28-day cementitious pastes would not be significantly affected in a temperature range of approximately 500–600 °C.

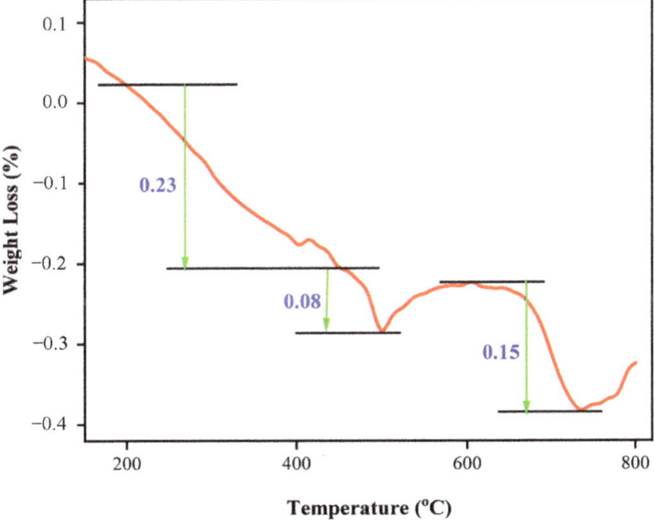

Figure 7. TGA curve of CDA10 after 28 days of curing.

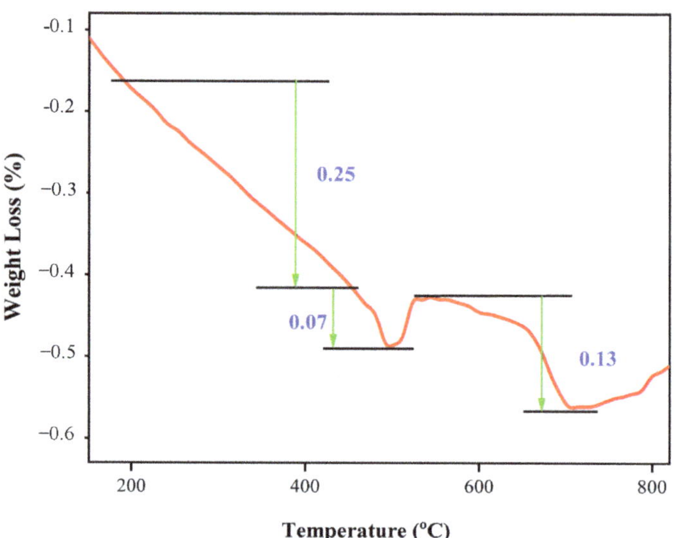

Figure 8. TGA curve of CDA20 after 28 days of curing.

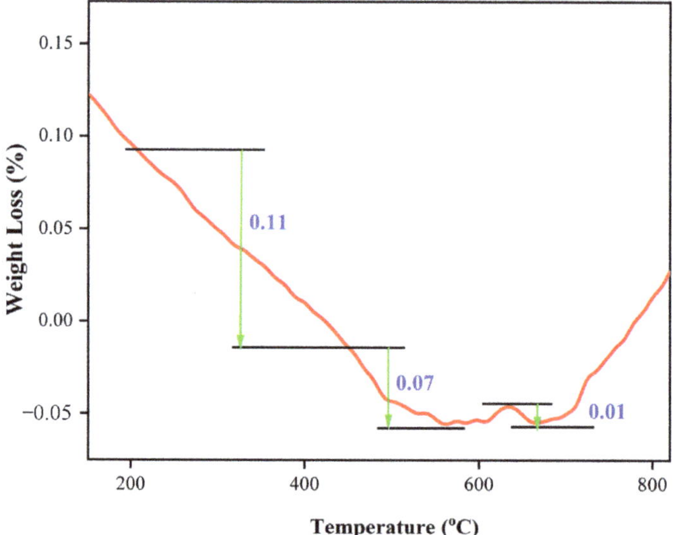

Figure 9. TGA curve of CDA30 after 28 days of curing.

3.3.2. Mineralogical Composition

The result of the Fourier-transform infrared spectroscopy (FTIR) analysis conducted on the mortar samples containing CDA content of 0%, 10%, and 30% after 7 and 28 days of curing is shown in Figures 10 and 11, respectively. The FTIR spectra for the 7-day samples, as shown in Figure 10, reveal major peaks at 3860–3640, 2930–2115, 1750–1345, and 1100–450 cm^{-1}. On the other hand, the spectra for the 28-day samples (Figure 11) exhibit major peaks at 3760–3640, 2930–2280, 2100–1300, 1016–620, and 490–420 cm^{-1}. The wavenumbers, curves, peaks, and valleys observed in the FTIR spectra of the selected mortar samples, i.e., CDA0, CDA10, and CDA30 were almost identical for both curing periods. This suggested that the CDA content had no significant effect on the hydration product during the development of the hydration process, despite minor differences in wavenumbers observed between the 7th and the 28th day samples. Figure 10 illustrates that unreacted particles were detected at the major peak wavenumbers of 744 cm^{-1} for CDA0 and CDA10 samples. However, for CDA30 samples, the wavenumbers increased and extended to the range of 813–681 cm^{-1}. The abundance of unreacted particles can impede the activation reaction, resulting in a delayed early-age strength development. Additionally, the production of a C–S–H gel was lower in the CDA10 and CDA30 mortars, while CDA0 exhibited superior C–S–H gel production compared to CDA10 and CDA30. The wavenumber band between 1644 and 1000 cm^{-1} indicates the presence of small amounts of carbon and the O–H–O extending and stretching vibration between 3460 and 1727 cm^{-1} suggests the presence of free water.

Based on the FTIR analysis depicted in Figures 10 and 11, the curve pattern observed for 28-day samples were comparable to that of the 7-day samples. Notably, unreacted particles were identified at identical wavenumbers within the range of 780–450 cm^{-1}, with a major peak observed at 700 cm^{-1} for CDA10. The formation of a C–S–H gel was evident from the peak observed at 1016–1115 cm^{-1}. The stretching vibration mode of Si–O–T (T: tetrahedral Si or Al) was assigned to the band observed between 950 and 1200 cm^{-1}. This specific frequency range of 1018–1045 cm^{-1} was noted to be characteristic of the presence of silicon tetrahedra (SiO$_4$) in the chain structure of the C–S–H gel [37]. Furthermore, both figures provided further insight into the formation of the C–S–H gel, with the frequency range of 1110–970 cm^{-1} indicating its formation. Unreacted particles were observed in the 600–400 cm^{-1} range, while O–H–O was detected in the 3440–2328 cm^{-1} range.

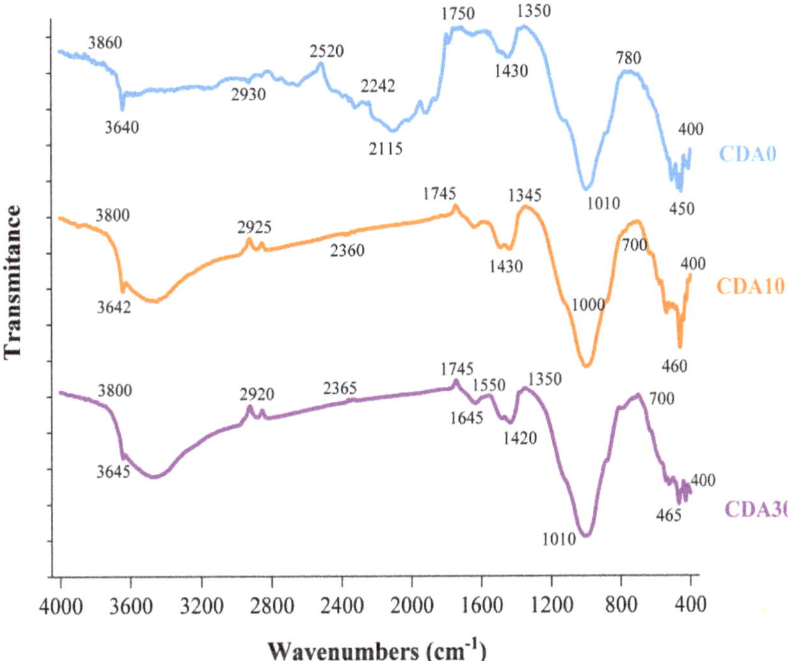

Figure 10. FTIR spectra of paste from a control and CDA-containing mortar cured for seven days.

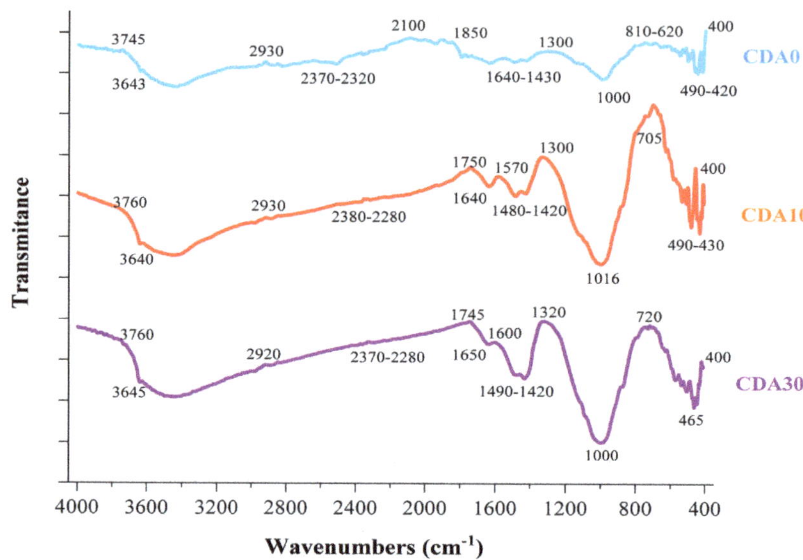

Figure 11. FTIR spectra of paste from a control and CDA-containing mortar cured for 28 days.

4. Conclusions

This research explored the potential use of cow dung ash (CDA) as a supplementary cementitious material by investigating its physical and chemical properties. The study also investigated how CDA in varying proportions (5%, 10%, 15%, 20%, 25%, and 30%) affects the workability, hardened, and microstructural properties of mortar. Based on the results obtained from the study, the following conclusions can be drawn:

- The oxide composition of CDA meets ASTM C150/C150M requirements for cement and is almost similar to that of ordinary Portland cement. However, OPC contains greater amounts of the major oxides of CaO and SiO_2 in comparison to CDA.
- As the percentage of CDA increases in the mortar mix, its workability decreases. Compared to the control mortar, the mortar mixes containing CDA exhibit reduced slump flow. The mortar containing 15% CDA satisfies the slump flow requirements outlined in the ASTM C1437 standard.
- When CDA was utilized beyond 5%, the water absorption of the mortar samples considerably increased due to the porous nature and significant presence of organic compounds that were not effectively removed. As a result, all of the density, homogeneity, and compressive strength of the mortar were negatively impacted. Although the use of CDA as a replacement for OPC resulted in a decrease in compressive strength, the reduction was not considered significant for up to a 15% substitution, as observed at the age of 56 days.
- The compressive strength of mortar specimens is not significantly affected by sulfate attack when containing up to 10% CDA, indicating that these specimens can withstand such an attack with only negligible reduction in strength.
- Mortar mixes containing 10%, 20%, and 30% CDA were found to exhibit thermal stability when exposed to temperatures ranging from 500°C to 600 °C. The FTIR analysis revealed the presence of unreacted particles and a wide-stretched C–S–H gel in the mortar samples.
- While the effect of CDA on the properties of mortar can differ across different tests, replacing OPC with up to 10% CDA can serve as a viable substitute for cementitious materials in mortar production. Adopting CDA as a supplementary cementitious material may offer a sustainable and eco-friendly approach that delivers advantages for both the environment and the construction sector.

- The study has demonstrated the potential of CDA as a promising option for partial replacement of cement. However, further investigations are needed to explore the possible synergistic effects of combining CDA with other supplementary cementitious materials (SCMs), with or without an alkali activator. Such investigations can help to identify the most effective combinations that can lead to improved mechanical properties and durability, as well as sustainable solutions in the cement industry.

Author Contributions: Conceptualization, M.A.W. and M.D.Y.; methodology, M.A.W.; software, M.A.W. and W.Z.T.; validation, M.D.Y., W.Z.T. and B.Z.H.; formal analysis, M.A.W.; investigation, M.A.W.; resources, M.D.Y.; data curation, W.Z.T.; writing—original draft preparation, M.A.W. and B.Z.H.; writing—review and editing, W.Z.T.; visualization, B.Z.H.; supervision, M.D.Y.; project administration, M.D.Y.; funding acquisition, M.D.Y. All authors have read and agreed to the published version of the manuscript.

Funding: This research received no external funding.

Informed Consent Statement: Not applicable.

Data Availability Statement: The data supporting the findings of this study are available upon request.

Conflicts of Interest: The authors declare no conflict of interest.

References

1. Global Cement and Concrete Association (GCCA). *Global Cement and Concrete Industry Announces Roadmap to Achieve Groundbreaking 'Net Zero' CO2 Emissions by 2050*; Global Cement and Concrete Association (GCCA): London, UK, 2021.
2. Lehne, J.; Preston, F. *Making Concrete Change Innovation in Low-Carbon Cement and Concrete*; Policy Commons: London, UK, 2018.
3. Cosentino, I.; Liendo, F.; Arduino, M.; Restuccia, L.; Bensaid, S.; Deorsola, F.; Ferro, G.A. Nano $CaCO_3$ particles in cement mortars towards developing a circular economy in the cement industry. *Procedia Struct. Integr.* **2020**, *26*, 155–165. [CrossRef]
4. Khankhaje, E.; Kim, T.; Jang, H.; Kim, C.-S.; Kim, J.; Rafieizonooz, M. Properties of pervious concrete incorporating fly ash as partial replacement of cement: A review. *Dev. Built Environ.* **2023**, *14*, 100130. [CrossRef]
5. Yehualaw, M.D.; Alemu, M.; Hailemariam, B.Z.; Vo, D.-H.; Taffese, W.Z. Aquatic Weed for Concrete Sustainability. *Sustainability* **2022**, *14*, 15501. [CrossRef]
6. Gedefaw, A.; Worku Yifru, B.; Endale, S.A.; Habtegebreal, B.T.; Yehualaw, M.D. Experimental Investigation on the Effects of Coffee Husk Ash as Partial Replacement of Cement on Concrete Properties. *Adv. Mater. Sci. Eng.* **2022**, *2022*, 4175460. [CrossRef]
7. Duvallet, T.Y.; Jewell, R.B. Recycling of bone ash from animal wastes and by-products in the production of novel cements. *J. Am. Ceram. Soc.* **2023**, *106*, 3720–3735. [CrossRef]
8. Yang, K.H.; Jung, Y.B.; Cho, M.S.; Tae, S.H. Effect of supplementary cementitious materials on reduction of CO_2 emissions from concrete. *J. Clean. Prod.* **2015**, *103*, 774–783. [CrossRef]
9. Endale, S.A.; Taffese, W.Z.; Vo, D.-H.; Yehualaw, M.D. Rice Husk Ash in Concrete. *Sustainability* **2022**, *15*, 137. [CrossRef]
10. Zeybek, Ö.; Özkılıç, Y.O.; Karalar, M.; Çelik, A.İ.; Qaidi, S.; Ahmad, J.; Burduhos-Nergis, D.D.; Burduhos-Nergis, D.P. Influence of Replacing Cement with Waste Glass on Mechanical Properties of Concrete. *Materials* **2022**, *15*, 7513. [CrossRef]
11. Acar, M.C.; Çelik, A.İ.; Kayabaşı, R.; Şener, A.; Özdöner, N.; Özkılıç, Y.O. Production of perlite-based-aerated geopolymer using hydrogen peroxide as eco-friendly material for energy-efficient buildings. *J. Mater. Res. Technol.* **2023**, *24*, 81–99. [CrossRef]
12. Beskopylny, A.N.; Shcherban', E.M.; Stel'makh, S.A.; Meskhi, B.; Shilov, A.A.; Varavka, V.; Evtushenko, A.; Özkılıç, Y.O.; Aksoylu, C.; Karalar, M. Composition Component Influence on Concrete Properties with the Additive of Rubber Tree Seed Shells. *Appl. Sci.* **2022**, *12*, 11744. [CrossRef]
13. Taffese, W.Z. Suitability Investigation of Recycled Concrete Aggregates for Concrete Production: An Experimental Case Study. *Adv. Civ. Eng.* **2018**, *2018*, 8368351. [CrossRef]
14. Basaran, B.; Kalkan, I.; Aksoylu, C.; Özkılıç, Y.O.; Sabri, M.M.S. Effects of Waste Powder, Fine and Coarse Marble Aggregates on Concrete Compressive Strength. *Sustainability* **2022**, *14*, 14388. [CrossRef]
15. Karalar, M.; Bilir, T.; Çavuşlu, M.; Özkiliç, Y.O.; Sabri Sabri, M.M. Use of recycled coal bottom ash in reinforced concrete beams as replacement for aggregate. *Front. Mater.* **2022**, *9*, 1064604. [CrossRef]
16. Shcherban', E.M.; Stel'makh, S.A.; Beskopylny, A.N.; Mailyan, L.R.; Meskhi, B.; Shilov, A.A.; Chernil'nik, A.; Özkılıç, Y.O.; Aksoylu, C. Normal-Weight Concrete with Improved Stress–Strain Characteristics Reinforced with Dispersed Coconut Fibers. *Appl. Sci.* **2022**, *12*, 11734. [CrossRef]
17. Çelik, A.İ.; Özkılıç, Y.O.; Zeybek, Ö.; Karalar, M.; Qaidi, S.; Ahmad, J.; Burduhos-Nergis, D.D.; Bejinariu, C. Mechanical Behavior of Crushed Waste Glass as Replacement of Aggregates. *Materials* **2022**, *15*, 8093. [CrossRef]
18. Karalar, M.; Özkılıç, Y.O.; Aksoylu, C.; Sabri, M.M.S. Flexural behavior of reinforced concrete beams using waste marble powder towards application of sustainable concrete. *Front. Mater.* **2022**, *9*, 1068791. [CrossRef]

19. Ige, J.A.; Olojo, F.O. Effect of Cow Dung Ash Calcined at Different Temperature on the Geotechnical Properties of Laterite Soil. *J. Archit. Civ. Eng.* **2021**, *6*, 38–47.
20. Ojedokun, O. Cow Dung Ash (CDA) as Partial Replacement of Cementing Material in the Production of Concrete. *Br. J. Appl. Sci. Technol.* **2014**, *4*, 3445–3454. [CrossRef]
21. Ekasila, S.; Vishali, B.; Anisuddin, S. An experimental investigation on compressive strength of concrete with rice husk and cow dunk ash using various curing methods. *Mater. Today Proc.* **2022**, *52*, 1182–1188. [CrossRef]
22. Ramachandran, D.; Vishwakarma, V.; Viswanathan, K. Detailed studies of cow dung ash modified concrete exposed in fresh water. *J. Build. Eng.* **2018**, *20*, 173–178. [CrossRef]
23. Pappis, I.; Sahlberg, A.; Walle, T.; Broad, O.; Eludoyin, E.; Howells, M.; Usher, W. Influence of Electrification Pathways in the Electricity Sector of Ethiopia—Policy Implications Linking Spatial Electrification Analysis and Medium to Long-Term Energy Planning. *Energies* **2021**, *14*, 1209. [CrossRef]
24. Taffese, W.Z.; Abegaz, K.A. Embodied Energy and CO_2 Emissions of Widely Used Building Materials: The Ethiopian Context. *Buildings* **2019**, *9*, 136. [CrossRef]
25. *ASTM C1084-19*; Standard Test Method for Portland-Cement Content of Hardened Hydraulic-Cement Concrete. ASTM International: West Conshohocken, PA, USA, 2020.
26. *ASTM C150/C150M-22*; Standard Specification for Portland Cement. ASTM International: West Conshohocken, PA, USA, 2022.
27. *ASTM C117-17*; Standard Test Method for Materials Finer than 75-μm (No. 200) Sieve in Mineral Aggregates by Washing. ASTM International: West Conshohocken, PA, USA, 2020.
28. Zhou, S.; Zhang, S.; Shen, J.; Guo, W. Effect of cattle manure ash's particle size on compression strength of concrete. *Case Stud. Constr. Mater.* **2019**, *10*, e00215. [CrossRef]
29. Venkatasubramanian, C.; Muthu, D.; Aswini, G.; Nandhini, G.; Muhilini, K. Experimental studies on effect of cow dung ash (pozzolanic binder) and coconut fiber on strength properties of concrete. *IOP Conf. Ser. Earth Environ. Sci.* **2017**, *80*, 012012. [CrossRef]
30. Sruthy, B.; Mathew, G.M.; Krishnan, A.G.; Raj, S.G. An Experimental Investigation on Strength of Concrete Made with Cow Dung Ash and Glass Fibre. *Int. J. Eng. Res.* **2017**, *6*, 492–495. [CrossRef]
31. Rawal, H.M.; Deshmukh, M.; Patil, J. Partial Replacement of Cow Dung with Cement. *Int. J. Sci. Res. Eng. Trends* **2019**, *5*, 518–521.
32. *ASTM C1585-20*; Standard Test Method for Measurement of Rate of Absorption of Water by Hydraulic-Cement Concretes. ASTM International: West Conshohocken, PA, USA, 2020.
33. Azhar, N.S.D.M.; Zainal, F.F.; Abdullah, M.M.A.B. *Effect of Geopolymer Paste on Compressive Strength, Water Absorption and Porosity*; AIP Publishing LLC: Melville, NY, USA, 2020. [CrossRef]
34. *IS 2250*; Code of Practice for Preparation and Use of Masonry Mortars. Bureau of Indian Standards: Delhi, India, 1981.
35. *IS 13311 (Part 1)*; Method of Non-Destructive Testing of Concrete, Part 1: Ultrasonic Pulse Velocity. Bureau of Indian Standards: Delhi, India, 1992.
36. Nawaz, M.A.; Ali, B.; Qureshi, L.A.; Aslam HM, U.; Hussain, I.; Masood, B.; Raza, S.S. Effect of sulfate activator on mechanical and durability properties of concrete incorporating low calcium fly ash. *Case Stud. Constr. Mater.* **2020**, *13*, e00407. [CrossRef]
37. Hwang, C.L.; Yehualaw, M.D.; Vo, D.H.; Huynh, T.; Largo, A. Performance evaluation of alkali activated mortar containing high volume of waste brick powder blended with ground granulated blast furnace slag cured at ambient temperature. *Constr. Build. Mater.* **2019**, *223*, 657–667. [CrossRef]

applied sciences

Article

The Use of De-Icing Salts in Post-Tensioned Concrete Slabs and Their Effects on the Life of the Structure

Tomás Luis Ripa Alonso [1,2], Noemí Corral Moraleda [2], Marcos García Alberti [1,*], Rubén Muñoz Pavón [1] and Jaime C. Gálvez [1]

[1] Departamento de Ingeniería Civil: Construcción, E.T.S. de Ingenieros de Caminos, Canales y Puertos, Universidad Politécnica de Madrid, 28040 Madrid, Spain; tomasluis.ripa@upm.es (T.L.R.A.); ruben.mpavon@upm.es (R.M.P.); jaime.galvez@upm.es (J.C.G.)
[2] LRA Infrastructures Consulting, Avda. Concha Espina, 28016 Madrid, Spain; noemicorral@lraingenieria.es
* Correspondence: marcos.garcia@upm.es; Tel.: +34-91-0674121

Abstract: This article expounds on the problem of the use of de-icing salts in the corrosion of steel rebars in bridge decks and their effect on post-tensioning elements. In particular, this paper focuses this problem on structures affected by an aggregate–alkali reaction and without any waterproof treatment using the example of one structure whose repair was carried out in 2020. In this structure, the internal stresses due to the aggregate–alkali reaction caused longitudinal cracks in the upper face of the deck, through which the penetration of chloride ions was concentrated, causing, finally, the brittle fracture of the steel bars and the corrosion of the prestressing elements. This article also explains some conclusions about the most probable mechanisms that resulted in the brittle fracture of the steel bars due to the extraordinary and unexpected nature of this phenomenon.

Keywords: de-icing salts; post-tensioned concrete slabs; corrosion fatigue; cracking; durability

Citation: Ripa Alonso, T.L.; Corral Moraleda, N.; García Alberti, M.; Pavón, R.M.; Gálvez, J.C. The Use of De-Icing Salts in Post-Tensioned Concrete Slabs and Their Effects on the Life of the Structure. *Appl. Sci.* **2023**, *13*, 6961. https://doi.org/10.3390/app13126961

Academic Editor: Syed Minhaj Saleem Kazmi

Received: 20 April 2023
Revised: 5 June 2023
Accepted: 6 June 2023
Published: 9 June 2023

1. Introduction

The use of de-icing salts during low-temperature periods in structures either without a waterproof treatment or a deficient one might result in the entrance of chloride ions through the upper face of their decks. Furthermore, this phenomenon may be aggravated by the presence of other previous relevant concrete pathologies such as pores, hollows or cracks inside the concrete which may ease the circulation of aggressive chemicals until steel reinforcements bars.

This article aims to show the role of de-icing salts in the corroding process of steel rebars and their effects on post-tensioned elements. Consequently, this article pays special attention to the brittle rupture phenomenon of transversal rebars for being extraordinary, as well as highlights the most likely mechanisms that gave rise to this pathology and emphasizing the important use of waterproofing treatments, not only to reduce brittle rupture failures of the steel rebars but also to prevent alkali–aggregate reactions of concrete. The article focuses on a representative case where this process has been observed. The Nudo de Colmenar Bridge, a motorway junction located in the north of Madrid Region with winter road maintenance activities based on the spreading of de-icing salts with concentrations varying from 10 to 15 g/m^2.

The science and engineering in corrosion of reinforced and post-tensioned concrete structures have been developed for more than half a century. The processes of study are slow and need decades to manifest and this means that continuous inspections are needed in order to detect the appearance of durability issues on the structure. Some maintenance tasks, such as the use of de-icing salts to avoid ice and snow in road structures, can be cheap and effective in the short term although they can be disastrous in the mid-term as shown in this study. In social terms, the tremendous financial burden caused by corrosion receives little attention in comparison with health or climate change even though they can be of similar or even greater impact in terms of costs [1]. Additionally, some educational challenges appear in this aspect, showing that civil engineers are generally not well trained to address these challenges This study shows a case study on the effects of the continuous use of de-icing salts in post-tensioned slabs and structures and the reduction of the life span, showing that the study of new alternatives to de-icing salts should be addressed for road maintenance tasks.

1.1. State of the Art

The need for the use of this type of salt, despite the negative chemical impact on concrete, is widespread throughout the world's infrastructure management. For this reason, many studies focus on the analysis of these salts, such as the one carried out by Sanchez Thomas et al., which analyzes the behavior of the main de-icing salts applied in infrastructure maintenance, modifying external conditions such as temperature or W/C ratio [2]. Other authors such as Guoju Ke et al. developed a decision procedure based on the analytic hierarchy process (AHP) to select the most suitable de-icing salt to be used [3]. Other outstanding studies detail the effect on the concrete using hardened cement paste modified with sodium silicate [4].

Regarding the study of post-tensioned concrete slabs, most published research focuses on a physical analysis of those elements, studying, for instance, the impact of loads on the structure. In this sense, research such as that of Youmn Al Rawi et al. stands out, as it analyzed the impact of loads produced by factors, such as rock fall during construction, comparing reinforced concrete slabs (RC) versus post-tensioned (PT) slabs [5]. Similar publications were presented by A. Jahami et al., which studied the load's impact on rehabilitated concrete slabs [6].

Focusing on a more economic aspect related to post-tensioned elements, there are also outstanding publications that detail an economic optimization of them, as is the case of the authors Yakov Zelickman and Oded Amir [7]. Such a study presents a methodology based on computational methods to save up to 50% of the budget and quantity of post-tensioned tendons [7].

However, over-cited studies outstand the ones related to analyzing de-icing salts and their effect on concrete. In this respect, authors such as Luping Tant et al. performed a long-term analysis of chloride penetration and reinforcement corrosion in highway concrete elements exposed since 1990 [8]. Other authors such as Meijie Xie et al. focused on the development of numerical models to predict chloride ingress in concrete subjected to atmospheric carbonation and de-icing salts usage [9]. In line with those investigations, authors such as Sara Al Haj Sleiman et al. study the performance of concrete in freeze–thaw filled exposure [10]. This study outlined the lack of normative methods related to concrete characterization in the presence of de-icing salts [10]. Authors such as Aref Ebrahimi et al. or Jan Deja focused on the concrete properties variation considering freezing and de-icing effects on them [11,12]. Special infrastructure types with outstanding exposure conditions are analyzed by authors such as Mullapudi, R., who focused on typical parking structure problems [13].

The current state of the art houses significant research focused on scientific aspects, as has been detailed previously. However, the novelty of this paper and its contribution to the state of the art relies on the engineering perspective that it provides. The reality of infrastructure management demands quick decision making based on inspections and expertise, with a lack of laboratory data analysis. In this sense, the paper aims to provide an overview of it and complement the research field, which is mostly focused on chemical or physical analysis.

1.2. Previous Experience

In the case of post-tensioned concrete slabs, chemical agents may gain entry into their sheaths, thus corroding their strands. This phenomenon has been observed by the investigation authors in several structures whose repair process was carried out in 2020 and was designed with concrete post-tensioned slabs without any waterproof treatment. In addition, these structures may also suffer an aggregate–alkali reaction causing the deterioration of the concrete.

In these structures, the most relevant damage was several longitudinal cracks observed on the upper face of their decks with numerous transversal broken rebars aligned with the mentioned cracks. Having studied the type of rupture of these rebars, apparently, without any previous elongation, it can be said that the brittle fracture was caused by a corrosion fatigue mechanism. Indeed, this process can be explained because of the combination of the rebars whose cross sections have been previously fragilized by the entrance of chloride ions through cracks with other mechanical concomitant loads (cyclic traffic loads and additional tensile load from concrete expansion) that together may produce the brittle fracture of the entire rebar cross-section.

2. Methodology

This paper focuses on the role of de-icing salt and its effects in a specific infrastructure located in Madrid. In order to archive this goal, various phases were followed. First, a complete description of the infrastructure must be done. Geometric and cross-section properties were defined, highlighting the structural typology and tendons characteristics. This information was relevant for the engineering team, in charge of the visual and special inspections, which are the next phases to commit.

Once the bridge was defined, the team proceeded with an initial visual inspection where the main damages were identified, including, for example, the ones related to the aggregate–alkali reaction. After that, a special inspection was carried out. In this sense, specific aspects such as the assessment of concrete mechanical characteristics, the microscopy analysis of concrete or the connectivity between venting pipes and prestressing sheaths were studied. The aforementioned phases are described as follows, including outstanding aspects that are necessary to obtain results underpinning the decision making.

The importance of periodic inspection phases is highlighted in this study. In this sense, there are different regulations that emphasize the definition of such inspections; for example, the guide to basic inspections of road works for the state road network of Spain or the guide for carrying out main inspections of crossing works on the state road network, both developed by the Ministry of Transport, Mobility and Urban Agenda (MITMA) [14,15]. These guides show three types of inspections: (1) basic, which could be assimilated as a visual inspection; (2) main inspection, or detailed visual inspection; and (3) special inspection. According to the referred guideline, maintenance operations and ordinary conservation activities are executed considering the results obtained from previous inspections [14,15]. Moreover, specific sheets for road maintenance are provided by the cited guide, securing the correct data collection during the inspection phases.

Infrastructure management is progressively gaining importance compared with design. Guides for basic inspection of road works or main inspections of crossing works and flyover bridges [14,15] were released in 2012, whereas the first version of the standard 3.1-IC of road layout design was released in 1939 [16]. The existence of specific regulations for infrastructure management denotes its growing importance in the civil engineering sector, which traditionally focused on the design and construction of new projects. The great social impact of maintenance has become as important as the design of the infrastructure itself. In this sense, the referred document shows the process followed to carry out maintenance in accordance with the indications of the existing regulations.

3. Infrastructure Inspection and Monitoring Carried Out

3.1. Infrastructure Description

The so-called Nudo de Colmenar Bridge (Madrid) connects the M-607 road (Madrid Way) and the M-40 ring road (A-1 Way). The structure is located in the Nudo de Colmenar detailed in Figure 1.

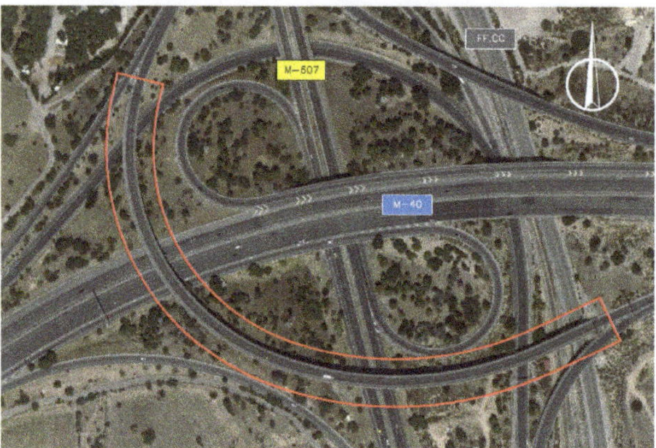

Figure 1. Structure location in the road junction Nudo de Colmenar.

The structure runs over several infrastructures, including the aforementioned highways M-40 ring road, M-607 road and connecting branches, as well as railway lines. It is a hyperstatic overpass whose typology is a continuous post-tensioning concrete slab. Its ground plan is curved. The structure has a total length of 561 m distributed in 17 spans with the following lengths: 20.5–32.2–35.6–37.9–56.4–32.9–36.7–39.0–28.8–29.6–36.5–28.6–30.3–32.9–33.1–28.4–21.6 m. These lengths were taken from the topographic survey carried out during the repair works made on the structure.

The cross-section is constant along the entire length of the deck. The slab has curved geometry with a maximum thickness of 1.40 m at the deck axis that is reduced to 0.20 m at the edge of the deck using a circular segment. The section is composed of a set of six circular lightning elements of variable diameter between 0.65 m and 1.00 m, as can be seen in Figure 2.

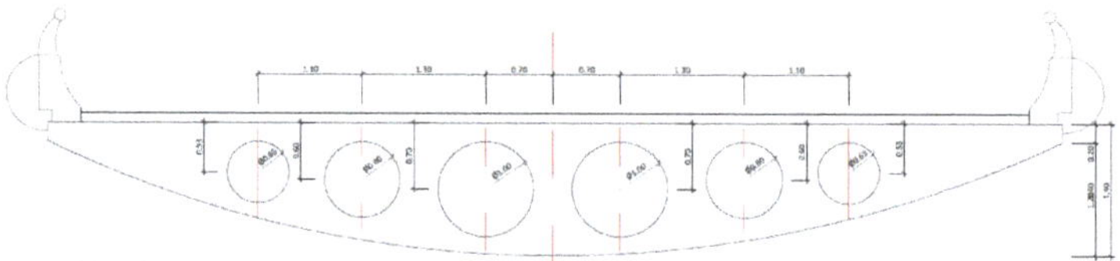

Figure 2. Cross-section of Nudo de Colmenar Structure.

The prestressing in the spans is not constant due to their different lengths, changing from 9 tendons of 15 Ø 0.6″ in the short spans to 16 tendons of 19 Ø 0.6″ in the longer spans.

3.2. Pathologies Detected in the Visual Inspection Initial Phase

The evaluation of the conservation status of a structure and the early detection of damage requires the realization of programmed and systematized inspections, which make it possible to know the structure's performance level at a specific time. Therefore, it is essential to define a conservation strategy that, in addition to other aspects, establishes the inspection levels and frequency with which they are carried out. With this purpose, different inspection levels can be defined according to the personnel that carry them out, the material resources used and the frequency with which they are executed.

The aim of this article is not to define the management of the structure's conservation. However, is important to highlight the importance of carrying out scheduled and systematized inspections, considering that they are a fundamental aspect of the early detection of pathologies and the starting point for the subsequent phases of study and evaluation.

In the case of the structure under study, the first alarm about the conservation status was detected within these programmed inspection works. During the main inspection (visual inspection by expert structural technicians, without special means of access and carried out every 4 years) excessive deflections in some spans and abundant cracking in the deck were noticed. In this way, the detection of the above damages within the programmed inspection works constitutes the first alarm about the state of conservation and gives rise to the subsequent evaluation phase. In this evaluation phase, a detailed visual inspection was carried out by engineers specializing in structural pathology, and a campaign of auscultation and testing was performed. The goal of these tasks was to determine the real scope of the pathologies detected and to assess their implications in the structural safety rate.

Thanks to the previously cited scheduled and systematized inspections, the identified damages revealed the existence of a serious pathological process were, fundamentally, the presence of a concrete degradation process by an aggregate–alkali reaction. It manifested itself in a series of particular damages, including the cut-off in the transverse reinforcing bars of the deck upper face and injection defects of the post-tensioning sheaths.

The process of concrete degradation by an aggregate–alkali reaction was manifested in the following observable damage:

Abundant longitudinal cracking on the bottom surface of the section, oriented parallel to the main prestressing compressions, can be seen in Figure 3.

Figure 3. Longitudinal cracking on the bottom surface of the deck.

Vertical marked deflections, especially in three spans: span # 5 (longest span over the M-40) and spans # 8 and # 11. The deformation of these spans pointed to a loss of concrete stiffness because of aggregate–alkali degradation. This deformation can be observed in Figure 4.

Figure 4. Marked deflections in spans # 5 and # 8 of Nudo de Colmenar structure.

Longitudinal cracks in the deck's upper face coincide with the direction of the prestressing. These cracks, detailed in Figure 5, were observed after milling the pavement.

Figure 5. Longitudinal cracks in the deck's upper face.

Moreover, in addition to the cracks observed on the deck's upper face, cores were extracted in the alignment of the cracks with internal fracturing of the concrete mass. In the visual examination of the cores extracted in the alignment of the fissures, the presence of a white halo around some aggregates was observed, as can be seen in Figure 6. It was compatible with the formation of silica gel of expansive nature, as a result of the aggregate–alkali reaction.

Figure 6. White halos around aggregates detected in the cores without microscopy needed.

Furthermore, sectional cuts in the transverse reinforcement bars were observed with a crack alignment coinciding with some of the longitudinal cracks and with a macroscopic appearance of brittle fracture, according to Figure 7. For the observation of the reinforcement, small test cuts were made in the alignment of the deck cracks.

Figure 7. Sectional cuts in the transverse reinforcement bars.

This type of brittle fracture is probably the most relevant damage observed in the structure under study, due to the extraordinary and unexpected nature of the phenomenon. As can be seen in Figure 7, necking is not observed in the breakage of the bars. In this sense, this fact points directly to a brittle fracture compatible with a corrosion fatigue phenomenon, which suggests the existence of a localized corrosion attack by chlorides coming from the winter maintenance treatments. Chlorides can easily penetrate inside the concrete mass through the existing cracks in the deck's upper face until they reach the reinforcement bars. Multiple factors such as the width or depth crack affect that penetration. However, the paper does not focus on internal chloride diffusion. Consequently, the modeling of chloride penetration is out of the scope of this paper, and we direct the reader to the listed references [17,18].

Lastly, signs of possible injection defects of the post-stressed sheaths were observed by examining the uncovered venting pipes in the milled area, by inserting a tape measure or a screwdriver inside them to check if they were clogged, as detailed in Figure 8. In this preliminary check it was observed that several of the venting pipes were not grouted up to the top face of the slab and that, therefore, it would be likely that there could be an inadequate injection of the prestressing sheaths. This led to the subsequent inspection with a borescope in the auscultation and testing campaign carried out in the structures.

Figure 8. Venting pipes check to verify the clogged conditions (see the handle of the screwdriver inserted into the hole).

In addition to the above pathologies, it should be noted that the presence of any waterproof treatment on the upper face of the deck was not observed. It certainly contributes to aggravating and accelerating any degradation process of the structure. Thus, the above damages, detected in the initial study phase of the state of conservation of the structure, pointed to the existence of a concrete process degradation by an aggregate–alkali reaction, which gave rise to several longitudinal cracks. Additionally, significant vertical deflections appeared because of the loss of stiffness of the concrete characteristic due to advanced stages of the process of arid–alkali degradation.

Additionally, and as will be seen below, the appearance of brittle failure of the upper transverse reinforcing bars with a failure alignment coinciding with the longitudinal cracks was observed. It suggests that the bar's failure is the consequence of a corrosion fatigue phenomenon. This phenomenon occurred as a result of chloride from the frequent use of de-icing salts that go into the slab cracks and the presence of cyclic traffic loads. In addition, the presence of wetting and drying cycles can affect the concrete internal movements of chlorides toward the surface [19].

The use of de-icing salts in post-stressed slabs without waterproof treatment could contribute to the progression of serious damages in the prestressing such as corrosion or even wires breakage. Although these defects were not observable in the initial inspection phases, they could be detected using borescope inspection in the following auscultation phases, as detailed in the following sections.

3.3. Special Inspection

The identification of the previous damage during the visual inspection gives rise to research on the structure to determine the extent of the damage and its real effect on the structural safety index. The research was based on the following process.

Determination of the concrete mechanical characteristics (modulus of elasticity and strength): analyzing its composition and making an electron microscopy observation, providing pictures like the ones shown in Figure 9. In both cases, the results confirmed the existence of an aggregate–alkali reaction in all the extracted cores.

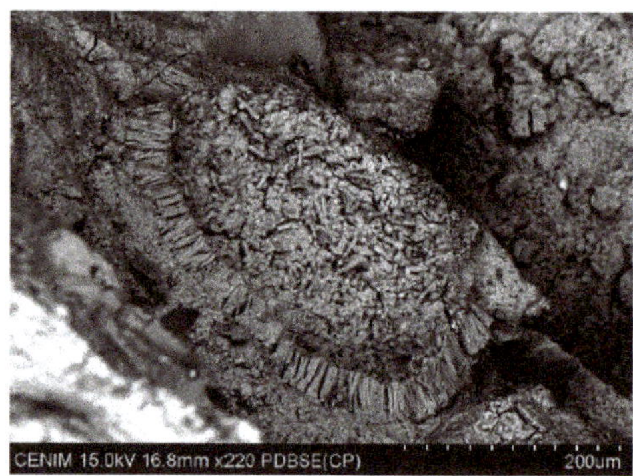

Figure 9. Appearance of the gels detected by microscopy.

In the case of the Colmenar Bridge, all the cores extracted showed, to a greater or lesser extent, fractured aggregates partially altered by chemical degradation processes, with the presence of pores and aggregate–lime interfaces clogged with whitish deposits. Likewise, a 54% average loss of elastic modulus was determined in the cores extracted.

Determination of the chloride content and other processes of a chemical attack on the concrete, observing a very high chloride content in some of the cores extracted (1.57% referred to concrete compared with 0.2% admissible according to the Spanish Structural Concrete Standard EHE-08).

In this sense, the analysis of chloride penetration is particularly important in the case of post-stressed concrete slabs because it can cause the corrosion of the post-tensioning strands, especially when these have injection deficiencies and other execution defects, as will be seen later on. The lack of an adequate waterproof treatment of the deck is also a determining factor in these cases.

In addition to the venting pipes check carried out in the initial phase and visual examination of some areas of prestressing elements, an inspection of the post-tensioning strands was carried out by using a borescope to evaluate the slurry and post-tensioning strand's condition.

The main damage detected in this inspection was some bad connections between the venting pipes and the sheaths (lack of drill holes in the sheath), incomplete filling of the sheath to a greater or lesser degree depending on the areas, leaving unprotected strands both in high points and adjacent sections, corroded and even broken strands due to a stress–corrosion process, slurry exudation and others. Figure 10 shows some examples of the previously mentioned damages. It was also observed that the layout elevations in high points were different from the design elevations, with a vertical displacement of the tendon axis with respect to the theoretical elevation up to 220 mm.

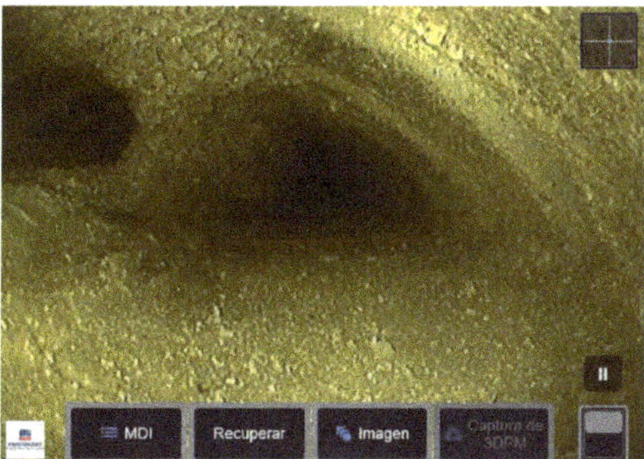

Figure 10. On the left, corroded cords and broken wires; on the right, unusual grout consistency with foam formation.

Defects that revealed malpractice in the execution of the prestressing, such as the presence of venting pipes simply supported on the sheath without any connection hole, were also observed, as shown in Figure 11.

Figure 11. Lack of connection between venting pipe and prestressing sheath (Colmenar Bridge).

4. Results and Evaluation

The detailed visual inspection carried out in conjunction with the research allowed for detecting and ratifying the existence of serious damages in the structure, which were concomitant with other execution defects. In this sense, the most relevant damage detected in the structure means a serious problem in post-stressed concrete slabs due to their implications on structural safety, are the following:

- Concrete degradation process due to aggregate–alkali reaction, chloride contamination because of the use of fluxing salts and lack of waterproofing of the deck;
- Systematic breakage of the transverse reinforcement due to a corrosion–fatigue process;
- Damage to the prestressing strands with a stress corrosion phenomenon and, consequently, the breakage of some wires.

The above damages mean a significant reduction of the safety conditions of the structure. However, although the structure under study had other minor defects, the previous ones have been highlighted because of the relation between them and the extraordinary and unexpected phenomenon of brittle fracture in transverse steel bars, which is described below.

Thus, the deterioration process of the structure began with a concrete sensitive to the aggregate–alkali reaction, that is, concrete with potentially reactive aggregates (reactive silica) that, in the presence of water and the alkalis from the cement paste, gave rise to the formation of an expansive gel that produces internal stresses in the concrete mass. In this sense, the lack of a waterproof treatment played a determining role in the activation of the aggregate–alkali reaction, since for this to occur it is necessary to provide water (usually from an external source); therefore, it is possible to slow down, or even stop this reaction by minimizing the entry of water into the structure, meaning with appropriate waterproofing.

The internal stresses that occur in the concrete as a result of the aggregate–alkali reaction, caused a high number of cracks, that in a massive element or idealized model would form a homogeneous and multidirectional cracking network, but in resistant elements such as bridge decks, the orientation of these cracks is conditioned by the geometry of the element and the orientation of the stresses. In the case of the structure presented in this article, cracks showed a parallel direction to the compression isostatics lines.

The longitudinal cracks observed with large crack openings (over normative limits) together with the lack of an adequate waterproof treatment cause the entry of chlorides from the de-icing salts used during winter into the structure, easing the corrosion of the passive bars and post-tensioning strands.

The bars of the transverse reinforcement showed full-section ruptures with a brittle appearance. In this sense, although the breakage of steel elements by a stress corrosion process is certainly a common phenomenon, in the case of the Colmenar Bridge, these breaks are completely different from the usual forms of breakage due to chloride contamination, which appears as local corrosion that progresses inwards the bar in an inverted cone form.

In the structure covered by the paper, the bars showed a macroscopic appearance of brittle fracture that pointed to the presence of mechanical actions concomitant with chloride attack. Although the breaks detected in the bars are still being analyzed using tests in specialized laboratories, the most likely hypothesis is that it is a corrosion–fatigue phenomenon, in which the steel bars breakage was caused by the simultaneous action of cyclic stress and the presence of an aggressive medium that caused localized corrosion of the steel. Indeed, the presence of cyclic loads associated with traffic on the structure would lead to the crack progression in the pitting of the steel bars caused by the chlorides. In this case, unlike stress corrosion processes, in which the presence of a permanent—or static—stress is necessary for material rupture to occur, in corrosion–fatigue processes this stress must be cyclic. In the case studied, it does not seem reasonable to attribute the rupture of the bars to a stress corrosion phenomenon, since they are not bars with a high level of stress and, in this case, the rupture plane would have the appearance of a quasi-fragile rupture.

Finally, the entry of chlorides through the longitudinal cracks, aggravated by the lack of a waterproof treatment and the injection defects of the post-tensioning sheaths, caused corrosion to reach the strands in certain sections, even causing the break of some of them due to stress corrosion cracking. It should also be noted that in any prestressing inspection, there is a high degree of uncertainty about the reliability of the preservation state since it is an element that is not accessible for inspection at 100% of its length and, therefore, it is necessary to consider this uncertainty in the assessment of the structure preservation state and the final repair proposal.

5. Discussion: Strategies for Prevention and Repair

The above pathologies are a serious problem in post-stressed slab bridge decks. However, for these pathologies to occur, a confluence of several factors is necessary, as is detailed in Figure 12.

The absence of adequate waterproofing to prevent the entry of water into the structure and the penetration of chlorides from the use of de-icing salts played a determining role in the progress of the above pathologies. In this sense, in addition to the presence of reactive silica and alkalis, the presence of water coming from an external source was necessary for the aggregate–alkali reaction to occur. The reaction would cease when one of the above reagents is consumed.

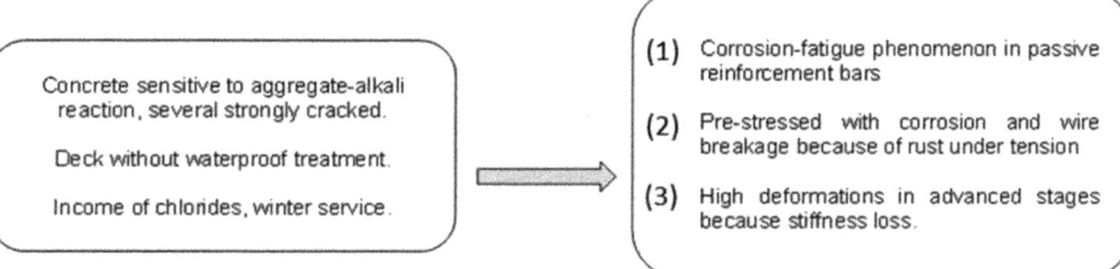

Figure 12. Damage and causal factors.

Additionally, the corrosion processes of the passive reinforcement and post-tensioning strands need the presence of an aggressive medium to prosper. In this case, the absence of a waterproofing treatment on the deck facilitated the penetration of chloride ions into the structure, aggravating this process in the previously cracked areas of the deck, where the corrosion rate increased due to the high percentage of chloride entering through the cracks.

Considering the above, it is simple to deduce that an adequate execution of waterproofing in the structure execution phase is a determining factor to avoid the progression of the above pathologies. However, in existing structures with a lack of or deficient waterproofing in which the damage has already progressed and led to pathological evolution, repair strategies based on the restitution of the affected areas and the complete encapsulation of the structure should be chosen. In this case, an appropriate repair strategy would consist of the following elements or phases:

- Hydro-demolition of the deck until the affected transverse reinforcement is revealed. Replacement of the reinforcement and reconstruction with fine aggregate concrete up to the initial level;
- Reinjection of prestressing sheaths by vacuum injection using a fluid slurry with corrosion inhibitor improves their durability significantly;
- Execution of an ultra-high performance concrete layer (UHPCFR) for the encapsulation of the structure of approximately 3 cm to improve structure performance in terms of durability due to its high compactness and the almost total absence of pores. In addition, it provides a further improvement of the structure's performance against transverse bending due to the high performance (up to 150 MPa compression and 12 MPa flexural bending) and multidirectional strength behavior.

The case of the Nudo de Colmenar Bridge is detailed in the present paper. As a professional project, the paper outlines a decision-making process based on engineering aspects rather than pure scientific data or laboratory results. In the practice of such projects, the decision-making process is commonly done without reduced scientific data support due to a lack of time. This fact is reflected as a limitation of the present paper, where aspects such as brittle fracture are determined by previous experience. Further studies would support brittle fracture or arid–alkali degradation with laboratory analysis results.

Due to the extension and evolution of the pathologies detected, and also the serious damage in the prestressing, it was concluded that the most reasonable alternative was the complete demolition of the bridge and subsequent reconstruction with a new prefabricated concrete deck. Real images of the described process are shown in Figure 13.

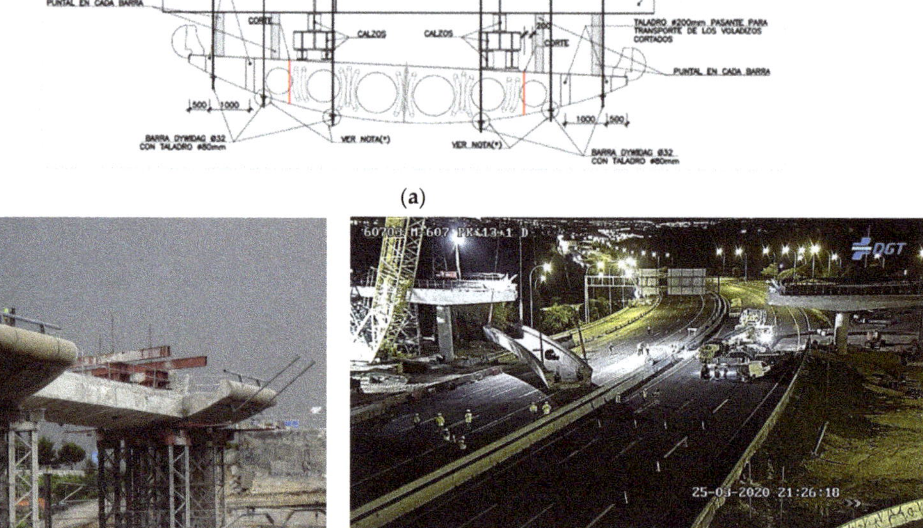

Figure 13. Dismantling of existing deck and assembly of new deck over M 40 road. (**a**) Section of the bridge, (**b**) real image of bridge, and (**c**) real image of operation process.

6. Conclusions

The pathologies observed in the structure constitute a serious problem that has progressed due to the confluence of several factors. The lack of waterproof treatment of the structure facilitated the chloride ions penetration from the use of de-icing salts as well as the degradation of the concrete by the aggregate–alkali reaction and the corrosion processes of the passive reinforcement and the prestressing.

In this degradation process, the internal stresses caused by the aggregate–alkali reaction led to the appearance of longitudinal cracks on the upper face of the deck. De-icing salts concentrated and went into these cracks, making easier the corrosion processes of the passive reinforcement and the prestressing cables. The propagation of this aggressive medium together with the concomitant mechanical loads caused brittle breaks in the transversal reinforcement bars. In this respect, although research is currently being carried out in specialized laboratories to determine the most probable cracking mechanisms, everything points to a corrosion fatigue phenomenon, in which the cyclic traffic loads were what induce the propagation of the cracking front with the added tensional load of the expansion of the concrete due to the aggregate–alkali reaction.

Furthermore, although this pathology has only recently been observed in the structure presented in the article, it is not an unusual phenomenon, but it has been observed in other bridges of a similar typology as well (as a result of identifications made in the Colmenar Bridge).

The above pathologies are the result of the confluence of several factors. However, water presence plays a determining role in the deterioration process. In this sense, an appropriate conservation strategy involves minimizing water entry into the structure throughout its service life. For this reason, the execution of high-performance waterproof treatments during the construction phases is a key aspect to avoid the progression of the previous pathologies and their effects on structural safety which, in cases such as the present Colmenar Bridge case, can be irreversible. Similarly, in existing structures without waterproofing in which these pathologies have already progressed, it is possible to slow down the deterioration rate by adopting measures aimed at restoring the deteriorated areas and encapsulating the structure using ultra-high-performance concretes. However, in this case, it will be essential to carry out a research and testing campaign to determine the magnitude, area and evolution degree of the pathologies detected. It will be used to assess the uncertainty about the reliability of the conservation status of the entire length of the prestressing since it is an essential element for the structural safety of the bridge but inaccessible for the complete inspection.

Author Contributions: Conceptualization, T.L.R.A., M.G.A. and N.C.M.; methodology, M.G.A., J.C.G. and R.M.P.; software, M.G.A., J.C.G. and R.M.P.; validation, M.G.A., J.C.G. and R.M.P.; formal analysis, M.G.A., J.C.G. and R.M.P.; investigation, T.L.R.A. and N.C.M.; resources, M.G.A., J.C.G. and R.M.P.; data curation, T.L.R.A. and N.C.M.; writing—original draft preparation, T.L.R.A. and N.C.M.; writing—review and editing, M.G.A., J.C.G. and R.M.P.; visualization, M.G.A., J.C.G. and R.M.P.; supervision, M.G.A., J.C.G. and R.M.P.; project administration, T.L.R.A. and N.C.M.; funding acquisition, T.L.R.A., M.G.A. and J.C.G. All authors have read and agreed to the published version of the manuscript.

Funding: This research received no external funding.

Institutional Review Board Statement: Not applicable.

Informed Consent Statement: Not applicable.

Data Availability Statement: The data presented in this study are available on request from the corresponding author.

Conflicts of Interest: The authors declared no conflict of interest.

References

1. Angst, U.M. Challenges and opportunities in corrosion of steel in concrete. *Mater. Struct.* **2018**, *51*, 4. [CrossRef]
2. Sanchez, T.; Conciatori, D.; Keserle, G.C. Influence of the type of the de-icing salt on its diffusion properties in cementitious materials at different temperatures. *Cem. Concr. Compos.* **2022**, *128*, 104439. [CrossRef]
3. Ke, G.; Zhang, J.; Tian, B. Evaluation and selection of de-icing salt based on multi-factor. *Materials* **2019**, *12*, 912. [CrossRef] [PubMed]
4. Skripkiūnas, G.; Nagrockienė, D.; Girskas, G.; Janavičius, E. Resistance of modified hardened cement paste to frost and de-icing salts. *Balt. J. Road Bridge Eng.* **2012**, *7*, 269–276. [CrossRef]
5. Al Rawi, Y.; Temsah, Y.; Baalbaki, O.; Jahami, A.; Darwiche, M. Experimental investigation on the effect of impact loading on behavior of post-tensioned concrete slabs. *J. Build. Eng.* **2020**, *31*, 101207. [CrossRef]
6. Jahami, A.; Temsah, Y.; Khatib, J.; Baalbaki, O.; Darwiche, M.; Chaaban, S. Impact behavior of rehabilitated post-tensioned slabs previously damaged by impact loading. *Mag. Civ. Eng.* **2020**, *93*, 134–146.
7. Zelickman, Y.; Amir, O. Optimization of post-tensioned concrete slabs for minimum cost. *Eng. Struct.* **2022**, *259*, 114132. [CrossRef]
8. Tang, L.; Boubitsas, D.; Huang, L. Long-term performance of reinforced concrete under a de-icing road environment. *Cem. Concr. Res.* **2023**, *164*, 107039. [CrossRef]
9. Xie, M.; Dangla, P.; Li, K. Reactive transport modelling of concrete subject to de-icing salts and atmospheric carbonation. *Mater. Struct.* **2021**, *54*, 240. [CrossRef]
10. Al Haj Sleiman, S.; Izoret, L.; Alam, S.Y.; Grondin, F.; Loukili, A. Freeze–thaw field exposure and testing the reliability of performance test temperature cycle for concrete scaling in presence of de-icing salts. *Mater. Struct.* **2022**, *55*, 2. [CrossRef]
11. Besheli, A.E.; Samimi, K.; Nejad, F.M.; Darvishan, E. Improving concrete pavement performance in relation to combined effects of freeze-thaw cycles and de-icing salt. *Constr. Build. Mater.* **2021**, *277*, 122273. [CrossRef]
12. Deja, J. Freezing and de-icing salt resistance of blast furnace slag concretes. *Cem. Concr. Compos.* **2003**, *25*, 357–361. [CrossRef]
13. Mullapudi, R. Typical Parking Structure Problems, Repairs, and Cost Assessment. In Proceedings of the Ninth Congress on Forensic Engineering, Denver, CO, USA, 4–7 November 2022; pp. 960–968.
14. Guía de Inspecciones Básicas de Obras de Paso. Available online: https://cdn.fomento.gob.es/portal-web-drupal/inspecciones_obras_paso.pdf (accessed on 1 May 2023).
15. Guía Para la Realización de Inspecciones Principales de Obras de Paso en la Red de Carreteras del Estado. Available online: https://www.mitma.gob.es/recursos_mfom/0870250.pdf (accessed on 1 May 2023).
16. Norma 3.1-IC de la Instrucción de Carreteras. Trazado. Available online: https://apps.fomento.gob.es/CVP/handlers/pdfhandler.ashx?idpub=ICW050 (accessed on 1 May 2023).
17. Guzmán, S.; Gálvez, J.C.; Sancho, J.M. Modelling of chloride ingress into concrete through a single-ion approach. Application to an idealized surface crack pattern. *Int. J. Numer. Anal. Methods Geomech.* **2014**, *38*, 1683–1706. [CrossRef]

18. Guzmán, S.; Gálvez, J.C.; Sancho, J.M. Cover cracking of reinforced concrete due to rebar corrosion induced by chloride penetration. *Cem. Concr. Res.* **2011**, *41*, 893–902. [CrossRef]
19. Bernal, J.; Fenaux, M.; Moragues, A.; Reyes, E.; Gálvez, J.C. Study of chloride penetration in concretes exposed to high-mountain weather conditions with presence of deicing salts. *Constr. Build. Mater.* **2016**, *127*, 971–983. [CrossRef]

applied sciences

Article

Exploring Design Optimization of Self-Compacting Mortars with Response Surface Methodology

Stéphanie Rocha [1], Guilherme Ascensão [2,*] and Lino Maia [1,3]

[1] CONSTRUCT-LABEST—Laboratory for the Concrete Technology and Structural Behaviour, Faculdade de Engenharia, Universidade do Porto, Rua Dr. Roberto Frias, 4200-465 Porto, Portugal; up202010607@g.uporto.pt (S.R.); linomaia@fe.up.pt (L.M.)
[2] RISCO, Department of Civil Engineering, University of Aveiro, 3810-193 Aveiro, Portugal
[3] Faculty of Exact Sciences and Engineering, University of Madeira, Campus da Penteada, 9020-105 Funchal, Portugal
* Correspondence: guilhermeascensao@ua.pt; Tel.: +351-966944248

Abstract: The ever-evolving construction sector demands technological developments to provide consumers with products that meet stringent technical, environmental, and economic requirements. Self-compacting cementitious mixtures have garnered significance in the construction market due to their enhanced compaction, workability, fluidity, and mechanical properties. This study aimed to harness the potential of statistical response surface methodology (RSM) to optimize the fresh properties and strength development of self-compacting mortars. A self-compacting mortar repository was used to build meaningful and robust models describing D-Flow and T-Funnel results, as well as the compressive strength development after 24 h (CS24h) and 28 days (CS28d) of curing. The quantitative input factors considered were A (water/cement), B (superplasticizer/powder), C (water/powder), and D (sand/mortar), and the output variables were Y1 (D-Flow), Y2 (T-Funnel), Y3 (CS24h), and Y4 (CS28d). The results found adjusted response models, with significant R^2 values of 87.4% for the D-Flow, 93.3% for the T-Funnel, and 79.1% for the CS24h. However, for the CS28d model, a low R^2 of 39.9% was found. Variable A had the greatest influence on the response models. The best correlations found were between inputs A and C and outputs Y1 and Y2, as well as input factors A and D for responses Y3 and Y4. The resulting model was enhanced, thereby resulting in a global desirability of approximately 60%, which showcases the potential for the further refinement and optimization of RSM models applied to self-compacting mortars.

Keywords: self-compacting mortars; design of experiments; fresh properties; compressive strength; ANOVA

1. Introduction

Concrete is one of the main elements used in civil construction, due to its variability in use, which only intensifies with urban development. However, concrete production heavily relies on natural resource consumption, including the fundamental constituents for cement production and the extraction of coarse and fine aggregates, thereby causing deleterious environmental impacts [1,2].

Self-compacting concrete (SCC) was developed to improve concrete fluidity and self- consolidation properties, thus obviating the requirement for external compaction elements while also fostering enhanced mechanical properties [3–6]. When producing self-compacting cement-based products, it is recommended to reduce the proportion of coarse aggregates and to increase the amount of fine aggregates and cement to reduce the risk of segregation and increasing properties such as viscosity, void filling, and stability, in addition to mechanical properties [7]. However, such a mix design approach may be prone to cracking and shrinkage, in addition to increased production costs [5,8].

The use of mineral and chemical additives in SCC aims to increase durability and workability while reducing production costs [9,10]. Moreover, contemporary architecture

Citation: Rocha, S.; Ascensão, G.; Maia, L. Exploring Design Optimization of Self-Compacting Mortars with Response Surface Methodology. *Appl. Sci.* **2023**, *13*, 10428. https://doi.org/10.3390/app131810428

Academic Editors: Francisco B. Varona Moya and Mariella Diaferio

Received: 28 August 2023
Revised: 13 September 2023
Accepted: 16 September 2023
Published: 18 September 2023

demands slender structures of highly intricate and complex geometry, which increase the demand for innovative self-leveling mortars and concrete with a high strength development [10–12]. However, the design of high-strength cement-based materials with self-compacting attributes is challenges due to inherent contradictions in mixture design requirements. For instance, achieving a high early strength typically requires low water/binder ratios (w/b), which, conversely, reduce the mixture's self-compacting ability. Therefore, traditional mixed design approaches may prove inadequate when addressing the demanding requirements of self-compacting mortars. Data-driven mixed design methodologies may offer a promising alternative method to gain valuable insights on how to enhance the fresh properties of self-compacting mortars while maximize their strength development. Design of experiments (DoE) is a statistical tool that has been used in the optimization of materials, and its advantages include correlating the investigated variables, finding an optimal response within an investigation region, reducing the number of required experimental trials, and even defining optimized models according to predefined parameters [13,14]. The response surface methodology (RSM) can be used to find relationships between input and output variables and to define the optimization criteria among variables [15–18].

In recent decades, investigations using statistical tools in concrete and mortar mix design have been intensified. Research has applied the RSM to maximize the replacement of cement or aggregates with—among others—pumice stone [1], foundry sand [18], plastic waste and silica fume [19], and hybrid steel fibers [20]. The authors developed high-significance models, thereby making it possible to optimize the performance of cementitious mixtures while reducing the use of prime raw materials. Other studies were undertaken with the primary objective of improving the environmental and economic performance of cementitious materials [21–38].

More recently, the use of statistical methodologies has gained attention in the realm of self-compacting cementitious materials. Its application extends to various domains, thus encompassing the assessments of rheological effects and mechanical properties, as well as the mitigation of early-stage cracking. Li et al. [39] employed regression models, utilizing central composite design (CCD), to examine the influence of four mixed design variables (fly ash content (FA), silica fume content (SF), sand–binder ratio (s/b), and water–binder ratio (w/b)) on the rheological and mechanical properties of concrete. The responses measured included mini slump flow, mini V-funnel results, and compressive strength after 28 days of curing. The authors applied multiresponse optimization to determine the optimal ranges for each compositional variable (FA: 10–20 wt%, SF: 6–10 wt%, s/b: 1.1–1.2, w/b: 0.35–0.36). Similarly, Safhi et al. [40] investigated the feasibility of using treated marine sediments as cement replacements in self-compacting mortars. The authors' findings indicated that treated marine sediments led to reductions in the mechanical performance (the elastic modulus and compressive strength) while having no significant deleterious effects on the rheological properties. The authors prescribed a sediment-to-cement ratio ranging from 0.0703 to 0.3462 to maintain acceptable levels of mechanical performance loss while benefiting from the reduced environmental impact and production costs.

Conversely, Matos et al. [28] focused on addressing the early-stage cracking proclivity of self-compacting ternary white mortars. The authors found that quadratic models were suitable for adequately describing the mortar's properties, and subsequent mix design optimization steps allowed for reductions in segregation (the T-Funnel time increased from 8.5 to 9.8 s) and yearly shrinkage (which reduced from 558×10^{-6} to 540×10^{-6}) that significantly minimized the risk of cracking.

In a prior publication [41], the authors comprehensively characterized the performance of self-compacting mortars, which they made publicly available in a substantial dataset encompassing a wide array of rheological and mechanical properties.

The objective of the study herein is to advance the data analysis leveraging the statistical tool of RSM through a central composite design to define, adjust, and optimize models for high-performance self-compacting mortars a based upon the aforementioned experimental dataset.

2. Materials and Methods

The performance of self-compacting mortars was evaluated using published datasets, which were collected during the authors' previous experimental studies [41].

The mortar specimens were produced using CEM I 42.5 R cement, limestone filler, normalized sand, and a polycarboxylate-based superplasticizer. The cement was purchased from Secil Portugal (Outão, Portugal) and was compliant with EN 197-1 specifications. Limestone filler with specific gravity of 2.68 g/cm^3 was provided by Omya S.A under the commercial reference of Betocarb 10 HP—OU. Normalized sand compliant with EN 196-1 was used in all experiments, with a water absorption of 0.30% and specific gravity of 2.63 g/cm^3. The selected superplasticizer (ViscoCrete®-20 HE, from SIKA, Vila Nova de Gaia, Portugal) was characterized by a density of 1.08 g/cm^3 and a solid content of 40%.

The dataset includes a total of thirty formulations of self-compacting mortars with four quantitative input variables. A central composite design was defined using Design–Expert software (Stat-Ease, Inc., Minneapolis, MN 55413-2561, USA—Design–Expert® Software, version 13.0.9.0 64-bit; Serial Number 0964-0841-3719-3394) consisting of a factorial design of four factors in two levels (2^4) representing 8 axial and 6 central realizations. The four input variables analyzed included the following—A: water-to-cement ratio (w/c); B: superplasticizer-to-powder (CEM I 42.5 R + limestone filler) ratio (Sp/p); C: water-to-powder (CEM I 42.5 R + limestone filler) ratio (w/p); and D: sand-to-mortar ratio (s/m). All ratios refer to volumetric relations. The evaluation levels were $-\infty$, -1, 0, $+1$, and $+\infty$, with ∞ being equal to 2. Table 1 shows the relation between coded points and real values.

Table 1. Equivalence of coded and real values (reproduced from [41]).

Levels	A: w/c	B: Sp/p	C: w/p	D: s/m
-2	0.78741	0.02069	0.46929	0.42240
-1	0.84110	0.02210	0.50129	0.45120
0	0.89478	0.02351	0.53328	0.48000
$+1$	0.94847	0.02492	0.56528	0.50880
$+2$	1.00216	0.02633	0.59728	0.53760

Workability and strength development were evaluated as outputs. Four response variables were defined and examined, namely, these included the following: Y1: D-Flow results (in mm); Y2: T-Funnel (in seconds); and compressive strength measured after 1 and 28 days of curing (Y3 and Y4, respectively, both expressed in MPa). The D-Flow and T-Funnel testing were conducted following EFNARC specifications and guidelines for self-compacting concrete, whereas compressive strength was determined as per EN 196-1. The detailed description of the testing protocols can be found elsewhere [41]. For compressive strength measurements, a minimum of four specimens was tested with respect to formulation and curing age. The average strength values were considered representative, and only those values are reported here. All experiments have been conducted in randomized order. Table 2 shows the coded input values (A–D) and the average result of each individual response (Y1–Y4).

Table 2. Coded input variables and results for response factors (adapted from [41]).

Std	Run	Coded Values				Results			
		A	B	C	D	Y1	Y2	Y3	Y4
1	11	−1	−1	−1	−1	325	22	62.7	115.8
2	29	1	−1	−1	−1	341	18	55.7	108.0
3	21	−1	1	−1	−1	325	21	62.7	112.1
4	26	1	1	−1	−1	359	15	57.7	114.6
5	9	−1	−1	1	−1	361	16	60.6	114.5
6	1	1	−1	1	−1	377	12	53.4	109.0
7	27	−1	1	1	−1	368	13	58.2	115.8
8	23	1	1	1	−1	370	13	55.6	105.1
9	22	−1	−1	−1	1	229	108	61.4	104.1
10	18	1	−1	−1	1	318	31	55.5	106.4
11	6	−1	1	−1	1	309	162	62.9	111.2
12	30	1	1	−1	1	308	30	53.9	101.8
13	2	−1	−1	1	1	304	29	62.0	112.5
14	5	1	−1	1	1	344	19	54.6	106.9
15	17	−1	1	1	1	328	24	61.7	111.4
16	8	1	1	1	1	342	18	56.9	112.2
17	25	−2	0	0	0	270	45	62.9	103.9
18	20	2	0	0	0	345	16	51.3	106.0
19	10	0	−2	0	0	329	21	58.4	117.7
20	24	0	2	0	0	349	17	56.7	115.4
21	15	0	0	−2	0	306	40	60.4	113.5
22	16	0	0	2	0	358	13	57.4	114.0
23	14	0	0	0	−2	370	12	57.5	114.6
24	19	0	0	0	2	282	52	58.8	110.0
25	4	0	0	0	0	342	18	60.2	108.2
26	7	0	0	0	0	339	21	58.7	107.2
27	13	0	0	0	0	332	19	58.5	111.7
28	12	0	0	0	0	348	17	56.1	113.8
29	3	0	0	0	0	338	19	59.2	117.7
30	28	0	0	0	0	338	24	61.7	*

*—Not possible to be measured.

3. Results

3.1. Preliminary Data Analysis

Table 3 shows the minimum, maximum, mean, standard deviation, and coefficient of variation values of the investigated response variables. The collected data was homogenous and presented low standard deviation and coefficients of variation that indicate concentration around the mean values. However, this pattern was not observed in the flow results obtained through the T-Funnel test, which displayed a more dispersed distribution, which was reflected through a higher coefficient of variability (CV = 106%).

Table 3. Responses of 30 self-compacting concrete mixes.

Levels	Y_1	Y_2	Y_3	Y_4
	D-Flow (mm)	T-Funnel (s)	CS24h * (MPa)	CS28d * (MPa)
Minimum	229	12	51.3	101.8
Maximum	377	162	62.9	117.7
Mean	332	29	58.4	110.9
Std. Dev.	32	31	3.1	4.4
CV (%)	10	106	5.3	4.0

* CS24h and CS28d stand for compressive strength after 1 and 28 days of curing, respectively.

Figure 1 shows the heatmap of the correlations between the input and output variables. The degree of interaction between the variables is shown according to color. Positive correlations are indicated in shades of red, whereas negative correlations are indicated in shades of blue. Strong correlations can be observed between A and Y3 (−0.878), Y1 and Y2 (0.642), and D and Y2 (−0.618). However, a significant number of moderate positive correlations, ranging between ±0.363 and ±0.465, primarily existed and comprised Y1 and

Y2 variables. It is crucial to assess whether these correlations have an impact on the behavior of the self-compacting mortars, and therefore should be reflected in response models.

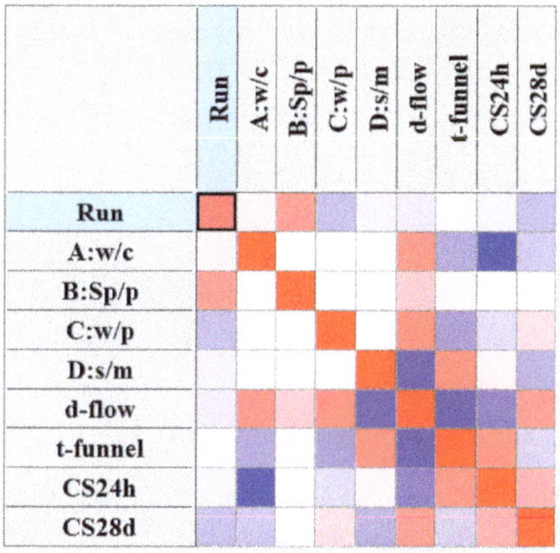

Figure 1. Heatmap of correlations between input and output variables. Darker colors indicate stronger correlations, with red and blue differentiating positive and negative relationships, respectively.

3.2. Fresh Properties of Self-Compacting Mortars

Regression Model for D-Flow

The regression analysis has shown that the quadratic models exhibited a superior fit. Notably, they presented an adjusted R^2 of 84.2% and a predicted R^2 of 76.0%, both of which had discrepancy values of less than 20%. However, the analysis of variance (ANOVA) revealed a "Lack of fit F-value" of 7.04, along with an associated *p*-value of 0.02. Consequently, there was only a 2% calculated likelihood that the lack of fit F-value could be attributed to random noise; it was more likely a result of the overall model's shortcomings. Hence, these findings highlighted the necessity for refinement of the polynomial model to ensure its soundness and validity. Figure 2 shows the normal probability of the residuals (a) and the residuals versus the run plot (b) as part of the data diagnostic analysis. These graphical representations provide valuable insights into the behavior of the residuals from the regression model. Notably, an evident outlier was observed in data point Std 9, Run 22, which exhibited significant deviations in the Y1 results compared to the remaining realizations. Similar outlier patterns were also noticed in relation to the Y2 output variable. In addition, Cook's distance, DFFITS, and DFBETAS, which are commonly used measures to assess the influence of individual data points on regression models (not shown here for the sake of brevity), indicate that this particular realization significantly departed from the majority of data points, thereby demonstrating its status as an outlier. The exact cause for such behavior remains unknown; however, it is most likely attributed to experimental variability during the preparation or testing of the mortar specimens. As a consequence, to ensure the robustness of the analysis, this outlier data point was excluded from the dataset, and a subsequent run was conducted that considered the revised dataset.

After excluding the outlier data point, a second run was conducted that demonstrated that the linear regression models presented better fitting than the quadratic models.

The analysis of variance (ANOVA) yielded adjusted R^2 values of 78.6% and a predicted R^2 of 71.3%, which can be considered to be fairly reasonable results, with differences being smaller than 20%. Despite its significance, the linear model F-value (6.16) indicated a suboptimal fitting. Therefore, the data diagnostic analysis was repeated to ensure the reliability of the analysis and identify any remaining outliers.

Figure 3 displays that the Std 17, Run 25 data point exhibited an outlier profile similar to the previously excluded Std 9 data point. The Cook's distance, DFFITS, and DFBETAS plots further confirmed the classification of Std 17, Run 25 as an outlier, although they are not shown here for the sake of brevity.

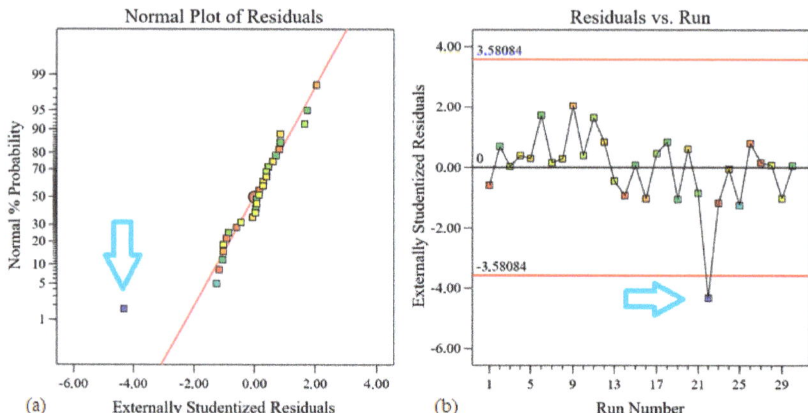

Figure 2. Normal probability of residuals (**a**) and Residuals versus Run plot (**b**) for D-Flow model. Arrows included to highlight outlier data point.

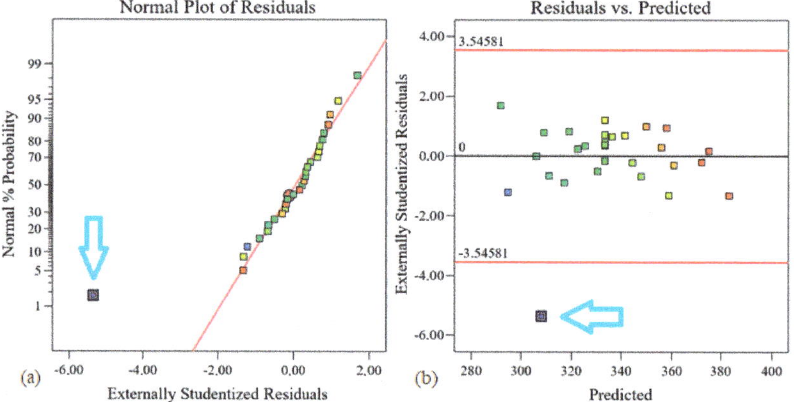

Figure 3. Plot of the normal probability of residuals for the D-Flow model (**a**) and Residuals versus Predicted plot for the D-Flow model (**b**); both plotted after the initial fitting.

Analogously to the previous case, the outlier was also excluded, and a new analysis was conducted, wherein the linear model yielded satisfactory results. The key performance indicators of the model are presented in Table 4. The model proved to be statistically significant, as evidenced by a *p*-value of less than 0.0001 and an F-value of 47.70. Furthermore, terms A, B, C, and D were also significant, with values lower than 0.1. Particularly noteworthy are the superior outcomes observed for terms C and D, which corroborates the model's nonreduction. The F-value was 2.74, thus implying that the associated error was not statistically significant. The probability of such an error being attributed to noise surpassed 13.45%, thereby confirming the model's suitability for use.

The model's statistical fitting demonstrated enhanced performance, with the predicted R^2 value reaching 83.15% and the adjusted R^2 reaching 87.37%. The adequate precision value was found to be 24.03 (>>4.0), thus further indicating a robust relationship between the signal and noise and the model discrimination. In addition, the standard deviation was reduced to 8.14, which was accompanied by a coefficient of variation of 2.41%. Therefore, the model effectively described the design space, and Table 5 shows the estimated coefficients that were obtained. These coefficients were derived from fits around the overall average of all the runs in an orthogonal design, and the variance inflation factor (VIF) exceeding 1.0 demonstrates that the factors A, B, C, and D were found to be multicollinear.

Table 4. Fitting results for D-Flow fitted models.

Source	Sum of Squares	Mean Square	F-Value	*p*-Value
Model	12,633.5	3158.4	47.7	<0.0001
A-w/c	1266.9	1266.9	19.1	0.0002
B-Sp/p	314.3	314.3	4.8	0.0399
C-w/p	4079.9	4079.9	61.6	<0.0001
D-s/m	8303.5	8303.5	125.4	<0.0001
Residual	1523.0	66.2		
Lack of Fit	1382.7	76.8	2.74	0.135
Pure Error	140.4	28.1		
Cor Total	14,156.5	-		

Table 5. Coefficients of coded factors—D-Flow.

Factor	Intercept	A—w/c	B—Sp/p	C—w/p	D—s/m
Coefficient Estimate	335.32	8.27	3.71	13.38	−19.09
Standard Error	1.56	1.89	1.70	1.70	1.70
95% CI Low	332.10	4.36	0.19	9.85	−22.62
95% CI High	338.54	12.19	7.24	16.91	−15.56
VIF	-	1.01	1.01	1.01	1.01

Figure 4 shows the normal probability of the residuals (a), the Residuals versus Run plot (b), and the Predicted versus Actual plot (c). The normal probability of the residuals exhibited a linear trend, thereby suggesting a favorable model fit. In addition, a desirable random dispersion pattern is observed, with all data points falling within the specified limits (±3.55) (Figure 4b). A strong correlation can be observed between the predicted and actual values, with the data points exhibiting a clear tendency to align in a preferential direction. Also located are the excluded points (Std 9, Run 15 and Std 17, Run 25) in the RSM (Figure 4c).

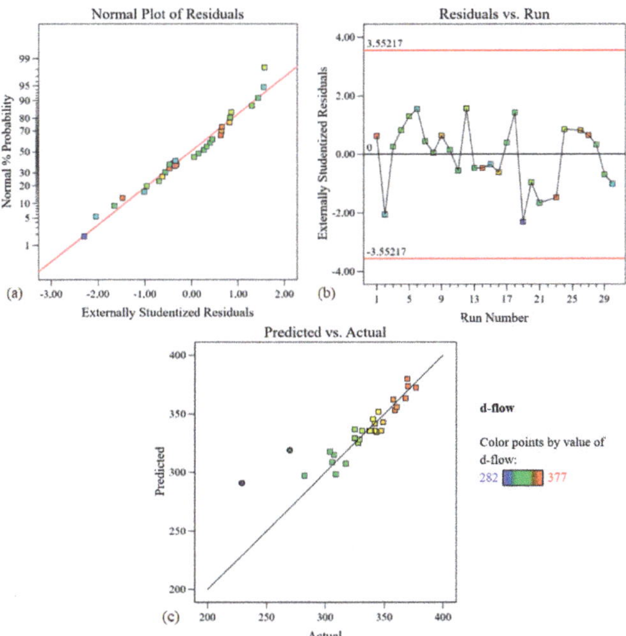

Figure 4. Graphs for D-Flow model: (**a**) Normal Plot of Residuals; (**b**) Residuals versus Run; and (**c**) Predicted versus Actual.

3.3. Regression Model for T-Funnel

The best response model for the T-Funnel was the 2FI (two-factor interaction), which obtained a predicted R^2 of 15.58% and an adjusted R^2 of 69.72%, which are values that are considered significant. The percentage difference between the predicted and adjusted values was 54.14% (a value well above 20%), which is not recommended, as the errors are expected to be significant.

The model, despite being considered significant in its best response, had a significant lack of fit, which is not ideal, as there are high chances of errors. An inverse transform was recommended by the Box–Cox plot diagnostic to improve the overall model fit. A lambda (λ) equal to -1 was considered within the 95% confidence interval, with its limits ranging between of -1.55 and -0.93.

After applying the inverse transform ($\lambda = -1$), the linear model yielded the most significant fit (Table 6). The F-value of 101.78 indicates that there is only a 0.01% probability of such a result being attributed to random noise. Furthermore, the terms A, B, C, and D were statistically significant, as their p-values were less than 0.05. The lack of fit was not significant, thus further demonstrating the well-fitting property of the model. The F-value for the lack of fit was 0.77, and there was a 69.49% chance that the lack of fit could be attributed to noise. The predicted R^2 was 91.79%, while the adjusted R^2 was slightly higher at 93.29% (ANOVA analysis). These results show a strong correlation between the predicted and actual values. The low standard deviation of 0.0054 reinforces the model's significance. A precision value of 36.02 further supports the model's efficacy. Table 7 shows the estimated coefficients for the T-Funnel linear model. All of the input variables exhibited a standard error of 0.0011, thereby indicating the precision of the estimates. Moreover, the variance inflation factor (VIF) has been calculated to be 1.00, thus corroborating the orthogonality of the factors studied here.

Table 6. Fitting results for T-Funnel linear model (inverse).

Source	Sum of Squares	Mean Square	F-Value	*p*-Value
Model	0.0117	0.0029	101.78	<0.0001
A-w/c	0.0019	0.0019	65.19	<0.0001
B-Sp/p	0.0001	0.0001	4.70	0.0398
C-w/p	0.0037	0.0037	128.64	<0.0001
D-s/m	0.0060	0.0060	208.57	<0.0001
Residual	0.0007	0.0000		
Lack of Fit	0.0005	0.0000	0.7718	0.6949
Pure Error	0.0002	0.0000		
Cor Total	0.0124	-		

Table 7. Coefficients in terms of coded factors—D-Flow.

Factor	Intercept	A–w/c	B–Sp/p	C–w/p	D–s/m
Coefficient Estimate	0.0499	0.0088	0.0024	0.0124	−0.0158
Standard Error	0.0010	0.0011	0.0011	0.0011	0.0011
95% CI Low	0.0479	0.0066	0.0001	0.0102	−0.0181
95% CI High	0.0519	0.0111	0.0046	0.0147	−0.0136
VIF		1.00	1.00	1.00	1.00

The residual normal probability plot in Figure 5a reveals that the data points were closely aligned with the straight line, despite adopting an S-shaped pattern, which indicates a satisfactory distribution of the residuals. The Residuals versus Run plot demonstrates random dispersion, thus lacking any discernible trend. Notably, the scattered data points fell well within the specified limits of ± 3.54047. Figure 5c shows lambda (λ) values of -1, which lie within the confidence interval (CI) of -1.55 to -0.93. Figure 5d illustrates the relationship between the predicted and actual points, thereby providing an intuitive visual insight into the model's good predictive performance and accuracy. Complementary analyses, such as the Residuals versus Predicted, Residuals versus Water/Cement, Cook's Distance, Leverage, DFFITS, and DFBETAS, yielded results within the established limits in good agreement with such observations.

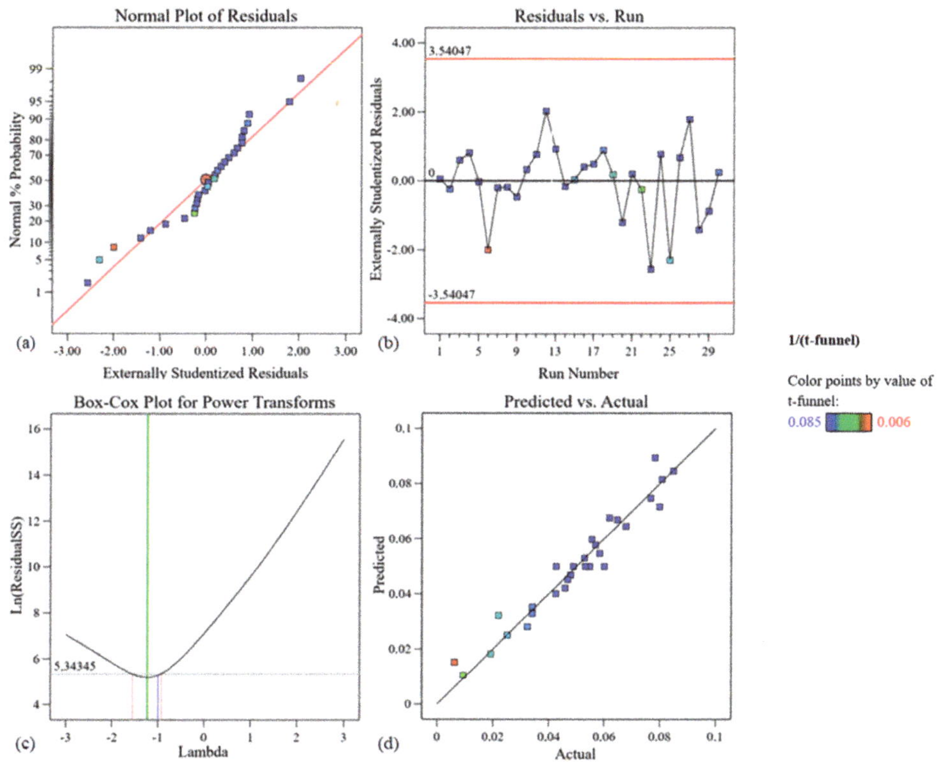

Figure 5. T-Funnel linear model: (**a**) Normal Plot of Residuals; (**b**) Residuals versus Run; (**c**) Box–Cox for Power Transforms; and (**d**) Predicted versus Actual.

3.4. Regression Model for 24-Hour Compressive Strength

For the 24 h compressive strength (CS24h) measurements, regression analysis was performed in order to define the polynomial model. The reduced linear model (removing factor B) was the one that best presented satisfactory results. The B term was removed from the model due to its considerably high p-value (0.9491), which aimed at enhancing the model performance. Due to the linear nature of the model, the removal of the B term did not influence the other terms. Table 8 shows the model's F-value of 37.54 and p-values below 0.05, thereby providing evidence of the model's statistical significance after removing term B. The terms A and C were also deemed to be statistically significant, as was supported by their high sum of squares values and p-value less than 0.05 (0.001 for A and 0.0351 for C). The F-value was calculated to be 0.4778, thereby surpassing the significance level of 0.05 and indicating that the lack of fit was not statistically significant concerning pure error. The possibility of a misfit F-value occurring due to noise only stood at approximately 89.34%.

The results of the regression model revealed a predicted R^2 value of 0.7607, which agreed well with an adjusted R^2 value of 0.7908. The difference between the predicted and adjusted R^2 values being less than 20% indicates the model's statistical significance. Moreover, the model demonstrated an adequate precision of 23.1254, thereby showcasing its ability to accurately describe the modeled data space. The standard deviation of the regression model was calculated to be 1.42, with a coefficient of variation (CV) of 2.43%.

Table 8. Fitting results for compressive strength results after 24 h of curing.

Source	Sum of Squares	Mean Square	F-Value	*p*-Value
Model	227.36	75.79	37.54	<0.0001
A—w/c	215.95	215.95	106.96	<0.0001
C—w/p	10.42	10.42	5.16	0.0316
D—s/m	0.9928	0.9928	0.4917	0.4894
Residual	52.49	2.02		
Lack of Fit	35.04	1.67	0.4778	0.8934
Pure Error	17.46	3.49		
Cor Total	279.86	-		

Table 9 presents the coefficients of the coded factors for a 95% confidence interval and are considered fits around the average response of all the runs. The standard errors for factors A, C, and D were 0.29, with VIFs equal to 1, thus confirming the orthogonality of the factors. For the VIF, values less than 10 were deemed acceptable.

Table 9. Coefficients in terms of coded factors—compressive strength after 24 h of curing.

Factor	Intercept	A—w/c	C—w/p	D—s/m
Coefficient Estimate	58.44	−3.00	−0.6589	0.2034
Standard Error	0.2594	0.290	0.2900	0.2900
95% CI Low	57.91	−3.60	−1.26	−0.3928
95% CI High	58.97	−2.40	−0.0628	0.7996
VIF	-	1	1	1

Figure 6 presents the normal plot of the studentized residuals (a) and the relationship between the predicted and actual values (b). Notably, the data points exhibited a close-to-linear dispersion. However, the adjusted coefficient of determination (R^2) was determined to be 79.08%, and some exceptions can be seen in Figure 6b.

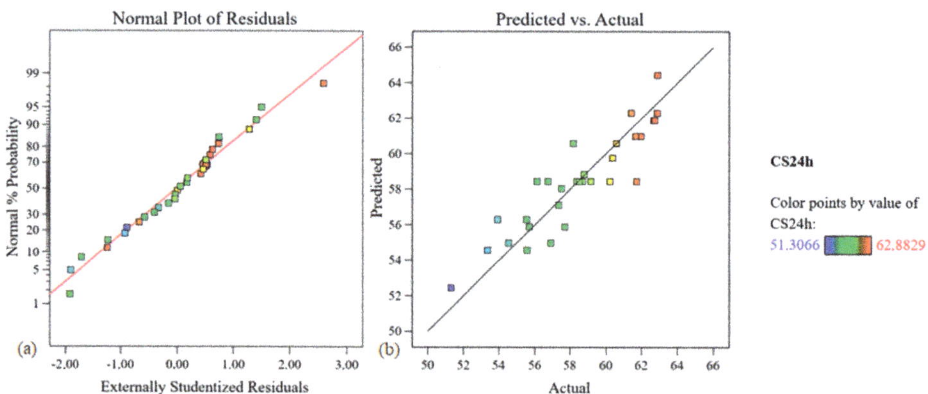

Figure 6. Regression model for compressive strength after 24 h of curing: (**a**) Normal Plot of Residuals and (**b**) Predicted versus Actual.

3.5. Regression Model for 28-Day Compressive Strength

The regression analysis seemed to suggest that quadratic models were the most promising candidates. However, upon ANOVA analysis, the quadratic model was found to be nonsignificant, as the *p*-value was 0.2889 and therefore under the recommended threshold of 0.05. Among the terms examined, only A^2 exhibited statistical significance, with a *p*-value of 0.0252, while all the other terms (A, B, C, D, AB, AD, BC, BD, CD, B^2, C^2, and D^2) were found to be not significant. Moreover, the predicted R^2 was found to be -0.8825, with an adjusted R^2 of 0.1506. This substantial discrepancy suggests that the quadratic model's efficacy was low. Graphic diagnosis was conducted and revealed a potential outlier at Std 18, Run 20, which was clearly shown by the distribution in the Predicted versus Actual plot, the Residual versus Predicted plot, the Residual versus Factor plot, and the Cook's distance relationships (not show here for the sake clarity). This data point exceeded the DFFITS limit and was consequently excluded from subsequent analysis. Still, these results suggested that the quadratic model could be reduced (by removing some terms) to enhance its significance. Table 10 shows the fitting results for the regression model compressive strength after removing the terms B, C, AB, AC, AD, BC, BD, CD, B^2, C^2, and D^2.

Table 10. Fitting results for compressive strength results after 28 days of curing.

Source	Sum of Squares	Mean Square	F-Value	*p*-Value
Model	239.43	79.81	6.99	<0.0015
A—w/c	72.33	72.33	6.33	<0.0189
D—s/m	59.12	59.12	5.18	0.0321
A^2	161.61	161.61	14.15	0.0010
Residual	274.10	11.42		
Lack of Fit	201.81	10.09	0.5584	0.8297
Pure Error	72.29	18.07		
Cor Total	513.53	-		

The likelihood of such a high F-value occurring due to noise was only 0.15%. Therefore, the quadratic model was directionally reduced to address the presence of insignificant terms and to ensure statistical significance. The significant terms A, D, and A^2 were retained. The newly built model had a nonsignificant lack of fit, with an 82.97% probability that misfit could be attributable to noise. The predicted R^2 value of 30.89% exhibited reasonable agreement with the adjusted R^2 value of 39.95%, with a difference of less than 20%. Adequate precision was measured, with a value of 9.366, which indicates that the signal was sufficiently strong and the model was well-suited to describe the design space. Table 11 presents the estimated coefficients for compressive strength after 28 days of curing, which all lay within the 95% confidence interval. The variance inflation factor (VIF) for the factors A and A^2 was 1.14, thereby indicating multicollinearity, while the variable D had a VIF of 1.0, thereby demonstrating orthogonality.

Table 11. Coefficients in terms of coded factors—compressive strength after 28 days of curing.

Factor	Intercept	A-w/c	D-s/m	A^2
Coefficient Estimate	113.19	−2.04	−1.57	−3.22
Standard Error	0.8726	0.8097	0.6898	0.8573
95% CI Low	111.39	−3.71	−2.99	−4.99
95% CI High	114.99	−0.3665	−0.1457	−1.46
VIF	-	1.14	1.00	1.14

Figure 7a shows the normal plot of the residuals for the compressive strength model after excluding the data point Std 18, Run 20. One can observe that the data points appear distributed across a line, but a fully linear distribution is not entirely perceived. Figure 7b shows the Predicted versus Actual relationship plot. The data point that was excluded is clearly noticeable in the lower quadrant of Figure 7b. In Figure 7b, one can also observe that the data points tended to distribute along a straight line; however, they exhibited some dispersion due to the moderate/low coefficient of determination (R^2).

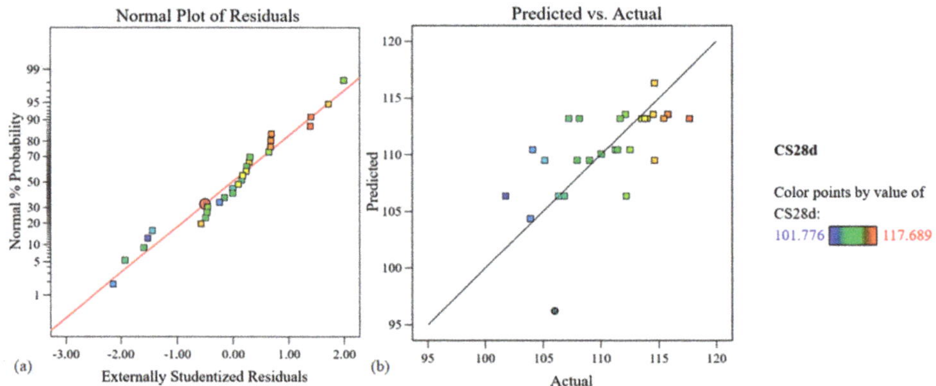

Figure 7. Graphs for CS28d model: (**a**) Normal Plot of Residuals and (**b**) Predicted versus Actual.

4. Discussion

4.1. Model Optimization

After finding the ideal models for each of the self-compacting mortar variables, through a central composite design, the model was optimized in order to find the best solutions based on certain criteria. Targets were set for each input and output variable, thereby encompassing maximization, minimization, achieving specific values, or falling within delimited intervals. The established limits for each variable were based on the database comprising 30 mixtures. In addition, weights ranging from 0.1 to 10 were defined (with the default being 1), along with a degree of importance ranging from 1 plus (+) to 5 plus (+++++). This approach adheres to the methodology proposed by Myers, Montgomery, and Anderson-Cook [42], who aimed to maximize global desirability across a set of output variables.

This study sought to identify mortar compositions that met the following criteria: (i) developed a higher compressive strength at 28 days with the lowest possible cement content to reduce production costs and environmental impacts and (ii) featured a high workability in order to facilitate the flow and ease of consolidation of the mixtures in slender structures and/or in highly reinforced arrangements. Table 12 presents the criteria for optimizing high-strength self-compacting mortars. For the input variable A (w/c), the objective was to minimize cement consumption; hence, a maximization goal was defined, with a weight of one and a maximum importance (factor 5 = +++++). For the input variable C (w/p), the objective was to maximize the water content over the powder content, and it was also assigned the same importance as factor A but with a weight of one. As for the variable D (s/m), maximizing it allowed for reducing the consumption of mortar in relation to sand, thus further contributing to reducing costs and the environmental impacts.

To increase the flowability of the self-compacting mortars and to optimize their performance during the concreting of slender elements with high reinforcement rates, the response variable Y1 (D-Flow) needed to be maximized and was ascribed a degree of importance of four. However, the maximum degree of importance was attributed to variable Y4, thereby aiming to maximize the compressive strength after 28 days with a minimum acceptance criterion of 110 MPa. The establishment of a minimum compressive strength criterion was envisioned to attain values surpassing the sample's average. Variables B (Sp/p), Y2 (T-Funnel), and Y3 (compressive strength after 24 h) were deemed to remain within acceptable value ranges, with a moderate degree of importance being attributed. After computing, a comprehensive set of 85 candidate solutions was found. Table 13 summarizes the top five optimized mixtures ranked based on their global desirability scores.

Table 12. Determined criteria to optimize the mix.

Factor	Goal	Lower Limit	Upper Limit	Weight	Importance
A: w/c	To be maximized	0.8411	0.9485	1	+++++
B: Sp/p	To be in range	0.0221	0.0249	1	+++
C: w/p	To be maximized	0.5013	0.5653	1	+++++
D: s/m	To be maximized	0.4512	0.5088	1	+++++
Y1: D-Flow (mm)	To be maximized	282	377	1	++++
Y2: T-Funnel (s)	To be in range	11.78	162.22	1	+++
Y3: CS24h (MPa)	To be in range	51.3066	62.88	1	+++
Y4: CS28d (MPa)	To be maximized	110.00	117.69	1	+++++

Table 13. The top five optimized mortar formulations according to global desirability scores.

Factor	1	2	3	4	5
A: w/c	0.897	0.904	0.889	0.889	0.878
B: Sp/p	0.024	0.023	0.024	0.024	0.023
C: w/p	0.564	0.563	0.561	0.555	0.564
D: s/m	0.481	0.4894	0.499	0.495	0.485
Y1: D-Flow (mm)	350.05	342.67	336.12	334.94	341.24
Y2: T-Funnel (s)	15.95	17.46	19.79	20.22	18.10
Y3: CS24h (MPa)	57.68	57.37	58.32	58.44	58.80
Y4: CS28d (MPa)	113.06	112.27	112.35	112.57	113.27
Desirability	0.591	0.586	0.570	0.555	0.549

Figure 8 shows the desirability ramps of the different factors and responses for the candidate solution 1. It is observed that the input factors A, C, and D can be maximized without compromising the main objective of increasing the output variable Y4 (CS28d). Variable Y1 (D-Flow) attained its highest value of 343.3 mm, yet it still remained within the specified range of 282 to 377 mm. Factor B and output variables Y2 and Y3 adhered to the designated intervals, thereby aligning with the predefined criteria.

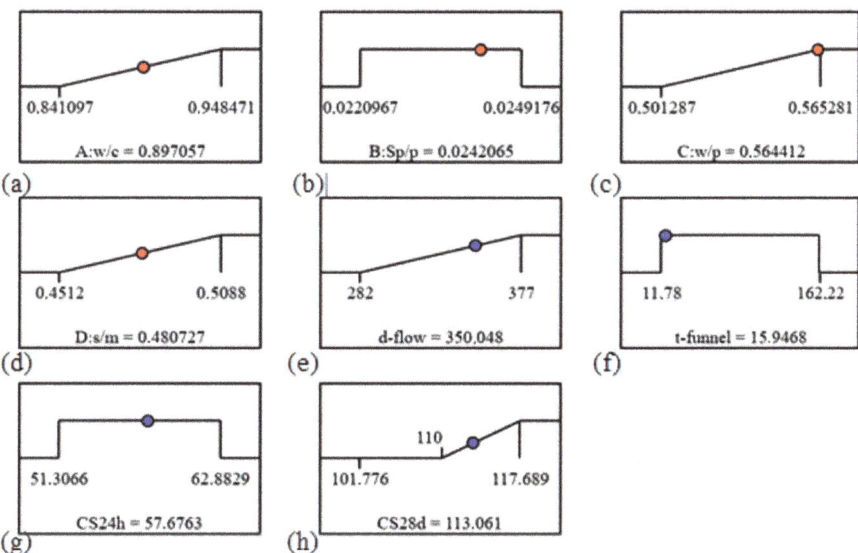

Figure 8. Desirability ramps for numerical optimization of different input and output variables of the best candidate solution found: (**a**) A: w/c; (**b**) B: Sp/p; (**c**) C: w/p; (**d**) D: s/m; (**e**) Y1: D-Flow; (**f**) Y2: T-Funnel; (**g**) Y3: CS24h; and (**h**) Y4: CS28d.

4.2. Optimization of Self-Compacting Mortars Using Response Surface Methodology

Response surface methodology (RSM) is a valuable tool that enables the establishment of relationships between independent variables and corresponding responses, thereby facilitating the optimization of mixed designs to achieve envisioned target features. Through RSM, plots are generated to visually illustrate the influence of each variable, which are represented by slopes or curvatures. Figure 9 shows the relationship between each output response and two of its key governing input variables.

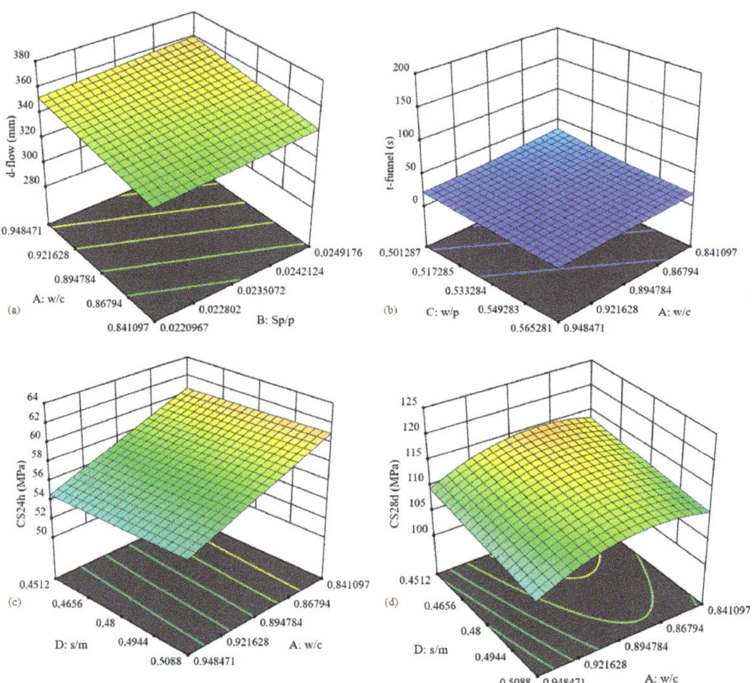

Figure 9. Rheological properties and strength development as a function of different variables: (**a**) D-Flow; (**b**) T-Funnel; compressive strength at 24 h (**c**); compressive strength after 28 days of curing (**d**).

Figure 9a shows the influence of A (w/c) and B (Sp/p) on Y1 (D-Flow). It is clear that increasing the w/c and Sp/p led to higher fluidity. This can be attributed to the higher amount of water in the system and the lower cement content when variable A increased, in addition to the greater amount of superplasticizer used in relation to powder, thus contributing to greater fluidity. Figure 9b shows how variables A (w/c) and D (s/m) are related to the response variable Y2 (T-Funnel). The reduction of the w/c factor (A) increased the funnel time due to the decrease in the amount of water in relation to the cement. In addition, with the increase in the D factor (s/m), the amount of sand increased in relation to the mortar, thus increasing the T-Funnel.

Figure 9c shows Y3 as a function of A and D. A strong inverse relationship can be observed between variable A and response Y3, thereby demonstrating that decreasing the w/c enhanced the compressive strength at 24 h of curing. However, for variable D, there was a minor increase in the s/m factor when the response variable was increased. Nevertheless, only variables A and D were shown to be related; the other variables promoted the following influences: regardless of the variation in B (Sp/p), there were insignificant changes in the Y3 response (CS24h), while, as there was an increase in the variable C (s/m), the CS24h also increased.

The temporal evolution of the strength development as a function of A and D can be depicted by observing the difference between Figures 9c and 9d. The previously straight surface became curved. The curvature occurred mainly in A, as decreasing the w/c resulted in an increased compressive strength at 28 days (Y4). However, the observed growth was not continuous, as the highest values of the CS28d were present in the third quadrant, which is shown in Figure 9 (between 0.86794 and 0.894784). In addition, an upward slope can be observed on the D axis, thereby indicating that decreasing the variable D resulted in a higher compressive strength at 28 days, especially at low w/c ratios.

These results suggest that, as curing time increases, the relationship between the two input variables A (w/c) and D (s/m) with the output Y4 (CS28d) also increases, most notably with variable D (s/m). Variables B (Sp/p) and C (w/p) were not tested here, since they did not show relevant interference for the response factor Y4 (CS28d).

5. Conclusions

The studies conducted in this work aimed to identify significant models that were capable of optimizing the performance of self-compacting mortars and to evaluate the influence of different model parameters on the prediction accuracy. Response surface methodology (RSM) and central composite design were employed as effective tools for modeling the fresh state properties and strength development, thereby resulting in the following conclusions:

Following refinements, the D-Flow linear model emerged as a commendably well-fitted model, which was characterized by an elevated adjusted determination coefficient of 87.4%. In addition, the residual normal and predicted versus actual plots revealed good correlations, thereby demonstrating the model's statistical significance.

Of note, an inverse transform was applied to the T-Funnel model, which yielded an enhanced adjusted determination coefficient of 93.3%. The coherence between the predicted and empirical outcomes testifies to the model's precision and validity.

The outlier exclusion yielded a well-fitted CS24h model with an adjusted R^2 of 79.1%, a low standard deviation, and good correction coefficients. The ideal model was found to be a linear reduction with the omission of variable B, which was dictated by its negligible significance. The CS28d model was found to be statistically significant despite a moderated adjusted R^2 of 39.9%, thus showcasing the potential actions for future improvements. The expeditions in model determination and material optimization have shown the pivotal roles of the variables A and C in shaping the D-Flow and T-Funnel responses, thus further demonstrating the relevance of the w/c ratio, which is an influencer that transcended the response categories.

This pioneering research advances the comprehension of the statistical methodologies within the scope of civil engineering and construction materials, and it is particularly relevant in the development of advanced self-compacting cement-based products. The successful exploration of the proposed models further elucidates a roadmap for forthcoming explorations, with the potential to reduce production costs, curtail environmental impacts, and increase the technical performance of advanced self-compacting mortars.

Author Contributions: Conceptualization, L.M.; Methodology, L.M.; Validation, L.M. and S.R.; Formal analysis, S.R.; Investigation, S.R.; Resources L.M.; Data curation, S.R.; Writing—original draft preparation, S.R.; Writing—review and editing, G.A. and L.M.; Visualization, S.R., G.A., and L.M.; Supervision: L.M. Funding acquisition, G.A. and L.M. All authors have read and agreed to the published version of the manuscript.

Funding: The research work of Stéphanie Rocha and Lino Maia was financially supported by Base Funding—UIDB/04708/2020 of CONSTRUCT—Instituto de I&D em Estruturas e Construções—which was funded by national funds through the FCT/MCTES (PIDDAC) and by national funds through FCT—Fundação para a Ciência e a Tecnologia, I.P., under the Scientific Employment Stimulus—Institutional Call—CEECINST/00049/2018. Guilherme Ascensão's work was supported by the Foundation for Science and Technology (FCT)—Aveiro Research Center for Risks and Sustainability in Construction (RISCO), Universidade de Aveiro, Portugal [FCT/UIDB/ECI/04450/2020]. This publication only reflects the authors' views, thus exempting the funding agency from any liability.

Institutional Review Board Statement: Not applicable.

Informed Consent Statement: Not applicable.

Data Availability Statement: The data presented in this study are available upon request from the corresponding author.

Conflicts of Interest: The authors declare no conflict of interest. The funders had no role in the design of the study; in the collection, analyses, or interpretation of data; in the writing of the manuscript, or in the decision to publish the results.

References

1. Ali, M.; Kumar, A.; Yvaz, A.; Salah, B. Central composite design application in the optimization of the effect of pumice stone on lightweight concrete properties using RSM. *Case Stud. Constr. Mater.* **2023**, *18*, e01958. [CrossRef]
2. Rashid, K.; Farooq, S.; Mahmood, A.; Iftikhar, S.; Ahmad, A. Moving towards resource conservation by automated prioritization of concrete mix design. *Constr. Build. Mater.* **2020**, *236*, 117586. [CrossRef]
3. Okamura, H.; Ouchi, M. Self-compacting high performance concrete. *Prog. Struct. Eng. Mater.* **1998**, *1*, 378–383. [CrossRef]
4. Shi, C.; Wu, Z.; Lv, K.; Wu, L. A review on mixture design methods for self-compacting concrete. *Constr. Build. Mater.* **2015**, *84*, 387–398. [CrossRef]
5. De Schutter, G.; Bartos, P.J.; Domone, P.; Gibbs, J. *Self-Compacting Concrete (Vol. 288)*; Whittles Publishing: Caithness, The Netherlands, 2018; Available online: http://hdl.handle.net/1854/LU-689512 (accessed on 20 July 2023).
6. Okamura, H.; Ouchi, M. Self-Compacting Concrete. *J. Adv. Concr. Technol.* **2003**, *1*, 5–15. [CrossRef]
7. Aitcin, P.C. *Concreto de Alto Desempenho*, 1st ed.; Pini: São Paulo, Brazil, 2000.
8. Beersaerts, G.; Ascensão, G.; Pontikes, Y. Modifying the pore size distribution in Fe-rich inorganic polymer mortars: An effective shrinkage mitigation strategy. *Cem. Concr. Res.* **2021**, *141*, 106330. [CrossRef]
9. Petrou, A.K.M.F.; Ioannou, I. Durability performance of self-compacting concrete. *Constr. Build. Mater.* **2012**, *37*, 320–325.
10. Nuruzzaman; Ahmad, T.; Sarker, P.K.; Shaikh, F.U.A. Rheological behaviour, hydration, and microstructure of self-compacting concrete incorporating ground ferronickel slag as partial cement replacement. *J. Build. Eng.* **2023**, *68*, 106127. [CrossRef]
11. Zerbino, R.; Barragán, B.; Garcia, T.; Agulló, L.; Gettu, R. Workability tests and rheological parameters in self-compacting con-crete. *Mater. Struct. Constr.* **2009**, *42*, 947–960. [CrossRef]
12. Ascensão, G.; Marchi, M.; Segata, M.; Faleschini, F.; Pontikes, Y. Reaction kinetics and structural analysis of alkali activated Fe–Si–Ca rich materials. *J. Clean. Prod.* **2020**, *246*, 119065. [CrossRef]
13. Upasani, R.S.; Banga, A.K. Response Surface Methodology to Investigate the Iontophoretic Delivery of Tacrine Hydrochloride. *Pharm. Res.* **2004**, *21*, 2293–2299. [CrossRef] [PubMed]
14. Montgomery, D.C. *Design and Analysis of Experiments*; John Wiley: Hoboken, NJ, USA, 2017.
15. Box, K.B.; Wilson, G.E.P. On the experimental attainment of optimum conditions. *J. R. Stat. Soc.* **1951**, 1–38. [CrossRef]
16. Box, J.S.; Hunter, G.E.P.; Hunter, W.G. *Statistics for Experimenters: An Introduction to Design, Data Analysis and Model Building*; Wiley: New York, NY, USA, 1978.
17. Awolusi, T.; Oke, O.; Akinkurolere, O.; Sojobi, A. Application of response surface methodology: Predicting and optimizing the properties of concrete containing steel fibre extracted from waste tires with limestone powder as filler. *Case Stud. Constr. Mater.* **2019**, *10*, e00212. [CrossRef]
18. Ali, M.; Khan, M.I.; Masood, F.; Alsulami, B.T.; Bouallegue, B.; Nawaz, R.; Fediuk, R. Central composite design application in the optimization of the effect of waste foundry sand on concrete properties using RSM. *Structures* **2022**, *46*, 1581–1594. [CrossRef]
19. Khan, M.I.; Sutanto, M.H.; Bin Napiah, M.; Khan, K.; Rafiq, W. Design optimization and statistical modeling of cementitious grout containing irradiated plastic waste and silica fume using response surface methodology. *Constr. Build. Mater.* **2021**, *271*, 121504. [CrossRef]
20. Ghafari, E.; Costa, H.; Júlio, E. RSM-based model to predict the performance of self-compacting UHPC reinforced with hybrid steel micro-fibers. *Constr. Build. Mater.* **2014**, *66*, 375–383. [CrossRef]
21. Dahish, H.A.; Almutairi, A.D. Effect of elevated temperatures on the compressive strength of nano-silica and nano-clay modified concretes using response surface methodology. *Case Stud. Constr. Mater.* **2023**, *18*, e02032. [CrossRef]
22. Mohammed, B.S.; Fang, O.C.; Hossain, K.M.A.; Lachemi, M. Mix proportioning of concrete containing paper mill residuals using response surface methodology. *Constr. Build. Mater.* **2012**, *35*, 63–68. [CrossRef]
23. Nunes, S.; Matos, A.M.; Duarte, T.; Figueiras, H.; Sousa-Coutinho, J. Mixture design of self-compacting glass mortar. *Cem. Concr. Compos.* **2013**, *43*, 1–11. [CrossRef]
24. Keleştemur, O.; Yildiz, S.; Gökçer, B.; Arici, E. Statistical analysis for freeze–thaw resistance of cement mortars containing marble dust and glass fiber. *Mater. Des.* **2014**, *60*, 548–555. [CrossRef]
25. Ferdosian, I.; Camões, A. Eco-efficient ultra-high performance concrete development by means of response surface methodology. *Cem. Concr. Compos.* **2017**, *84*, 146–156. [CrossRef]

26. Mermerdaş, K.; Algın, Z.; Oleiwi, S.M.; Nassani, D.E. Optimization of lightweight GGBFS and FA geopolymer mortars by response surface method. *Constr. Build. Mater.* **2017**, *139*, 159–171. [CrossRef]

27. Zahid, M.; Shafiq, N.; Isa, M.H.; Gil, L. Statistical modeling and mix design optimization of fly ash based engineered geopolymer composite using response surface methodology. *J. Clean. Prod.* **2018**, *194*, 483–498. [CrossRef]

28. Matos, A.M.; Maia, L.; Nunes, S.; Milheiro-Oliveira, P. Design of self-compacting high-performance concrete: Study of mortar phase. *Constr. Build. Mater.* **2018**, *167*, 617–630. [CrossRef]

29. Khan, M.I.; Sutanto, M.H.; Napiah, M.B.; Zahid, M.; Usman, A. Optimization of Cementitious Grouts for Semi-Flexible Pavement Surfaces Using Response Surface Methodology. *IOP Conf. Ser. Earth Environ. Sci.* **2020**, *498*, 012004. [CrossRef]

30. Al Salaheen, M.; Alaloul, W.S.; Malkawi, A.B.; de Brito, J.; Alzubi, K.M.; Al-Sabaeei, A.M.; Alnarabiji, M.S. Modelling and Optimization for Mortar Compressive Strength Incorporating Heat-Treated Fly Oil Shale Ash as an Effective Supplementary Cementitious Material Using Response Surface Methodology. *Materials* **2022**, *15*, 6538. [CrossRef]

31. Shi, X.; Zhang, C.; Wang, X.; Zhang, T.; Wang, Q. Response surface methodology for multi-objective optimization of fly ash-GGBS based geopolymer mortar. *Constr. Build. Mater.* **2022**, *315*, 125644. [CrossRef]

32. Anurag; Singh, S. Estimation of the impacts of adding recycled demolition waste and steel fibers on different mechanical properties of self-compacting concrete. *Mater. Today Proc.* **2022**, *48*, 1723–1730. [CrossRef]

33. Waqar, A.; Bheel, N.; Almujibah, H.R.; Benjeddou, O.; Alwetaishi, M.; Ahmad, M.; Sabri, M.M.S. Effect of Coir Fibre Ash (CFA) on the strengths, modulus of elasticity and embodied carbon of concrete using response surface methodology (RSM) and optimization. *Results Eng.* **2023**, *17*, 100883. [CrossRef]

34. Khavari, B.C.; Shekarriz, M.; Aminnejad, B.; Lork, A.; Vahdani, S. Laboratory evaluation and optimization of mechanical properties of sulfur concrete reinforced with micro and macro steel fibers via response surface methodology. *Constr. Build. Mater.* **2023**, *384*, 131434. [CrossRef]

35. Hamada, H.M.; Al-Attar, A.A.; Tayeh, B.; Yahaya, F.B.M.; Abu Aisheh, Y.I. Optimising concrete containing palm oil clinker and palm oil fuel ash using response surface method. *Ain Shams Eng. J.* **2023**, *14*, 102150. [CrossRef]

36. Mita, A.F.; Ray, S.; Haque, M.; Saikat, H. Prediction and optimization of properties of concrete containing crushed stone dust and nylon fiber using response surface methodology. *Heliyon* **2023**, *9*, e14436. [CrossRef]

37. Sinkhonde, D. Generating response surface models for optimisation of CO2 emission and properties of concrete modified with waste materials. *Clean. Mater.* **2022**, *6*, 100146. [CrossRef]

38. Adamu, M.; Trabanpruek, P.; Jongvivatsakul, P.; Likitlersuang, S.; Iwanami, M. Mechanical performance and optimization of high-volume fly ash concrete containing plastic wastes and graphene nanoplatelets using response surface methodology. *Constr. Build. Mater.* **2021**, *308*, 125085. [CrossRef]

39. Li, L.; Wan, Y.; Chen, S.; Tian, W.; Long, W.; Song, J. Prediction of optimal ranges of mix ratio of self-compacting mortars (SCMs) based on response surface method (RSM). *Constr. Build. Mater.* **2022**, *319*, 126043. [CrossRef]

40. Safhi, A.e.M.; Benzerzour, M.; Rivard, P.; Abriak, N.-E.; Ennahal, I. Development of self-compacting mortars based on treated marine sediments. *J. Build. Eng.* **2019**, *22*, 252–261. [CrossRef]

41. Maia, L. Experimental dataset from a central composite design with two qualitative independent variables to develop high strength mortars with self-compacting properties. *Data Brief* **2021**, *40*, 107738. [CrossRef] [PubMed]

42. Myers, R.H.; Montgomery, D.C.; Anderson-Cook, C.M. *Response Surface Methodology*, 3rd ed.; Wiley: New York, NY, USA, 2009.

applied
sciences

Article

Developing a New Procedural Binary Particle Swarm Optimization Algorithm to Estimate Some Properties of Local Concrete Mixtures

Fatima Alsaleh [1], Mohammad Bassam Hammami [2], George Wardeh [3,*] and Feras Al Adday [4]

[1] Faculty of Civil Engineering, University of Aleppo, Aleppo 12212, Syria; alsaleh-f@alepuniv.edu.sy
[2] Artificial Intelligence Engineering, University of Aleppo, Aleppo 12212, Syria; bassamoo7@alepuniv.edu.sy
[3] Civil Engineering Mechanics and Materials Laboratory (L2MGC), CY Cergy-Paris University, 95031 Neuville-sur-Oise, France
[4] Faculty of Civil and Architectural Engineering, University of Kalamoon, Damascus Countryside, Damascus-Homs International Highway, Deirattiah P.O. Box 222, Syria; feras.aladday@uok.edu.sy
* Correspondence: george.wardeh@cyu.fr

Abstract: Artificial intelligence techniques have lately been used to estimate the mechanical properties of concrete to reduce time and financial expenses, but these techniques differ in their processing time and accuracy. This research aims to develop a new procedural binary particle swarm optimization algorithm (NPBPSO) by making some modifications to the binary particle swarm optimization algorithm (BPSO). The new software has been created based on some fresh state properties (slump, temperature, and grade of cement) obtained from several ready-mix concrete plants located in Aleppo, Syria to predict the density and compressive strength of the regional concrete mixtures. The numerical results obtained from NPBPSO have been compared with the results from BPSO and artificial neural network ANN. It has been found that BPSO and NPBPSO are both predicting the compressive strength of concrete with less number of iterations and more accuracy than ANN (0.992 and 0.998 correlation coefficient in BPSO and NPBPSO successively and 0.875 in ANN). In addition, NPBPSO is better than BPSO as it prevents the algorithm from falling into the problem of local solutions and reaches the desired optimal solution faster than BPSO. Moreover, NPBPSO improves the accuracy of obtained compressive strength values and density by 30% and 50% successively.

Keywords: concrete; binary particle swarm algorithm; artificial neural networks; compressive strength; density

Citation: Alsaleh, F.; Hammami, M.B.; Wardeh, G.; Al Adday, F. Developing a New Procedural Binary Particle Swarm Optimization Algorithm to Estimate Some Properties of Local Concrete Mixtures. *Appl. Sci.* **2023**, *13*, 10588. https://doi.org/10.3390/app131910588

Academic Editors: Francisco B. Varona Moya and Mariella Diaferio

Received: 19 June 2023
Revised: 20 September 2023
Accepted: 20 September 2023
Published: 22 September 2023

1. Introduction

To determine the strength of concrete mixtures using the traditional approach, the following steps are required: (1) Identify the components of the mixture, such as the types and amounts of sand, gravel, cement, and auxiliary materials. (2) Determine the correct amount of water to add to the mix, considering local evaporation factors. (3) Implement the mix using appropriate procedures and steps, considering ambient temperature and humidity. (4) Fill molds with the mixture according to the prescribed shapes and dimensions. (5) Allow the mixture to harden and form a concrete base by leaving it inside the mold for 7 to 28 days. (6) Extract the concrete beam from the mold and expose it to external forces using special testing devices to determine its compressive strength. (7) After completing the experiments, remove any waste resulting from the examination process [1–5].

Due to the speed of artificial intelligence (AI) techniques in solving engineering problems, there has been a tendency to use these techniques in various fields of civil engineering, including designing construction materials (concrete mixtures for example) or estimating their properties. As it is hard to predict the compressive strength of concrete due to the different nonlinearities inherent in the mixture designs, various concrete companies are continuously looking to use new methods and technologies to predict the compressive strength. Such methods include numerical modelling and artificial intelligence due to their advantages. These methods are efficient and environmentally friendly, as there is no waste in the testing process. In addition, they are more economical since there is no need for test means, test materials, or even laboratory employees. Moreover, these methods are flexible since many parameters can be taken into consideration, along with the speed of their implementation. It is crucial to accurately predict and evaluate the compressive strength of concrete mixtures, as it is one of the most important features of concrete [6,7].

Several recent studies focused on evaluating compressive strength using machine learning (ML) practices [8] which involve regular, big, and complete information. However, collecting this information is restricted due to the lack of data corresponding to the diverse input characteristics [9]. The concept of utilizing particle swarm optimization (PSO) begins by resetting particles within the search space randomly. Then, the particles construct upon their previous successful attempts and those of their neighbors to discover the optimal particle state. This is achieved by resetting the particle's location and updating its velocity [10]. Furthermore, the parameters of PSO can be easily modified, making it suitable for a wide range of practical problems [11]. In simpler terms, particle velocity in each cycle is determined by three factors: (1) the particle's current location, (2) the best location it has ever been, and (3) the best location within the entire group. This concept is explained in greater detail in reference [12]. PSO is a widely used procedure in the field of Swarm Intelligence that relies on optimization [13]. The goal of the optimization process is to find the best possible solutions to specific problems while taking into account any relevant constraints [14].

For this paper, the binary particle swarm optimization (BPSO) algorithm was selected due to its high level of adaptability and simplicity. Throughout the research, this algorithm has faced some challenges, stemming from local solutions and time constraints. In order to solve these issues, the BPSO algorithm was modified, and a new procedural binary particle swarm (NPBPSO) algorithm was obtained, which was able to overcome the obstacle of local solutions and is able to achieve a reduction in the required time to reach the optimal solution.

2. Literature Review

Concrete has several advantageous characteristics, including high wear resistance, low water permeability, and good compressive strength, and it is widely used in civil structures [15–18]. To maintain resident safety and structure durability, construction engineers are mainly concerned with the quality of building materials, notably the compressive strength of concrete. One typical method of evaluating the concrete's other physical and mechanical characteristics is measuring its compressive strength, which acts as a significant and trustworthy indicator of whether or not a concrete mixture conforms with engineering design criteria [16,17]. The process of precisely measuring the compressive strength of concrete mixtures is difficult, time-consuming, and associated with multiple problems [18–20]. Although statistical and experimental models incorporate a lot of data from laboratory tests, the results' accuracy is still poor [20].

Artificial intelligence models (AIMs) have been proposed as an alternative method to address the challenges of compressive strength prediction connected to the impact of various mixed design parameters [21–26]. By predicting the compressive strength of the concrete, a project's time and expense can be reduced. As a result, AIMs can be used to identify this important characteristic [27].

A mathematical model for estimating the compressive strength of concretes with additives was developed by Kandiri et al. [28] using an artificial neural network (ANN) technique. In the testing phase, the proposed model showed acceptable accuracy with a mean absolute percentage error of 11.10%. Ngo et al. [29] used artificial neural networks (ANNs), support vector regression (SVR), linear regression (LR), and M5P techniques for the prediction of axial strength in circular steel tube confined concrete columns. The authors outlined key distinctions between the techniques and concluded that the M5P was the best artificial intelligence (AI) model for predicting experimental results when compared to others. Goutham and Singh [30] used support vector regression (SVR) to predict the compressive strength of concrete. By comparing the analytical results with those of a non-destructive test, the authors concluded that SVR can be successfully used to predict the compressive strength of concrete.

Using four artificial intelligence models (AIMs), namely ICA-XGBoost, AIM ICA-ANN, ICA-SVR, and ICA-ANFIS, Duan et al. [31] evaluated the compressive strength of concrete made by recycled aggregates. The ICA-XGBoost model is the best one for determining the compressive strength of concrete, according to the findings. According to the authors, the proposed method can be used to verify that recycled concrete has the required mechanical characteristics in structural engineering [31].

Another study by H. N. Muliauwan et al. [32] determined the most exact Input/Output (I/O) connections between the components of concrete mixtures by employing many AIMs. The three AIMs employed in this investigation were support vector machines, linear regression, and artificial neural networks. The simulation's results using roughly 1030 compressive strength test values demonstrated that AIMs can facilitate the development of precise predictive models for concrete properties without the need for substantial expenditures on costly laboratory experiments.

Using six different types of AIMs, Cihan [27] employed AI to forecast the compressive strength of concrete. The adopted techniques were linear regression, classification and regression trees, K-nearest neighbor and extreme learning machine, adaptive neuro-fuzzy inference system (ANFIS), random forest, and SVR. The correlation factor, absolute mean error, root mean squared, and mean were used as standards to evaluate the efficiency of these approaches. Comparative results showed that the ANFIS outperforms the competition as a prediction model. The findings of the random forest model were nearly identical to those of the ANFIS, while the classification and regression tree had the lowest level of correctness. To estimate compressive strengths, Nafees. et al. [33] used three models namely genetic programming (GEP), ANFIS, and MLPNN that is a form of ANN. The results of the study showed that GEP models for data predictions are more precise than machine learning (ML) and that a new mathematical formula might be created and utilized to estimate additional database properties. The strength of lightweight concrete was predicted by Kumar et al. [25] using six machine learning algorithms: GPR, EL, SVMR, enhanced SVMR and GPR, and ensemble learning (EL). The results of this study showed that the optimized GPR model had the greatest accuracy. Furthermore, the improved GPR and SVMR models showed excellent behavior. K. Nasrollahzadeh and E. Nouhi, 2016 [34], applied the fuzzy inference model to improve a new precise procedure and to evaluate the square concrete columns' strength and strain subjected to a vertical load strengthened by fiber polymer wraps. An experimental compressive strength of 261 and a crucial experimental strain of 112 were gathered from the previous studies. The outputs of the finally proposed (Takagi–Sugeno) fuzzy inference models were well agreed with the experimental data of both strain and strength [34].

In order to estimate the density and compressive strength of local concrete mixtures based on the specific properties of their constituents, this work aims to develop the binary particle swarm algorithm (BPSA) with a new procedure. Because the experimental data currently available may be regarded as discrete space, and since the binary particle swarm approach is quick to reach the best answer with fewer iterations than other algorithms, it was chosen for this investigation.

Moreover, artificial swarm intelligence (ASI) may considerably improve prediction accuracy and collective insights.

3. New Procedural Binary Particle Swarm Optimization (NPBPSO)

3.1. Binary Particle Swarm Optimization

Binary particle swarm optimization was devised by Eberhart and Kennedy in 1995 [35,36]. This technique produces a swarm of particles through a random process, each of which stands for a potential solution to the problem. Thus, to find the best particle, the approach must search iteratively using two equations [37]. The first is the particle velocity equation (speed equation)

$$V_{ij}^{t+1} = WV_{ij}^t + c_1 r_{1j}\left(pbest_i^t - X_{ij}^t\right) + c_2 r_{2j}\left(gbest_i^t - X_{ij}^t\right) \tag{1}$$

where i denotes the particle number; j: the number of elements inside the particle; V_{ij}^t: the velocity of the particle i in the previous instant t; V_{ij}^{t+1}: the velocity of the particle i in the next instant $t + 1$; $pbest_i^t$: the most appropriate value reached by the i-particle until the iteration t; $gbest_i^t$: the most appropriate value within the swarm has been reached up to repetition t,

$$gbest_i^t = max\left(pbest_1^t\ pbest_2^t \ldots pbest_{55}^t\right)$$

r_{1j} and r_{2j}: random values to ensure diversity of investigation and fall within the range [0, 1]; X_{ij}^t: the position of the particle i in the previous instant t; W: a variable representing a percentage of the particle velocity at the previous moment; c_1 and c_2 acceleration variables that control the speed of reaching the best solution. In applications that use this algorithm, the variables c_1 and c_2 and W are calibrated experimentally [38,39].

W's value is constrained to the range [0.3–0.9], while the fields for the two variables c_1 and c_2 are [0.4–2]. As a result, a series of experiments were carried out to arrive at the correct values. The probability of a change in the values of the constituent parts of the particle is determined by Equation (1). Equation (2) represents the particle's new state.

$$X_{ij}(t+1) = \begin{cases} 1 & \text{if} \quad u_{ij} < sig\ \left[v_{ij}\ (t+1)\right] \\ 0 & \text{if} \quad u_{ij} \geq sig\ \left[v_{ij}\ (t+1)\right] \end{cases} \tag{2}$$

u_{ij}: a random value within the interval [0, 1] generated according to an equal probability function at the beginning of each iteration.

$$sig\left(v_{ij}(t)\right) = \frac{1}{1 + e^{-v_{ij}(t)}} \tag{3}$$

Sigma function $sig(v)$ aims to narrow the numeric values into confined space [0, 1] in order to improve the performance of the algorithm [40]. Figure 1 shows a systematic diagram for working of the binary optimization particle swarm.

The algorithm used here operates differently than neural networks. It begins by creating a swarm of potential solutions, referred to as "particles." With the use of two mathematical equations, the speed equation and the new state equation, the algorithm continually enhances these solutions. Following each iteration, an evaluation function is used to assess the resulting solutions and identify the most optimal one. For the particle, after each iteration, all the resulting "particle" solutions are evaluated using a special function called the evaluation function, by which the best particle "optimal solution" is reached. The data in this algorithm are not divided into "training, checking, testing" groups as it is in the neural networks' algorithm. Rather, all the data are applied to the algorithm, and through the evaluation function, we can identify the best particle, the "best solution", and ensure that the algorithm is able to reach the optimal solution.

Even though the binary particle swarm optimization algorithm shows good accuracy regarding to other AI techniques such as ANN, it still suffers from the phenomenon of immature convergence (falling into a local solution) in which the search process may get stuck in a region that contains an optimal value, which results in a loss of diversity [41].

In addition, the new procedural binary particle swarm optimization algorithm NPBPSO provides the optimum solution with fewer iterations.

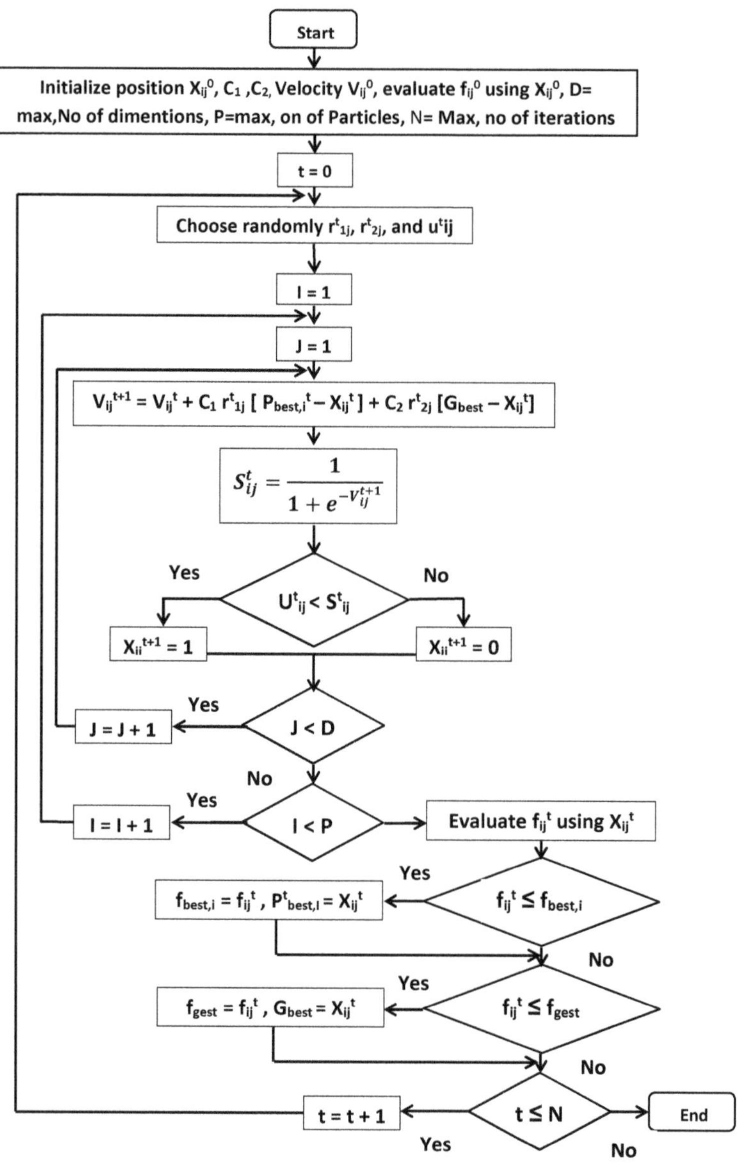

Figure 1. The flowchart of binary particle swarm optimization algorithm [42].

3.2. New Procedural Binary Particle Swarm Optimization (NPBPSO)

To address the issue of immature convergence in the binary particle swarm optimization (BPSO) algorithm, a new approach called NPBPSO has been proposed in this research. Immature convergence occurs when the search process becomes trapped in a local solution, leading to a loss of diversity and optimal value. Figure 2 depicts the modifications made to the procedural binary particle swarm optimization technique to prevent immature convergence. Figure 2 illustrates the modifications applied to the new procedural binary particle swarm optimization technique to avoid immature convergence.

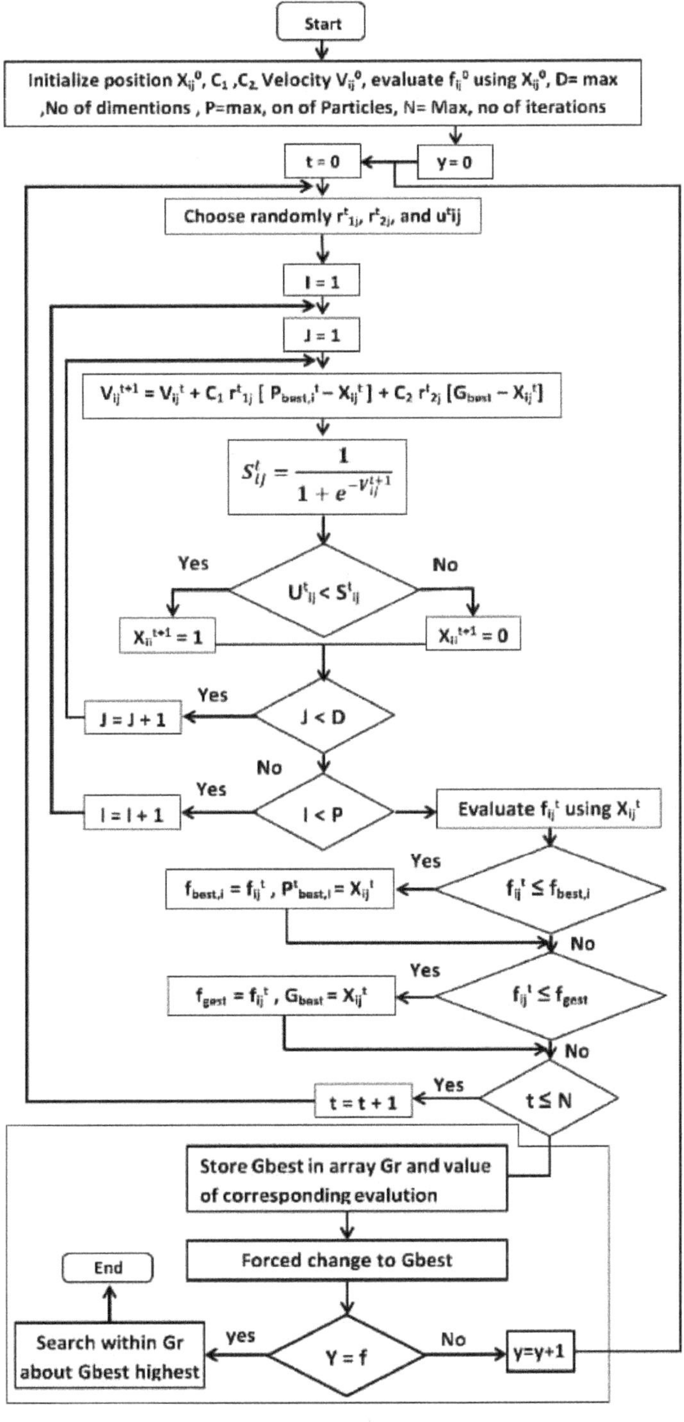

Figure 2. The flowchart of the new procedural binary particle swarm optimization algorithm.

3.3. Function of the Suggested Evaluation

The evaluation function was created using the target formula [43] since the constraints of the problem of the compressive strength of the mixture are independent of one another (slump, temperature, and cement grade). Whenever the value given by equation 4 is higher, the solution is considered to be better, and the particles on the next iteration will move towards that solution.

$$Fitness = \sum_{j=1}^{n} (Q + CL + T)j \qquad (4)$$

where Q, CL, and T denotes the grade of cement, the slump, and the temperature, respectively.

3.4. Piloting Tests

A total of 60 iterations of the particle swarms' program were completed before the mean value of those results was calculated and represented. Table 1 expresses the conventions used during program piloting.

Table 1. Conditions adopted during the piloting of programs.

Processor	1.7–1.73 GHzCore
memory	12 G Byte
operating system	Windows 7

4. Methodology and Materials

The database used for the investigation in this study contains 163 concrete mixtures. Some of them were from the experimental works of the laboratories of the faculty of civil engineering at the University of Aleppo, and the others were from the prefabricated concrete factory of the industrial region of Aleppo City in Syria. These data include the grade of cement, slump, temperature, density, and compressive strength at 7 and 28 days where the density was between 2350 and 2550 kg/m3, cement grade values were N20, N25, N32, and N40, fresh slump was within the range of 40–125 mm, the exterior temperature was varying from 10 to 30 °C, the compressive strength at 7 days was between 10 and 40 MPa, and the compressive strength at 28 days was between 22 and 57 MPa. The statistical distribution of the measurements is shown in Figure 3.

In the next step, the binary particle swarm optimization algorithm was developed using cement grade, slump, and temperature as input to obtain the density and compressive strength as output by means of the evaluation function.

Then, some modification to the binary particle swarm optimization algorithm was made to improve the procedure of having results by avoiding local solutions of BPSO and having the optimum solution with fewer iterations. The modified algorithm is named the new procedural binary particle swarm optimization algorithm NPBPSO.

Later, the NPBPSO was run using the same inputs to get the density and compressive strength at 7 and 28 days.

Finally, the results of BPSO and NPBPSO were compared with ANN and investigated in order to know the best one to be used for estimating the density and compressive strength of local concrete mixtures.

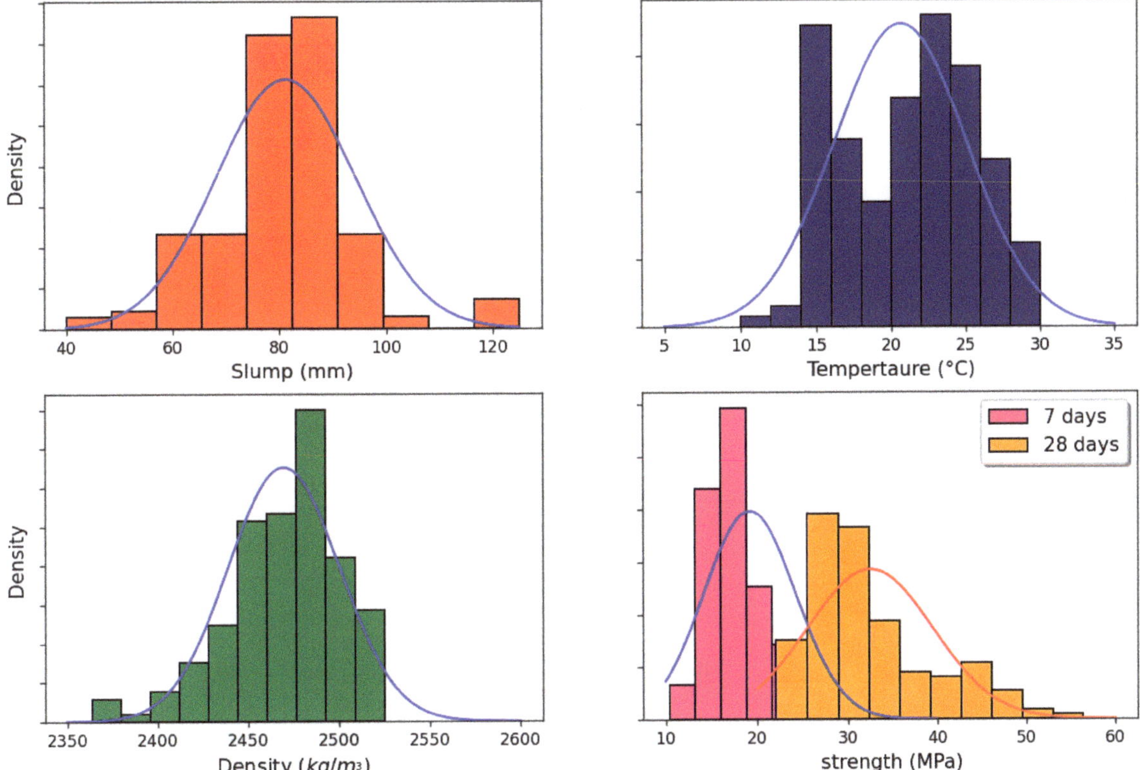

Figure 3. Statistical distribution of database values.

5. Results and Discussion

In the previous research [44], the neural network algorithm was used to estimate the compressive strength of cement concrete for each period of 28 days.

In that research, a neural network was designed consisting of three layers: the first layer was the "input layer", which contained twenty neurons; then, the second layer was the "hidden layer", which consisted of ten neurons; the third layer was the output layer, which consisted of five neurons. The algorithm used in training the neural network was the rapid deployment algorithm, and the number of training sessions for this network was 1000 training sessions, where in each training session, the weights of neurons within the network layers were adjusted in order to reach the optimal solution and reduce errors to the lowest possible extent.

The same data of experimental mixtures were entered into the network (163 mixtures from the experimental works of the laboratories of the faculty of civil engineering at the University of Aleppo and a prefabricated concrete factory of the industrial region of Aleppo City in Syria). The used input data of the neural network were temperature, slump, and cement grade, and the output was the density and the compressive strength at 7 and 28 days of concrete.

The data of experimental mixtures were divided into three groups: The first group, containing 111 mixtures of data, was used to train the network. The second group, containing 25 mixtures of data, was used to audit the training results. The third group, containing 25 mixtures of data, was used to test and to ensure that the neural network received sufficient training and was able to show satisfactory results.

As the correlation coefficient is quite low, $R^2 = 0.875$ (even though it is acceptable regarding other research [45–47]), the results presented in Figure 4 indicate that the selected

ANN technique does not fit the experimental data well in terms of compressive strength; there was a need to use different AI techniques, and as for the reason mentioned above, the authors selected the BPSO and then improved it to NPBPSO.

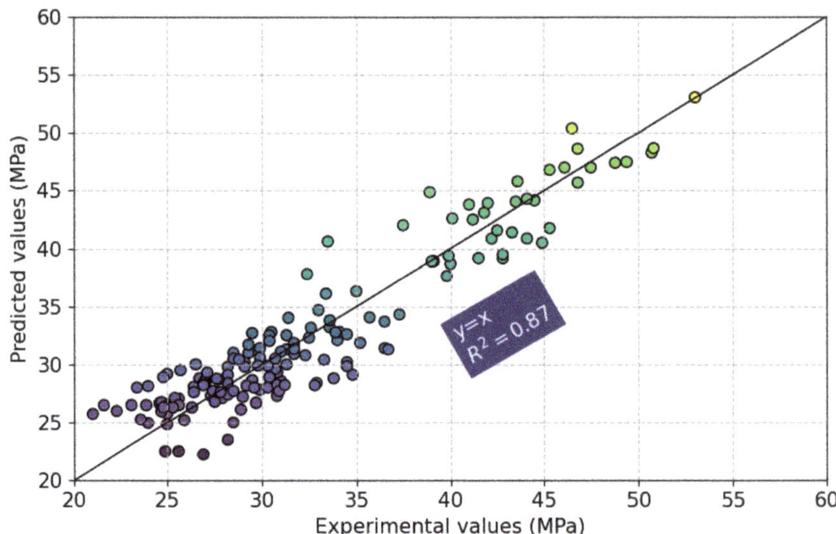

Figure 4. Comparison of the neural network results with the real values of the test data in the resistance condition after 28 days.

The results of the neural networks algorithm, related to estimating the resistance of concrete for 28 days, showed that there was a shift between the experimental values (in the laboratory) and the values generated by the neural networks with a capacity of (1.91 Mpa). We also notice that there is a large dispersion in the values generated by the neural network algorithm.

Figure 5 presents experimental compressive strength measurements and BPSO results obtained after 7 and 28 days, respectively. The predicted compressive strength values at the age of 7 and 28 days have been found to be extremely close to those experimentally achieved with a shift of (0.01112 MPa).

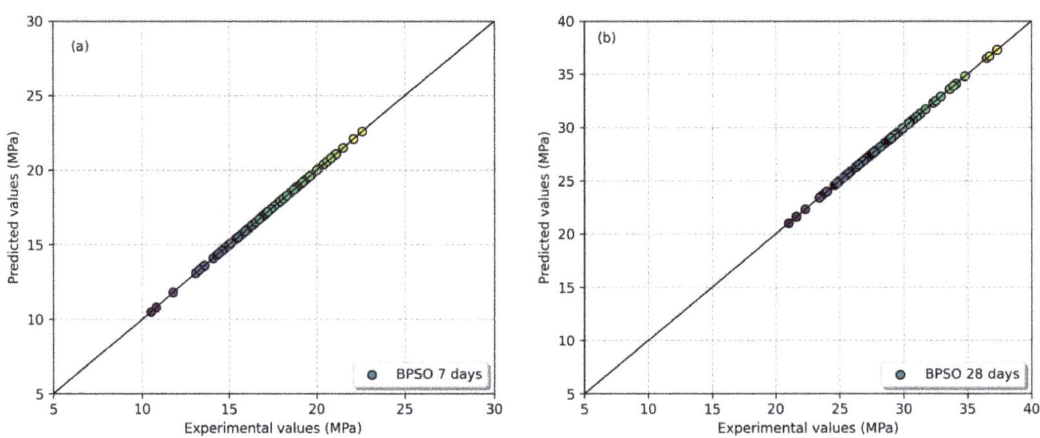

Figure 5. Experimental versus predicted compressive strength using BPSO (**a**) after 7 days, (**b**) after 28 days.

Figure 6 shows the differences between experimental compressive strength values and others resulting from applying the new procedural binary particle swarm optimization NPBPSO after 7 and 28 days, respectively, where it has been noticed that most of the values resulting from NPBPSO matched the experimental ones. As for the points that did not match, the shift ratio reached (0.00798 MPa); in other words, the accuracy of the values resulting from artificial intelligence has been improved by almost 30%.

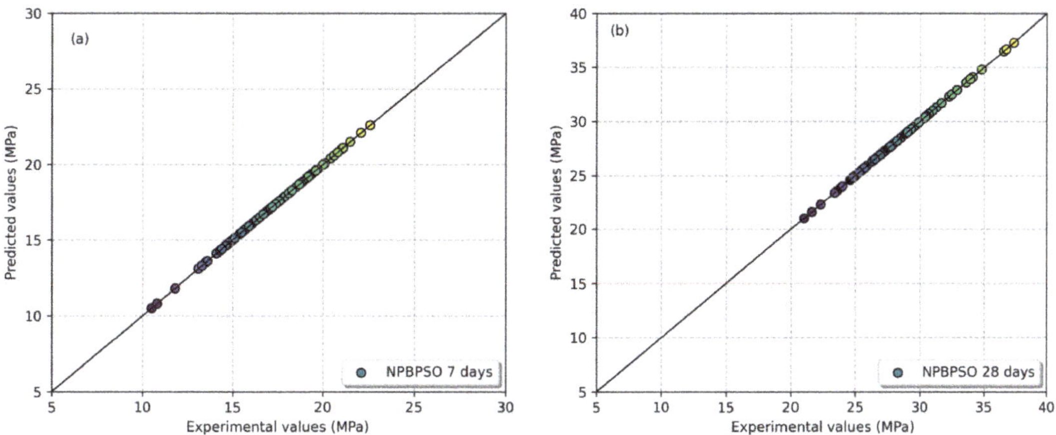

Figure 6. Experimental versus predicted compressive strength using NPBPSO (**a**) after 7 days, (**b**) after 28 days.

Finally, Table 2 shows an improvement in the results after using the improved version of the algorithm (NPBPSO), as shown in the last column of the table.

Figure 7 shows the differences between average density and others resulting from applying the new procedural binary particle swarm optimization NPBPSO after 28 days, where it has been noticed that most of the values resulting from NPBPSO matched the experimental ones. As for the points that did not match, the shift ratio reached (1 Kg/m^3),

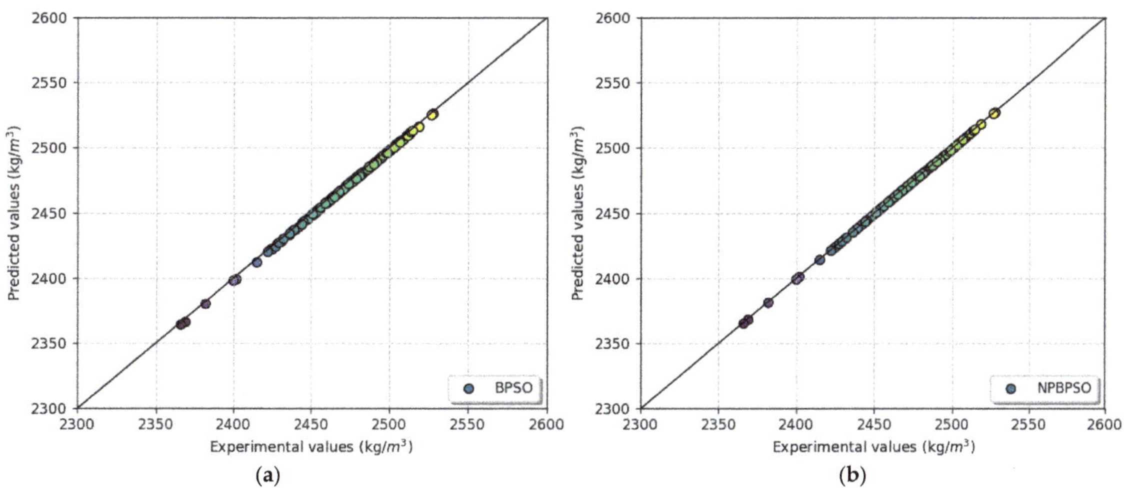

Figure 7. Experimental versus predicted values of the density at 28 days (**a**) using BPSO, (**b**) using NPBPSO.

Table 2. Statistics evaluation of compressive strength results.

		Standard Deviation	Variance	R^2	Experimental/ Predicted Average
7 days	BPSO	5.26	27.66	0.999451	1.0005
	NPBPSO	5.26	27.66	0.999423	1.0004
28 days	BPSO	7.24	52.52	0.999653	1.0003
	NPBPSO	7.24	52.52	0.999784	1.0002

Table 3 shows, through the fourth column, that the values resulting from the modified algorithm (NPBPSO) are close to the experimental values, and therefore, the proposed modification of the algorithm achieved the desired results, while the last column shows significant improvement in the time to reach the results

Table 3. Statistics evaluation of density results.

	Standard Deviation	Variance	R^2	Experimental/ Predicted Average	Execution Time
BPSO	28.77	828.12	0.992211	1.0010	60 iterations
NPBPSO	28.79	829.36	0.998454	1.0004	25 iterations

6. Conclusions

By comparing the results shown by the neural networks algorithm to obtain the compressive strength of concrete for 7 days and 28 days, respectively, (1.413 MPa—7 days and 1.91 MPa—28 days) and the results shown by the binary particle swarm algorithm BPSO (0.01112) MPa—7 and 28 days), the great superiority of the BPSO algorithm over the neural networks algorithm appears in two main important points: the first is the number of cycles or iterations that the algorithm needs to reach the results (i.e., the time to obtain the results), and the second point is the accuracy of the results that the algorithm shows.

The binary particle swarm algorithm was developed with a new procedure (NPBPSO) in order to estimate the strength of concrete and the compressive strength. Through this development, we were able to achieve two goals. One is to prevent the algorithm from falling into the problem of local solutions that the algorithm can fall into. The other is that the speed of the algorithm reaches the desired optimal solution.

After that, the new algorithm (NPBPSO) was converted into a computer program using a high-level programming language, where the data of the experimental compressive strength tests of the concrete mixtures that were conducted in the laboratories of the Faculty of Civil Engineering at the University of Aleppo and the ready-mixed concrete factory were entered.

BPSO and NPBDSO compressive strength and density outputs were obtained.

With an average shift (0.01112 MPa) in compressive strength, it was discovered that the solutions obtained using the binary particle swarm technique are close to the experimental values. However, the new procedural binary particle swarm algorithm (0.00798 MPa) shows a shift from the experimental values with approximately 30% improvement in accuracy.

The density results for BPSO and NBPSO show a shift rate of 2.311 kg/m^3 and 1.001 kg/m^3, respectively, with an accuracy improvement of about 50%. As a result of this research, it can be found that both BPSO and NBPSO techniques provide good results with better accuracy than NPBPSO because of the modification made to avoid local solutions and reduce the number of iterations.

The intensity results for both BPSO and NBPSO show a significant improvement in the speed of Al-Khwarizmi's access to optimal solutions, as the number of iterations needed to reach the optimal solution in the version (BPSO) decreased from (60 iterations) to (25 iterations) in the modified version of the algorithm (NPBPSO).

Finally, it can be found that BPSO and NBPSO AI techniques are good at predicting some properties of concrete such as compressive strength and density, which means that they can be highly recommended due to their speed, accuracy, and low cost.

Author Contributions: Conceptualization, F.A. and G.W.; methodology, M.B.H. and F.A.; software—Algorithm development, M.B.H.; validation, G.W. and M.B.H.; formal analysis, M.B.H.; investigation, F.A.A.; resources, F.A.; data curation, M.B.H.; writing—original draft preparation, F.A.A.; writing—review and editing, G.W.; visualization, M.B.H. and G.W.; supervision, F.A. All authors have read and agreed to the published version of the manuscript.

Funding: This research received no external funding.

Data Availability Statement: The data that support the findings of this study are available from the first author Fatima ALSALEH upon reasonable request.

Conflicts of Interest: The authors declare no conflict of interest.

References

1. *Concrete Mix Design Technician Study Guide*; North Carolina Division of Highways, Materials and Tests Unit: Raleigh, NC, USA, 2019.
2. Giatec Scientific Lnc. Concrete Mix Design Just Got Easier. Available online: www.Giatecscientific.com/author/admin/ (accessed on 20 June 2023).
3. Al Adday, F.; Awad, A. Pre-wetting of recycled concrete as alternative of chemical, natural and industrial waste additives. *Int. J. Innov. Technol. Explor. Eng.* **2019**, *8*, 743–748. [CrossRef]
4. Al Adday, F. Fresh and hardened properties of recycled concrete aggregate modified by iron powder and silica. *Int. J. Geomate* **2019**, *16*, 222–230. [CrossRef]
5. Awad, A.; Al Adday, F.; Qasem, A.; Al-Dulaimy, A. An Experimental Study on the Possibility of Demolition of Destroyed Concrete Buildings with Different Types of Acid. *Int. J. Eng. Res. Technol.* **2020**, *13*, 2297–2304. [CrossRef]
6. Li, H. Prediction of high-performance concrete compressive strength through novel structured neural network. *J. Intell. Fuzzy Syst.* **2023**, *45*, 1791–1803. [CrossRef]
7. Van Damme, H. Concrete material science: Past, present, and future innovations. *Cem. Concr. Res.* **2018**, *112*, 5–24. [CrossRef]
8. Golafshani, E.M.; Behnood, A.; Arashpour, M. Predicting the compressive strength of eco-friendly and normal concretes using hybridized fuzzy inference system and particle swarm optimization algorithm. *Artif. Intell. Rev.* **2022**, *56*, 7965–7984. [CrossRef]
9. Lyngdoh, G.A.; Zaki, M.; Krishnan, N.A.; Das, S. Prediction of concrete strengths enabled by missing data imputation and interpretable machine learning. *Cem. Concr. Compos.* **2022**, *128*, 104414. [CrossRef]
10. Azam, A.; Bardhan, A.; Kaloop, M.R.; Samui, P.; Alanazi, F.; Alzara, M.; Yosri, A.M. Modeling resilient modulus of subgrade soils using LSSVM optimized with swarm intelligence algorithms. *Sci. Rep.* **2022**, *12*, 14454. [CrossRef]
11. Zhu, J.; Liu, J.; Chen, Y.; Xue, X.; Sun, S. Binary Restructuring Particle Swarm Optimization and Its Application. *Biomimetics* **2023**, *8*, 266. [CrossRef]
12. Mahdavi, G.; Nasrollahzadeh, K.; Hariri-Ardebili, M.A. Optimal FRP Jacket Placement in RC Frame Structures Towards a Resilient Seismic Design. *Sustainability* **2019**, *11*, 6985. [CrossRef]
13. Mohd Yamin, M.N.; Ab. Aziz, K.; Siang, T.G.; Ab. Aziz, N.A. Particle Swarm Optimisation for Emotion Recognition Systems: A Decade Review of the Literature. *Appl. Sci.* **2023**, *13*, 7054. [CrossRef]
14. Van Zyl, J.-P.; Engelbrecht, A.P. Set-Based Particle Swarm Optimisation: A Review. *Mathematics* **2023**, *11*, 2980. [CrossRef]
15. Baron, J.; Olivier, J.-P. *Les Bétons: Bases et Données Pour Leur Formulation*; Eyrolles: Paris, France, 1996.
16. Ghali, A.; Favre, R.; Elbadry, M. *Concrete Structures: Stresses and Deformations: Analysis and Design for Serviceability*, 3rd ed.; Spon Press: London, UK, 2002.
17. Brandt, A.M. *Cement-Based Composites: Materials, Mechanical Properties and Performance*; CRC Press: Boca Raton, FL, USA, 2005.
18. Smarzewski, P. Mechanical and Microstructural Studies of High Performance Concrete with Condensed Silica Fume. *Appl. Sci.* **2023**, *13*, 2510. [CrossRef]
19. Beskopylny, A.N.; Stel'makh, S.A.; Shcherban', E.M.; Mailyan, L.R.; Meskhi, B.; Chernil'nik, A.; El'shaeva, D.; Pogrebnyak, A. Influence of Variotropy on the Change in Concrete Strength under the Impact of Wet–Dry Cycles. *Appl. Sci.* **2023**, *13*, 1745. [CrossRef]
20. Torrenti, J.M.; Reynouard, J.M.; Pijaudier-Cabot, G. *Mechanical Behavior of Concrete*; ISTE Ltd.: London, UK, 2010.
21. Xiao, Q.; Li, C.; Lei, S.; Han, X.; Chen, Q.; Qiu, Z.; Sun, B. Using Hybrid Artificial Intelligence Approaches to Predict the Fracture Energy of Concrete Beams. *Adv. Civ. Eng.* **2021**, *2021*, 6663767. [CrossRef]
22. Wang, Y.R.; Lu, Y.L.; Chiang, D.L. Adapting Artificial Intelligence to Improve In Situ Concrete Compressive Strength Estimations in Rebound Hammer Tests. *Front. Mater.* **2020**, *7*, 8870. [CrossRef]

23. Sharma, J.; Singhal, R.S. Comparative research on genetic algorithm, particle swarm optimization and hybrid GA-PSO. In Proceedings of the 2015 2nd International Conference on Computing for Sustainable Global Development (INDIACom), New Delhi, India, 11–13 March 2015; pp. 110–114.

24. Hassan, R.; Cohanim, B.; de Weck, O.; Venter, G. A Comparison of Particle Swarm Optimization and the Genetic Algorithm. In Proceedings of the 46th AIAA/ASME/ASCE/AHS/ASC Structures, Structural Dynamics and Materials Conference, Austin, TX, USA, 18–21 April 2005; American Institute of Aeronautics and Astronautics: Reston, VA, USA, 2005. [CrossRef]

25. Kumar, A.; Arora, H.C.; Kapoor, N.R.; Mohammed, M.A.; Kumar, K.; Majumdar, A.; Thinnukool, O. Compressive Strength Prediction of Lightweight Concrete: Machine Learning Models. *Sustainability* **2022**, *14*, 2404. [CrossRef]

26. Rezvan, S.; Moradi, M.J.; Dabiri, H.; Daneshvar, K.; Karakouzian, M.; Farhangi, V. Application of Machine Learning to Predict the Mechanical Characteristics of Concrete Containing Recycled Plastic-Based Materials. *Appl. Sci.* **2023**, *13*, 2033. [CrossRef]

27. Cihan, M.T. Comparison of artificial intelligence methods for predicting compressive strength of concrete. *Gradevinar* **2021**, *73*, 617–632.

28. Kandiri, A.; Golafshani, E.M.; Behnood, A. Estimation of the compressive strength of concretes containing ground granulated blast furnace slag using hybridized multi-objective ANN and salp swarm algorithm. *Constr. Build. Mater.* **2020**, *248*, 118676. [CrossRef]

29. Ngo, N.-T.; Pham, T.-P.-T.; An, L.H.; Nguyen, Q.-T.; Nguyen, T.-T.-N.; Huynh, V.-V. Prediction of axial strength in circular steel tube confined concrete columns using artificial intelligence. *J. Sci. Technol. Civ. Eng. HUCE* **2021**, *15*, 10. [CrossRef] [PubMed]

30. Goutham, S.J.; Singh, V.P. Artificial Intelligence for Compressive Strength Prediction of Concrete. *IOP Conf. Ser. Mater. Sci. Eng.* **2020**, *1004*, 12010. [CrossRef]

31. Duan, J.; Asteris, P.G.; Nguyen, H.; Bui, X.-N.; Moayedi, H. A novel artificial intelligence technique to predict compressive strength of recycled aggregate concrete using ICA-XGBoost model. *Eng. Comput.* **2021**, *37*, 3329–3346. [CrossRef]

32. Muliauwan, H.N.; Prayogo, D.; Gaby, G.; Harsono, K. Prediction of Concrete Compressive Strength Using Artificial Intelligence Methods. *J. Phys. Conf. Ser.* **2020**, *1625*, 12018. [CrossRef]

33. Nafees, A.; Javed, M.F.; Khan, S.; Nazir, K.; Farooq, F.; Aslam, F.; Musarat, M.A.; Vatin, N.I. Predictive Modeling of Mechanical Properties of Silica Fume-Based Green Concrete Using Artificial Intelligence Approaches: MLPNN, ANFIS, and GEP. *Materials* **2021**, *14*, 7531. [CrossRef]

34. Nasrollahzadeh, K.; Nouhi, E. Fuzzy inference system to formulate compressive strength and ultimate strain of square concrete columns wrapped with fiber-reinforced polymer. *Neural Comput. Applic.* **2018**, *30*, 69–86. [CrossRef]

35. Kennedy, J.; Eberhart, R. Particle swarm optimization. In Proceedings of the ICNN'95—International Conference on Neural Networks, Perth, Australia, 27 November–1 December 1995; Volume 4, pp. 1942–1948. [CrossRef]

36. Eberhart, R.; Kennedy, J. A new optimizer using particle swarm theory. In Proceedings of the Sixth International Symposium on Micro Machine and Human Science, MHS'95, Nagoya, Japan, 4–6 October 1995; pp. 39–43. [CrossRef]

37. Shen, Y.; Wang, G.; Liu, Q. Correlative Particle Swarm Optimization for Multi-objective Problems. In *Advances in Swarm Intelligence: Second International Conference, ICSI 2011, Chongqing, China, 12–15 June 2011*; Proceedings, Part II; Tan, Y., Shi, Y., Chai, Y., Wang, G., Eds.; Springer: Berlin Heidelberg, Germany, 2011; pp. 17–25.

38. Ajith, A.; Grosan, C.; Ramos, V. *Swarm Intelligence in Data Mining*; Springer: Berlin/Heidelberg, Germany, 2006.

39. Singla, S.; Kumar, D.; Rai, H.; Singla, P. A Hybrid PSO Approach to Automate Test Data Generation for Data Flow Coverage with Dominance Concepts. *Int. J. Adv. Sci. Technol.* **2011**, *37*, 15–26.

40. Chan, F.T.S.; Tiwari, M.K. (Eds.) *Swarm Intelligence*; I-Tech Education and Publishing: London, UK, 2007. [CrossRef]

41. EL-Gammal, D.; Badr, A.; El Azeim, M.A. New Binary Particle Swarm Optimization with Immunity-Clonal Algorithm. *J. Comput. Sci.* **2013**, *9*, 1542. [CrossRef]

42. Talukder, S. Mathematical Modelling and Applications of Particle Swarm Optimization by Satyobroto. Master's Thesis, Blekinge Institute of Technology, School of Engineering, Karlskrona, Sweden, 2011.

43. Zobolas, G.I.; Tarantilis, C.D.; Ioannou, G. Exact, Heuristic and Meta-heuristic Algorithms for Solving Shop Scheduling Problems. In *Metaheuristics for Scheduling in Industrial and Manufacturing Applications*; Xhafa, F., Abraham, A., Eds.; Springer: Berlin/Heidelberg, Germany, 2008; pp. 1–40. [CrossRef]

44. Alsaleh, F.F.; Hussein, R. Prediction of cement concrete resistance using industrial neural networks. *Al Furat Univ. J. Basic Sci. Ser.* **2022**, 53.

45. Amaratunga, V.; Wickramasinghe, L.; Perera, A.; Jayasinghe, J.; Rathnayake, U. Artificial neural network to estimate the paddy yield prediction using climatic data. *Math. Probl. Eng.* **2020**, *2020*, 8627824. [CrossRef]

46. Başyigit, C.; Akkurt, I.; Kilincarslan, S.; Beycioglu, A. Prediction of compressive strength of heavyweight concrete by ANN and FL models. *Neural Comput. Appl.* **2010**, *19*, 507–513. [CrossRef]

47. Tan, K. Predicting compressive strength of recycled concrete for construction 3D printing based on statistical analysis of various neural networks. *J. Build. Constr. Plan. Res.* **2018**, *6*, 71. [CrossRef]

Article

Safety Evaluation of Existing R.C. Buildings: Uncertainties Due to the Location of In Situ Tests

Vincenzo Sepe [1,2], **Mariella Diaferio** [3,*] **and Roberta Caraccio** [4]

1 Department of Engineering and Geology, University "G. d'Annunzio" of Chieti-Pescara, 65127 Pescara, Italy; vincenzo.sepe@unich.it
2 UdA-TechLab, Research Center, University "G. d'Annunzio" of Chieti-Pescara, 65127 Pescara, Italy
3 Department of Civil, Environmental, Land, Building Engineering and Chemistry, Polytechnic University of Bari, 70125 Bari, Italy
4 Independent Researcher, 00100 Rome, Italy; ing.caraccioroberta@gmail.com
* Correspondence: mariella.diaferio@poliba.it

Abstract: The paper aimed to investigate the influence, on the assessment of the structural safety level of an existing r.c building, of the different choices that the technician in charge of a structural evaluation (the "analyst") can make regarding the structural elements to be tested to obtain a prescribed level of knowledge. To this end, the case study of a reinforced concrete framed structure built in the 1960s in Italy was investigated by means of numerical analyses. The probability distribution of the estimated safety levels was evaluated in the paper by means of a Monte Carlo approach, considering the alternative selections of elements done by a large number of analysts, and the probability of unsuccessful safety estimations is discussed for the knowledge levels considered in the Italian technical codes and the Eurocodes.

Keywords: confidence factor; assessment of seismic safety index; Monte Carlo simulation; existing buildings

1. Introduction

Citation: Sepe, V.; Diaferio, M.; Caraccio, R. Safety Evaluation of Existing R.C. Buildings: Uncertainties Due to the Location of In Situ Tests. *Appl. Sci.* **2024**, *14*, 2749. https://doi.org/10.3390/app14072749

Academic Editor: Kang Su Kim

Received: 7 February 2024
Revised: 16 March 2024
Accepted: 18 March 2024
Published: 25 March 2024

The quantitative assessment of the safety of an existing building requires, as is known, the implementation of a numerical model that is able to describe the geometric and mechanical characteristics of the building. For a reinforced concrete (r.c.) building, according to the Italian technical codes [1,2], ASCE 41-17 [3], and the Eurocodes [4], this requires, in particular, knowledge of the construction details (e.g., the quantity and arrangement of reinforcements) and the strength of the materials (concrete and steel).

The formulation of a numerical model, in all its phases, is inevitably affected by epistemic uncertainties, e.g., those due to the possible alternatives in the choice of the representative model of the building [5–7] or of the actions to be considered, and by random uncertainties, e.g., those due to the intrinsic variability of the mechanical parameters throughout the structure, which can therefore be described only from a probabilistic point of view.

This is the consequence of the variability of these properties within the building, which may be due to the different characteristics of the materials used, corresponding to different days during which the concrete was poured, different casting conditions, etc.

It is worth noting that, in practical investigations, modern survey technologies can ensure the exact description of the geometrical features of the investigated building, while the design documentation, if available, may provide data regarding the dimensions and positions of the steel bars inside each structural element. Of course, the acquisition of a given level of knowledge of the investigated building depends on the number of tests that are conducted to assess the mechanical properties of the materials.

For this reason, the codes usually suggest a minimum number of tests to be performed, but even if the number of such tests and the experimental technique are the same, a huge number of combinations of investigated elements and of points within the same element may be chosen, depending on the technician's criteria and/or for logistic reasons.

This is an intrinsic source of uncertainty, even neglecting other aspects that can have effects on the results of experimental investigations, such as the environmental conditions in which the tests are performed, the instruments utilised, and several other elements depending on the chosen technique [8–10].

Therefore, for the same knowledge level, a huge number of different experimental results may be obtained, and, consequently, for each different combination of experimental data, this may correspond to a different value of the assessed safety level of the building with respect to seismic events.

For random uncertainties, usually, codes suggest discrete knowledge levels with the aim of describing the availability of information on the existing building examined. Based on the acquired knowledge level, two possible approaches are suggested by standards: the amplification of the force demand or a reduction in the materials' resistance. For example, FEMA-356 [11] prescribes the introduction of a knowledge factor, dependent on the acquired knowledge level, which amplifies the force demand.

However, for the random uncertainty of mechanical parameters—which is the focus of this paper—the Italian technical codes [1,2], ASCE 41-17 [3], and the Eurocodes [4] introduce a confidence factor (CF) which reduces the mean resistance of materials as provided by on-site tests, with the CF depending on the knowledge level (KL). A larger value of CF is suggested by such codes for a lower number—and therefore a lower representativeness—of the tests carried out. In detail, the codes define three knowledge levels based on the number of acquired data: KL1 is limited knowledge, KL2 is extended knowledge, and KL3 is full knowledge. It is worth noting that different values for the confidence factor correspond to the different knowledge levels. According to the Italian technical code [2] and the Eurocodes [4], the confidence factor related to limited knowledge is $CF_{KL1} = 1.35$, while the confidence factor corresponding to extended knowledge is $CF_{KL2} = 1.20$ and the confidence factor corresponding to full knowledge is $CF_{KL3} = 1.00$.

The possibility of taking into account the many uncertainties that arise in the procedure for the seismic safety assessment of existing buildings with a single confidence factor applied to material strengths has been, and is still, widely debated in the scientific literature [12–16]. In detail, starting from the approach proposed in [12], the present research explores the effects of uncertainties related to the choice of on-site test locations on the assessment of seismic safety indexes.

Specifically, this paper investigates, through the case study of a typical reinforced concrete structure of the 1960s built in Italy, the probability that a safety assessment based on a pre-defined knowledge level and the value consequently assigned to the confidence factor can actually lead to the technician in charge obtaining a precautionary estimate of the building's capacity to resist seismic actions.

To this end, the actual level of seismic safety of the building—evaluated by assuming perfect knowledge (never available in practice) of the mechanical properties of all the structural elements [12]—is compared with the safety level evaluated under the assumption that the tests were carried out, and thus that the mechanical properties were estimated, only for a small number of structural elements. In this case, as is usual, the safety index was assessed by introducing the elements' average strengths divided by a CF whose value was fixed according to the level of knowledge corresponding to the number of tests.

For a given knowledge level, the probability distribution of the estimated safety level was evaluated in the paper by means of a Monte Carlo approach, which allows phenomena with significant uncertainties to be modelled by considering randomly generated scenarios and by evaluating the outcomes of the investigated model for each of them. In this way, the mean value of an output, its distribution, and its minimum or maximum values can be estimated. This method is used widely to evaluate the impact of risk and uncertainties in the prediction of outcomes. In the case examined here, the Monte Carlo method was applied to simulate a large set of alternative selections of the structural elements to be tested, with the same number of such elements randomly picked out, and by estimating the safety index of the building for each scenario.

Moreover, to take into account that in real cases perfect knowledge of structural characteristics can never be attained, nine alternative realisations of the reference model—with equivalent statistical properties—were considered, and therefore a total number of 9×10^3 analyses were carried out for each one of the three knowledge levels prescribed by the codes.

This procedure was intended to evaluate the effect on safety estimation uncertainty of different possible choices of structural element locations that, *ceteris paribus*, can be made by the technician in charge (denoted as the analyst in the following, according to [12]), even when the number of such tests is assumed to be constant.

In this way, the paper deals with an intrinsic uncertainty in safety evaluation, starting from the observation that no strategy can assure that the different positions of the tests chosen by different analysts—even when they perform the same number of tests—will lead to unique "best" values of the average strengths of concrete and reinforcing steel. Thus, the paper explores how the unavoidable uncertainty of these values affects the evaluation of the safety index obtained by a given analyst (i.e., what the probability is for a given analyst of underestimating or overestimating the "true"—but never really knowable—safety index). The rest of the paper is organised in four parts: Section 2 describes the case study and the main assumptions of the performed analyses, Section 3 describes the numerical model implemented to assess building safety levels, Section 4 reports the statistical analyses and discusses the effects of the confidence factor values on the accuracy of the assessed vulnerability index, and Section 5 presents the conclusions.

2. Case Study

The sample case described in the manuscript is a five-floor r.c. structure representative of a widespread typology in the existing building stock built in Italy in the 1960s, with a structure designed for gravitational loads only. Although its geometrical and mechanical characteristics have been suggested by typical real situations, the mechanical model analysed here does not reproduce—and therefore cannot be referred to—any specific existing building. The building plane covers about 800 m², with global dimensions of 58 m × 13 m and a total height of 16 m.

The building is realised by infilled r.c. frames and hollow bricks, with floors which can be considered as rigid in their plane. In the 1960s, the main part of the Italian territory was not classified as a seismic-prone zone; thus, the technical codes prescribed the design of buildings that would be capable of resisting gravitational loads and static loads. Typical of many buildings at that time, the loads of the case study building are carried by frames in the longitudinal direction, while only a few of the columns are connected by beams in the transversal direction and belong, therefore, to frames that are able to resist horizontal actions. In detail, as shown in Figure 1a, four longitudinal frames and four transversal frames are present. The span lengths of the beams range between 2.10 m and 5.30 m, the inter-story height is 3.30 m, and the mass distribution is uniform in plan and in elevation.

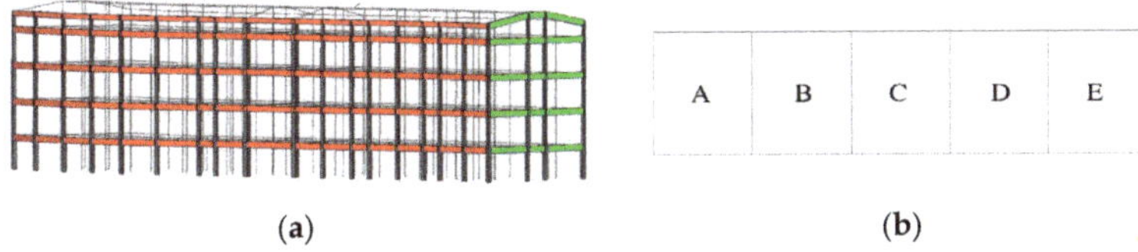

(a)   (b)

Figure 1. A schematic axonometric view of the FE model (**a**) and a plan view (**b**) of the building. A longitudinal and a transversal frame have been highlighted in the FE model (with beams shown in red and green colour, respectively). The letters in the plan view represent the different parts considered in the analyses, each one characterised at a given floor by constant mechanical properties of concrete and steel, as reported in Tables 1 and 2.

Table 1. Reference model M_0 (actual configuration): concrete compressive strength for each part of the building in Figure 1b.

	Concrete Strength [MPa]				
	A	**B**	**C**	**D**	**E**
First storey	16	28	16	13	10
Second storey	16	23	29	22	10
Third storey	28	22	22	20	17
Fourth storey	17	22	24	28	9
Fifth storey	22	18	21	28	10

Table 2. Reference model M_0 (actual configuration): steel strength for each part of the building in Figure 1b.

	Steel Strength [MPa]				
	A	**B**	**C**	**D**	**E**
First storey	360	440	360	440	460
Second storey	460	340	300	380	300
Third storey	360	380	480	500	400
Fourth storey	320	460	500	460	360
Fifth storey	340	460	340	380	320

The elements' cross-sections range between 40 cm × 30 cm and 25 cm × 25 cm for the columns and between 40 cm × 60 cm and 20 cm × 50 cm for the beams.

The geometry, load conditions, and design criteria of the case study are representative of typical residential r.c. buildings of the 1960s in Italy. In particular, the quantity and arrangement of steel reinforcements have been designed for gravitational loads according to the allowable stress approach.

Assuming a reference period (V_R) equal to 50 years, according to the Italian Building Code [1], a seismic hazard corresponding to medium–high seismic level was considered, with a peak ground acceleration (PGA) equal to 0.18 g at the Life-Safety Limit State (SLV).

3. Finite Element Model of the Building

A 3D finite element model of the building was implemented with beam elements and floors acting as rigid diaphragms, assuming the columns to be fixed at the base (Figure 1a).

In order to represent the possible variation in the mechanical properties, corresponding, for example, to different conditions of the preparation, casting, or curing of the concrete or to different batches of steel, the value of the concrete compressive strength (f_c) was considered to be in the range of 9 MPa to 29 MPa, while for the steel strength (f_s), the range of variability was considered to be between 300 MPa and 500 MPa.

On this basis, each floor of the building was divided into five parts (see Figure 1b). Each part was considered representative of a different casting day or batch of steel, and, for each part and for each floor, the values of the concrete and steel strengths were randomly chosen in the mentioned ranges, as summarised in Tables 1 and 2, respectively.

The aforementioned procedure aimed to fix the "real" distribution of strengths in the examined structure and, on that basis, to evaluate the "true" seismic safety index of the "real" structure, which represents the final scope of seismic vulnerability assessment.

For the "real" structure, the randomly extracted concrete strength values were characterised by a mean value of 19.6 MPa, a standard deviation equal to 6.1 MPa, and a coefficient of variation equal to 0.3, while the chosen steel strengths had a mean value of 396 MPa, a standard deviation of 63 MPa, and a coefficient of variation equal to 0.16. As it is reasonable to expect, the steel strength was characterised, therefore, by a variability lower than that of the concrete.

These values were assigned to the beams and columns of each of the different regions (Figure 1b) of the reference model of the structure, from now on denoted as M_0 (i.e., model 0) to distinguish it from alternative realisations (Section 4) of the same "true" model, and represent the actual mechanical strengths of these elements. The perfect knowledge of these—never attainable in practice, but here introduced as a hypothesis—allowed evaluation of the "true" level of seismic safety of the building once all the other geometrical and loading details were also known, as assumed in the following, in order to focus attention on the random variation of mechanical properties.

The dynamic analyses of the finite element model, obtained by introducing the values in Tables 1 and 2, were performed considering the spectral acceleration corresponding to the Life-Safety Limit State and assuming a behaviour factor (q) equal to 2, and the seismic safety index (α) of the model was evaluated as the ratio between the peak ground acceleration capacity (PGA_C) and the peak ground acceleration demand (PGA_D).

It is worth noting that the assessed seismic safety index was evaluated by considering "exact" strength values for each of the structural elements and not a mean value constant for the whole structure.

The obtained seismic safety index for the reference model was 0.315, and the analysis highlighted that the safety of the structure was mostly conditioned by the combined axial compression and bending moment capacity. Of course, the structural element capacity that conditions the safety of buildings in other cases studies may be different from the one considered here. However, the results discussed in Section 4 highlight some issues that are also expected to be observed in other possible structural typologies and/or different mechanisms of limit state attainments.

The subsequent step of the numerical investigation described here aimed at evaluating the impact that, during the acquisition of a fixed knowledge level, the alternative locations of the in situ tests selected by different analysts can have on the estimated safety level, all other conditions being equal. Different locations may in fact correspond to different experimentally acquired strengths and, consequently, to different average values of the strengths adopted by the technician in charge of the structural analyses on the basis of the available data. The obtained mean values and the corresponding confidence factors were then utilised to assess the seismic safety indexes.

To compare the evaluated indexes, it was assumed that the attainment of the Life-Safety Limit State is always conditioned by the combined axial compression and bending moment capacity of the columns. Without affecting the more general validity of the proposed approach, which will be further explored in the following part of this research, such an assumption allows, by means of the interpolating surfaces introduced in Section 4, a reduction in the computational effort required.

4. Effects of Uncertainties Related to Test Locations

According to the Italian technical codes [1,2], the seismic vulnerability assessment of an existing structure always requires the execution of in situ tests on the structural elements. Their number depends on the choice of the knowledge level, which in turn is related to the confidence factor (CF) (see Section 1) introduced to pass from the average values of test results to the design values of materials' strengths. All other factors being equal (the geometry of the building, the quantity and arrangement of the reinforcements, etc.), as assumed in this paper, the design strength of concrete and steel, on the other hand, affect the safety level of the building provided by the structural analysis.

However, even when the number of tests can be assumed as fixed (e.g., according to the indications of technical codes), the technician in charge may make alternative choices regarding the position of the tests. Therefore, for the same knowledge level, different analysts may estimate different mechanical properties for the same building.

As a consequence, even if the numerical model developed to assess the response of the investigated structure and the confidence factor are the same with respect to any other characteristic of the building, different analysts may estimate different values of the seismic safety index because of the variation in the mean values of the concrete and steel strengths that each analyst assumed to be representative of the existing building on the basis of the available in situ tests. Mean values of concrete and steel strength can therefore be viewed as samples for a statistical analysis of the seismic safety index, as discussed below.

The procedure here proposed is based on a Monte Carlo simulation to select the structural elements whose strengths are assumed to be measured in situ, with the aim of quantifying the uncertainty in the evaluated safety index as the location of the tests varies while keeping their number fixed.

In detail, the probability distribution of the safety indexes evaluated by 100 different analysts was obtained, i.e., considering the mean values of concrete and steel strengths corresponding to 100 different combinations of test locations.

To reduce the computational effort, however, for each given knowledge level, only 15 numerical FEM models were analysed, each one corresponding to a different combination of the mean strengths (f_c and f_s) virtually obtained by different analysts. As is usual in engineering practice, the obtained mean values were considered to be constant throughout the whole structure during the assessment of the safety index, which is different from the procedure (Section 3) for the evaluation of the actual index, where the knowledge of the local strength of each element was assumed and introduced in the model. The evaluated safety indexes for each f_c, f_s pair are listed in Table 3 for knowledge level KL2 and a confidence factor (CF_{KL2}) equal to 1.2 [2]. The safety indexes (α_{KL2}) corresponding to other combinations of f_c and f_s may be obtained by fitting the obtained value by means of a polynomial law (see Figure 2). Similarly, Table 4 shows the safety indexes (α_{KL1}) corresponding to knowledge level KL1 and $CF_{KL1} = 1.35$.

Table 3. Seismic safety index (α_{KL2}) for 15 different combinations of mean concrete and steel strength values corresponding to knowledge level KL2.

	300 MPa	340 MPa	380 MPa	420 MPa	460 MPa
15 MPa	0.242	0.268	0.283	0.293	0.315
20 MPa	0.253	0.278	0.288	0.302	0.323
25 MPa	0.258	0.283	0.297	0.310	0.327

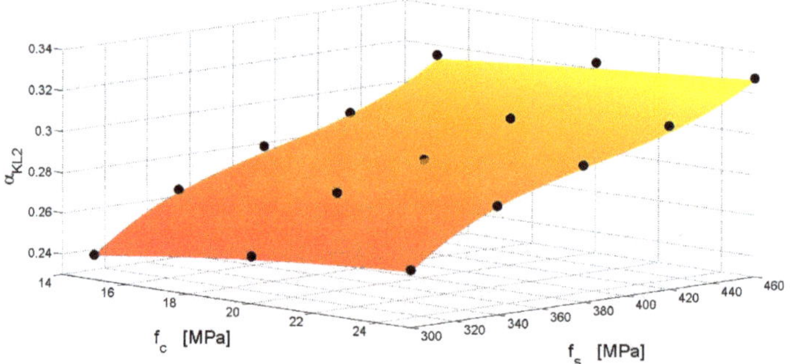

Figure 2. Polynomial surface (of order 3) fitting the values of the seismic safety indexes (α_{KL2}) in Table 3 (black dots). f_c and f_s are the mean concrete compressive strength and the mean steel strength, respectively.

At first the analyses with knowledge level KL2 are described below, assuming that no other information (e.g., results of tests during the building process) was available on the strength of materials. Under this hypothesis and with reference to the columns, i.e., the elements by which the safety level is conditioned in this case study (see Section 2), the Italian technical codes [1,2] suggest extended in situ testing that involves the testing of two samples of steel bars per storey and two concrete cores for each 300 m^2 of floor, and then six cores per storey for the investigated building (about 800 m^2 plan). For the five-floor building considered here, therefore, the strengths (f_c and f_s) to be considered in the safety evaluation were obtained for each analyst as the average of the strengths of 30 cores and 10 steel bars extracted by columns.

Table 4. Seismic safety index (α_{KL1}) for 15 different combinations of mean concrete and steel strength values corresponding to knowledge level KL1.

	300 MPa	340 MPa	380 MPa	420 MPa	460 MPa
15 MPa	0.230	0.247	0.263	0.283	0.293
20 MPa	0.236	0.253	0.268	0.288	0.297
25 MPa	0.242	0.258	0.273	0.293	0.302

A Monte Carlo approach was adopted to define 100 different cases. For each case, the procedure simulated the extraction of six values of concrete strength and two values of steel strength for each floor, corresponding to a random subset of the columns, and, based on such values, the mean values of the concrete compressive strength (f_c) and steel strength (f_s) were evaluated. The obtained values are plotted in Figure 3.

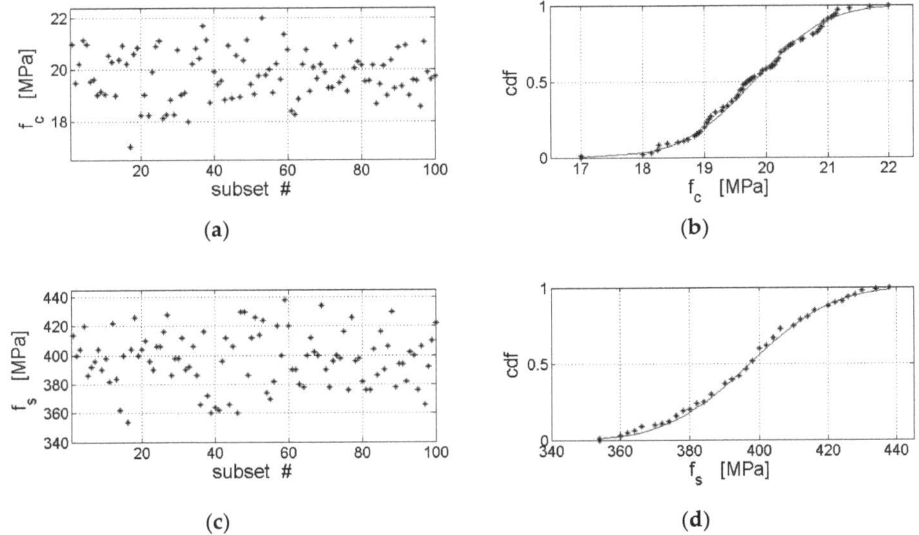

Figure 3. Mean concrete compressive strength (**a**) and mean steel strength (**c**) for 100 different subsets of structural elements. Cumulative distribution functions of concrete (**b**) and steel (**d**) mean strengths and their Gaussian interpolation (red lines).

By means of the interpolating surface in Figure 2, the seismic safety index was then evaluated for each combination of strength values in Figure 3, and their cumulative distribution is shown in Figure 4. Moreover, in Figure 4, the cumulative distribution of the Gaussian curve which fits the evaluated safety indexes is plotted.

The red broken line in Figure 4 corresponds to the safety index evaluated by means of the reference model (M_0) (Section 2), which assumed perfect knowledge (never attainable in practice) of the effective strength of each structural element. The safety index of the reference model can therefore be assumed as an ideal target, whose comparison with the indexes evaluated by different analysts may allow estimation of the reliability of the safety level attainable when a finite (and usually small) number of in situ tests are available, as in real cases.

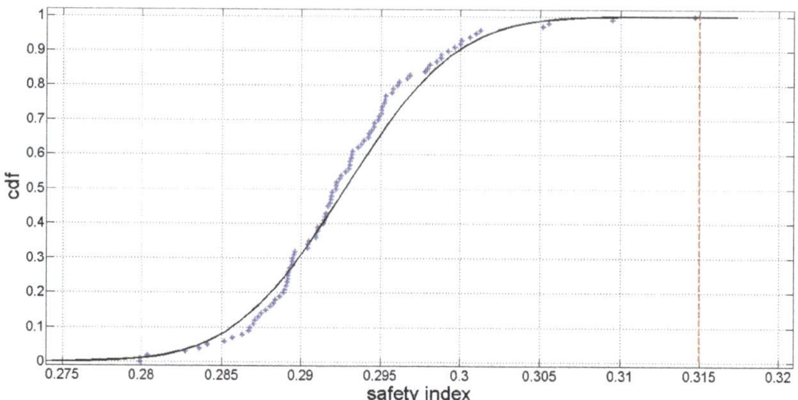

Figure 4. Cumulative distribution of seismic safety indexes obtained for knowledge level KL2 by performing a Monte Carlo simulation: the red broken line represents the actual safety index evaluated by means of the reference model (M_0); the blue stars represent the indexes corresponding to each of the 100 cases; the black line represents the Gaussian curve approximating the numerical data.

Figure 4 highlights that for the randomly selected locations of in situ tests corresponding to the average values of f_c and f_s in Figure 3, the probability of overestimating the seismic safety is negligible once the confidence factor $CF_{KL2} = 1.2$ is adopted in the analysis, as suggested by the Eurocodes [4] and the Italian codes [2] for the level of knowledge KL2.

To check with a wider set of analysts the probability of obtaining an unsafe evaluation of the safety index, the aforementioned procedure was repeated another nine times for different sets of 100 randomly selected values of concrete and steel strength, leading to another nine cumulative distribution functions of the safety index, which are shown in Figure 5 together with the curve in Figure 4. A total number of 10^3 combinations were therefore analysed to obtain the data plotted in Figure 5.

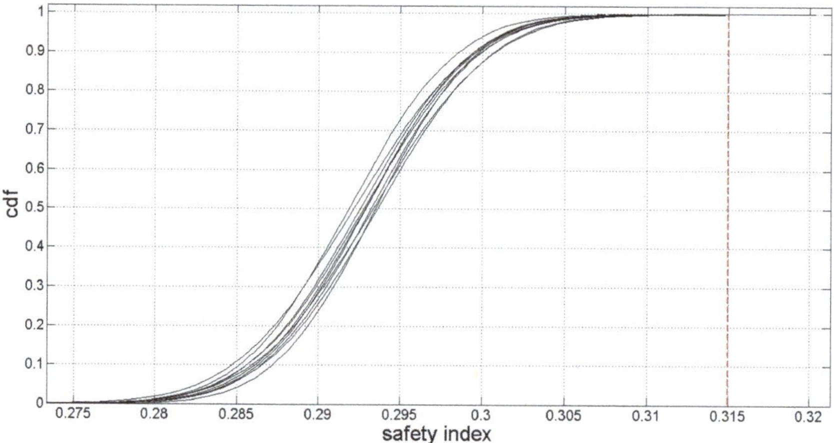

Figure 5. Cumulative distribution functions of seismic safety indexes obtained for knowledge level KL2 by performing the Monte Carlo simulation 10 times: the red broken line represents the actual safety index (α_0) of the reference model (M_0).

This confirms that for the reference model (M_0) corresponding to the concrete and steel resistance in Tables 1 and 2, this CF_{KL2} value is well calibrated because it effectively guarantees, in almost all cases, a slightly precautionary estimation of the structural capacity.

Due to the inhomogeneity of the mechanical properties in Tables 1 and 2, however, the capacity of the reference model (M_0) considered up to this point—and therefore its safety risk ($\alpha_0 = 0.315$)—was conditioned by the effective distribution of such characteristics in the different regions of each storey, which is never completely knowable in practice.

To evaluate the impact of such a distribution on the capacity of the model assumed as a reference to describe the real structure, therefore, eight alternative models (M_1–M_8) were considered to be representative of eight different realisations of the building described so far. These were obtained by randomly re-distributing the steel and the concrete properties of the elements between the different 25 regions of the building and thus obtaining different combinations of f_c, f_s without changing the parameters (i.e., the mean values and standard deviations) of their values. The procedure performed for the M_0 model was also carried out for each of the eight new models, leading to 10 cumulative distribution functions for each model, i.e., a total number of 9×10^3 analyses.

For each model, Figure 6 compares the cumulative distribution functions of the safety indexes evaluated by 10 different sets of analysts with the actual safety index of the model.

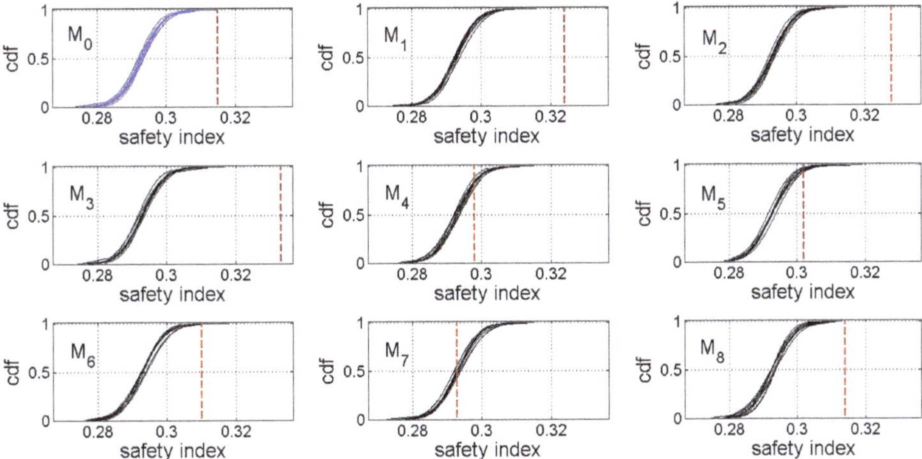

Figure 6. Cumulative distribution functions of seismic safety indexes obtained for knowledge level KL2 by performing the Monte Carlo simulation 10 times for nine different models (M_0–M_8) with the same geometrical features and mean values of material strengths. The red broken lines represent the actual safety indexes of each model.

Figure 7a shows, for each model and for each set of analysts, the probability of underestimation (p^*) of the actual safety index (blue circles) and the average $\mu_p{}^*$ (red crosses)—also reported in Table 5—as a function of the actual safety index of each model. Although for one of the models (model M_7) the mean probability of overestimating the actual safety index was about 50%, the overall mean value of the probability of underestimation (p^*) for the nine models was 92% (Table 5), thus confirming that the confidence factor CF_{KL2} allows a cautionary evaluation of the seismic safety index.

Based on the statistical distribution of the safety indexes of the nine models (M_0–M_8) and denoting their average values and standard deviations μ_α and σ_α, respectively, Figure 7a also reports, as a vertical green line, the average value ($\mu_\alpha = 0.313$), and it can be verified that, in this case, the probability of underestimation of the safety index by the analysts is almost equal to 100%. Figure 7a also shows with a green broken line the value of the safety index $\mu_\alpha - \sigma_\alpha = 0.299$ with an 84% probability of exceedance, and in this case the probability of underestimation of the safety index by the analysts is equal to 87%.

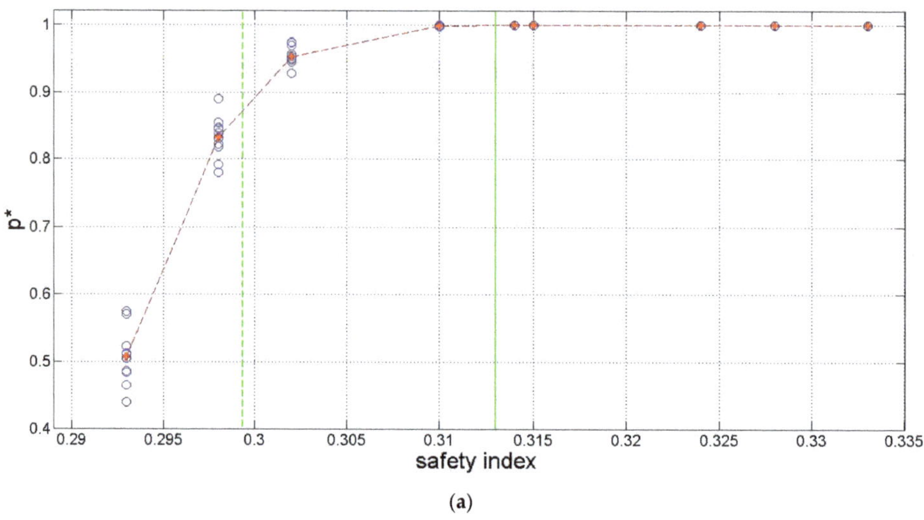

(a)

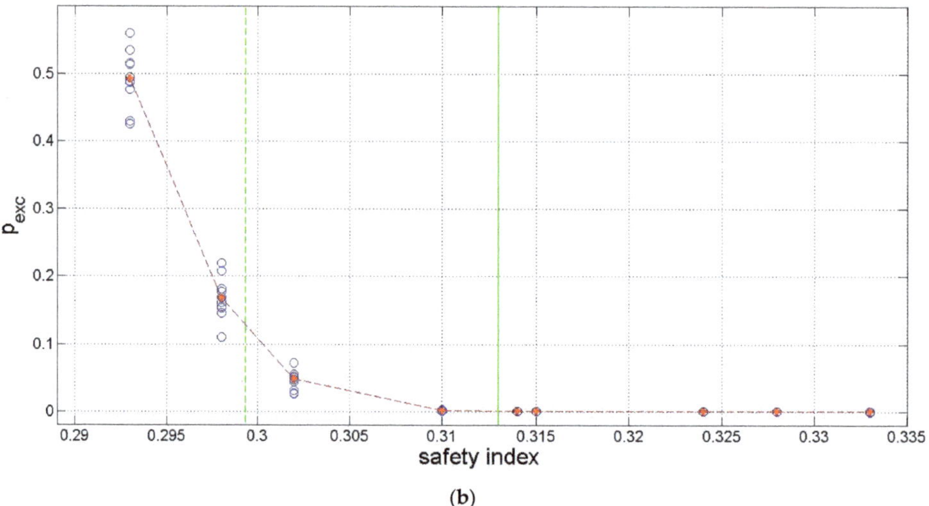

(b)

Figure 7. (**a**) Probability of underestimation (p*) vs. actual safety index and (**b**) probability of exceedance ($p_{exc} = 1 - p^*$) for knowledge level KL2. The green line represents the mean value (μ_α) of the actual safety indexes related to the nine models (M_0–M_8); the green broken line represents a safety index equal to μ_α-σ_α.

The same results can be shown by plotting the probability (p_{exc}) of exceeding the actual safety index, as reported in Figure 7b.

It is worth noting that for a major part of the analysed models (i.e., seven models out of the total number of nine), almost all the analysts underestimated the building's actual structural capacity with an average probability of at least 95% and that, considering all of the nine models, the mean value of the probability of underestimation was equal to 92%. Thus, adopting a confidence factor $CF_{KL2} = 1.2$, the probability of overestimating the real structural capacity of the sample case discussed here is limited to about 8%. This value can be considered acceptable when compared with other sources of uncertainties which are usually involved in structural capacity assessment, such as those related to the choice of software, the method adopted for the analysis (linear or non-linear), and those strictly connected to the experimental tests.

Table 5. Actual safety index (α) and the corresponding average probability of underestimation (μ_{p*}) for each of the nine models.

Model	Actual Safety Index	Average Probability of Underestimation (μ_{p*})
M_0	0.315	99%
M_1	0.324	100%
M_2	0.328	100%
M_3	0.333	100%
M_4	0.298	83%
M_5	0.302	95%
M_6	0.310	99%
M_7	0.293	51%
M_8	0.314	99%
Mean value	**0.313**	**92%**

To investigate the influence of the number of in situ tests, the procedure was repeated by assuming that three concrete cores and one steel bar were extracted at each of the five storeys, with a total number of 15 values of concrete strength and 5 values of steel strength randomly selected in a subset of the columns for each of the 100 simulated samples.

In this case, corresponding to the knowledge level KL1 of the Italian technical code, a confidence factor $CF_{KL1} = 1.35$ was suggested [2]. As shown in Figure 8, which reports the cumulative distribution functions of the seismic safety indexes obtained by means of an interpolating surface based on Table 4 (confidence factor $CF_{KL1} = 1.35$), the probability of an unsafe evaluation becomes negligible, as also shown in Figure 9, which was obtained by means of the same procedure adopted to evaluate Figure 7.

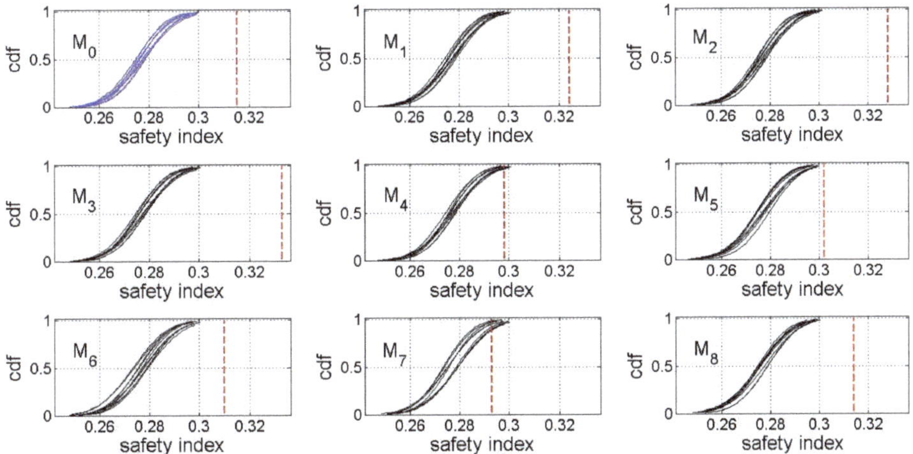

Figure 8. Cumulative distribution functions of seismic safety indexes obtained for knowledge level KL1 by performing the Monte Carlo simulation 10 times for nine different models (M_0–M_8) with the same geometrical features and mean values of material strengths. The red broken lines represent the actual safety indexes of each model.

This confirms that for the lowest level of knowledge allowed by the codes, the value of $CF_{KL1} = 1.35$ is well calibrated in the case study described here.

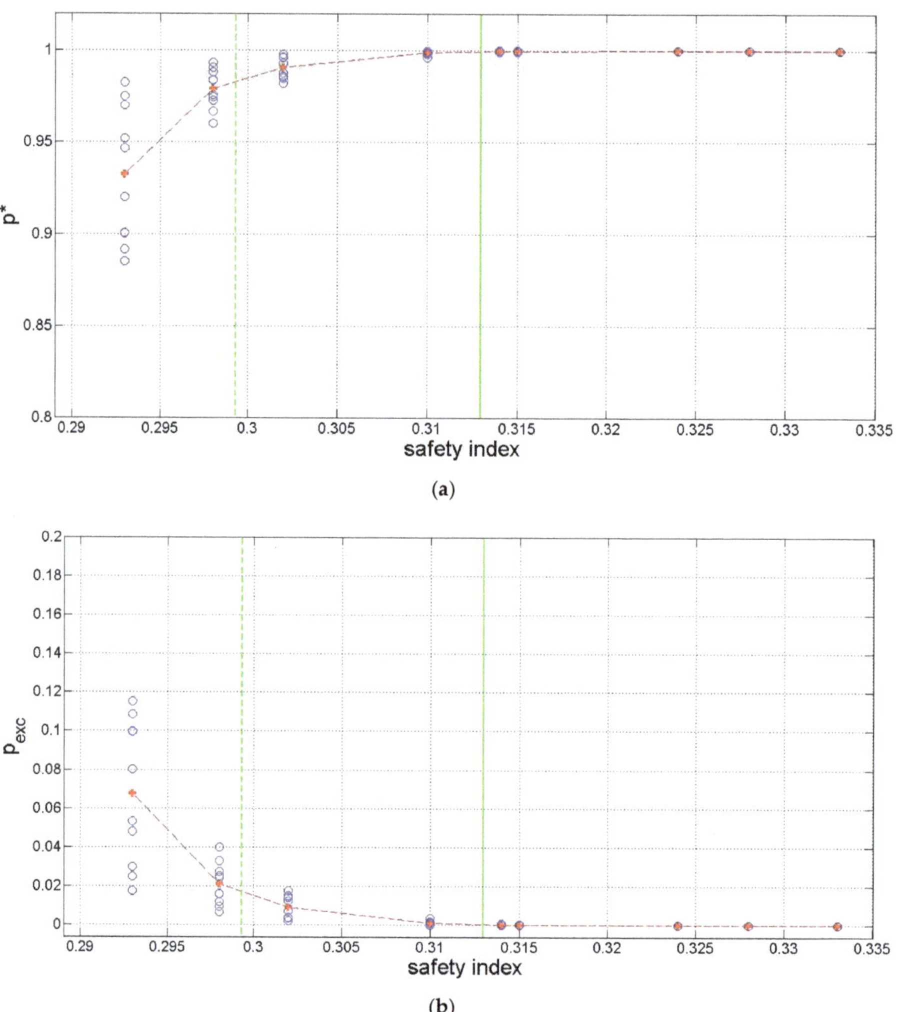

Figure 9. (**a**) Probability of underestimation (p*) vs. actual safety index and (**b**) probability of exceedance ($p_{exc} = 1 - p^*$) for knowledge level KL1. The green line represents the mean value (μ_α) of the actual safety indexes related to the nine models (M_0–M_8); the green broken line represents a safety index equal to μ_α-σ_α.

The above-described procedure was also carried out to investigate the case of the full-knowledge level, i.e., KL3, when the confidence factor is equal to 1.0 [2], and, in this case, the seismic safety index of the sample building discussed here was always overestimated.

Although this aspect deserves to be further explored in the future by examining different structures, what seems to emerge clearly from these circumstances is that the choice of the confidence factor cannot always and only be related to the number of available tests and that it should instead explicitly take into account their higher or lower homogeneity. Accordingly, it should eventually be assumed to be equal to 1 only when the dispersion of experimental results is particularly low, unlike the case reported here.

For the building considered here, it was in fact observed that, even if the nine alternative (and statistically equivalent) models M_0–M_8 are characterised by the same mean values of concrete and steel strength, i.e., $f_c^* = 19.64$ MPa and $f_s^* = 396$ MPa, their safety indexes vary with respect to their mean values (μ_α) in an interval ± 1.46 times their standard deviations σ_α. Moreover, the safety indexes of models M_0–M_8 (probably because they are conditioned by the structural elements with the lowest resistance) were generally lower than the "ideal" safety index ($\alpha^* = 0.337$), corresponding to the hypothesis of absolute homogeneity of the mechanical characteristics of concrete and steel, i.e., equal to f_c^* and f_s^* for each structural element, similarly to the results of [12]. As a consequence, assuming the average values of experimental tests as design strengths, as when CF = 1 is assigned, can lead to the overestimation of safety indexes in cases of significant material inhomogeneity, as in those of models M_0–M_8.

This suggests that, in such cases, the confidence factor should be precautionarily assumed to be greater than 1, independently of the number of in situ tests, in order to take into account that the structural safety of a building can be highly conditioned by the structural elements with the lowest capacities. The suggestion of adopting in such cases a confidence factor greater that 1 is in accordance with results in [12].

Future research will therefore be devoted to applying the above-described procedure to different probability distributions of mechanical properties, with the aim of exploring how and to what extent the calibration of the confidence factor should include the dispersion of experimental results, an issue that is only qualitatively considered in the Italian codes [2].

Extending the results discussed here, which were obtained by adopting the linear analysis with behaviour factor, the future steps of the research will also consider the effect of the introduction of non-linear analyses (for example, push-over analyses, etc.) for the assessment of the seismic safety indexes of buildings with mechanical parameter distributions similar to those examined here, with the aim of verifying the influence of their mean values and their variations on safety indexes.

5. Conclusions

In this paper, the case study of a framed reinforced concrete building designed to resist only gravitational and static loads, representative of a widespread typology of structures built in the 1960s in Italy, is examined. Indeed, buildings of this kind usually stand in seismic-prone zones, according to the current codes, and this requires improvements to their structural and building performances to guarantee acceptable levels of safety and habitability conditions. The actuality of this topic is also confirmed by the attention devoted by building codes to the analysis of existing buildings.

These codes prescribe, among other procedures, the execution of experimental tests, the numbers of which depend on the desired knowledge level.

Although this topic is still debated in the scientific literature [12–16], the Italian technical codes [1,2] and the Eurocodes [4] prescribe the introduction of a confidence factor, based on the acquired knowledge level, in the assessment of seismic safety indexes, which serves to reduce the mean values of experimentally evaluated mechanical strengths of structural materials.

It is worth noting that the positions in which the tests are performed depend on several factors, and, clearly, different analysts may choose different locations; consequently, they may assess different values of the mechanical properties and obtain different safety indexes.

In the present paper, starting from the approach in [12], the case study of a residential r.c. building built in the 1960s in Italy is discussed with the aim of investigating the implications of the different choices that the technician in charge (the "analyst") can make regarding the structural elements to be tested for obtaining a prescribed level of knowledge.

The case study of an r.c. building was examined, assuming, initially, perfect knowledge of the material mechanical properties of each structural element and, on this basis, evaluating the "actual" safety index.

Then, numerical analyses were performed by considering the three knowledge levels prescribed by the building technical codes and, for each level, numerically simulating the acquisition of experimental strengths.

In detail, a Monte Carlo approach was adopted to consider the effects of the different locations of the tests. A numerical simulation of 10^3 possible combinations of concrete and steel mechanical properties, as virtually assessed for each knowledge level suggested by the codes, was carried out, and the corresponding seismic safety indexes were evaluated by introducing the confidence factor suggested by technical codes for each knowledge level.

To evaluate the capacity of the model assumed as a reference to describe the real structure, the same procedure was applied to nine different realisations of the same building, and thus a total number of 9×10^3 combinations were analysed for each knowledge level.

The analyses performed for the sample case show that the confidence factors corresponding to KL1 and KL2 [2] could effectively provide a precautionary assessment of the structural capacity of the building. With the level of knowledge KL2, in fact, for the analysed models (representative of nine alternative realisations of the reference model with equivalent statistical properties), almost all the analysts underestimated the actual building's structural capacity with an average probability of 92%. Thus, adopting a confidence factor $CF_{KL2} = 1.2$, the probability of overestimating the real structural capacity of the sample case discussed here was limited to about 8%, while for the level of knowledge KL1 ($CF_{KL1} = 1.35$), the probability of an unsafe evaluation became negligible.

For the full-knowledge level, the safety index was instead overestimated when a value of $CF_{KL3} = 1$ was assumed, thus suggesting that the calibration of this parameter should not only depend on the number of available in situ tests but must consider their dispersion.

As is widespread in engineering practice, in the present paper a linear numerical analysis was performed. However, further studies must be performed to extend the findings to the assessment of safety indexes by means of non-linear approaches.

Author Contributions: Conceptualisation, V.S., M.D. and R.C.; methodology, V.S., M.D. and R.C.; software, V.S. and R.C.; validation, V.S. and M.D.; formal analysis, V.S. and M.D.; investigation, V.S., M.D. and R.C.; writing—original draft preparation, V.S. and M.D.; writing—review and editing, V.S., M.D. and R.C.; visualisation, V.S., M.D. and R.C.; supervision, V.S. and M.D. All authors have read and agreed to the published version of the manuscript.

Funding: This research received no external funding.

Institutional Review Board Statement: Not applicable.

Informed Consent Statement: Not applicable.

Data Availability Statement: The original contributions presented in the study are included in the article, further inquiries can be directed to the corresponding author.

Conflicts of Interest: The authors declare no conflicts of interest.

References

1. Italian Ministry for Infrastructures and Transportation. Aggiornamento delle Norme Tecniche per le Costruzioni (Italian Building Code). *Gazzetta Ufficiale*, 17 January 2018. (In Italian)
2. Italian Ministry for Infrastructures and Transportation, Circolare. Istruzioni per L'applicazione dell'Aggiornamento delle Norme Tecniche per le Costruzioni (Instructions for the Application of the Italian Building Code). *Gazzetta Ufficiale*, 21 January 2019. (In Italian)
3. *ASCE/SEI 41-17*; Seismic Evaluation and Upgrade of Existing Buildings. American Society of Civil Engineers: Reston, VA, USA, 2017.
4. *Eurocode 8*; Design of Structures for Earthquake Resistance—Part 3: Assessment and Retrofitting of Buildings and Bridges. European Committee for Standardization: Brussels, Belgium, 2005.
5. Sepe, V.; Spacone, E.; Raka, E.; Camata, G. Seismic analysis of masonry buildings: Equivalent frame approach with fiber beam elements. In Proceedings of the 9th International Conference on Structural Dynamics, EURODYN 2014, Porto, Portugal, 30 June–2 July 2014; pp. 237–244, ISBN 978-972-752-165-4.
6. Maracchini, G.; Clementi, F.; Quagliarini, E.; Lenci, S.; Monni, F. Preliminary study of the influence of different modelling choices and materials properties uncertainties on the seismic assessment of an existing RC school building. *AIP Conf. Proc.* **2017**, *1863*, 450011. [CrossRef]
7. Jalayer, F.; Iervolino, I.; Manfredi, G. Structural modeling uncertainties and their influence on seismic assessment of existing RC structures. *Struct. Saf.* **2010**, *32*, 220–228. [CrossRef]
8. Diaferio, M.; Vitti, M. Correlation curves to characterize concrete strength by means of UPV tests. *Lect. Notes Civ. Eng.* **2021**, *110*, 209–218.

9. Cristofaro, M.T.; Viti, S.; Tanganelli, M. New predictive models to evaluate concrete compressive strength using the SonReb method. *J. Build. Eng.* **2020**, *27*, 100962. [CrossRef]

10. Breysse, D. Nondestructive evaluation of concrete strength: An historical review and a new perspective by combining NDT methods. *Constr. Build. Mater.* **2012**, *33*, 139–163. [CrossRef]

11. FEMA. *356 Prestandard and Commentary for the Seismic Rehabilitation of Buildings*; Federal Emergency Management Agency: Washington, DC, USA, 2000.

12. Franchin, P.; Pinto, P.E.; Pathmanathan, R. Confidence factor? *J. Earthq. Eng.* **2010**, *14*, 989–1007. [CrossRef]

13. Rota, M.; Penna, A.; Magenes, G. A framework for the seismic assessment of existing masonry buildings accounting for different sources of uncertainty. *Earthq. Eng. Struct. Dyn.* **2014**, *43*, 1045–1066. [CrossRef]

14. Tondelli, M.; Rota, M.; Penna, A.; Magenes, G. Evaluation of Uncertanties in the Seismic Assesment of Existing Masonry Buildings. *J. Earthq. Eng.* **2012**, *12*, 36–64. [CrossRef]

15. Jalayer, F.; Elefante, L.; Iervolino, I.; Manfredi, G. Knowledge-based performance assessment of existing RC buildings. *J. Earthq. Eng.* **2011**, *15*, 362–389. [CrossRef]

16. Cattari, S.; Lagomarsino, S.; Bosiljkov, V.; D'Ayala, D. Sensitivity analysis for setting up the investigation protocol and defining proper confidence factors for masonry buildings. *Bull. Earthq. Eng.* **2015**, *13*, 129–151. [CrossRef]

MDPI

St. Alban-Anlage 66

4052 Basel

Switzerland

www.mdpi.com

Applied Sciences Editorial Office

E-mail: applsci@mdpi.com

www.mdpi.com/journal/applsci